Energy Efficiency Applications in Buildings

Energy Efficiency Applications in Buildings presents an investigation into energy use and measures to improve the energy efficiency of existing building stock in the UK. The aim of this research is to assess the domestic energy use of statistically representative residential buildings and their occupants' thermal comfort by considering the significant impact of overheating risks on energy consumption and occupants' well-being.

Divided into two volumes, these books present energy consumption and thermal comfort in the construction sector as a complex socio-technical problem that involves the analysis of an intrinsic interrelationship amongst dwellings, occupants, and the environment. Using case studies, the authors demonstrate the significance of improving energy efficiency and its impact on occupants during long-term heatwaves in the summer. Additionally, the volumes demonstrate how dynamic thermal energy simulation can be used as a learning laboratory for future trends in housing energy consumption reduction.

Volume 1: Theories, Methods, and Tools presents the background to the research and the assessment methods and tools adopted by the research team.

Volume 2: Performance Evaluation and Retrofitting Strategies describes the case studies and building performance and post-occupancy evaluations, before making strategic policy and retrofitting recommendations.

The research and roadmap presented in these volumes can be used as a guidance tool for building energy modelling and performance simulation, enabling architects, building engineers, and other practitioners to close the gap between the current understanding and the actual performance of existing building stocks.

Dr. Bertug Ozarisoy is a Lecturer in Architecture at the College of Technology and Environment, London South Bank University (LSBU), London, UK. His research centers on understanding the relationship between architecture and energy policy design, while also exploring the effects of passive cooling systems on domestic energy consumption and household thermal comfort.

Dr. Hasim Altan is a Professor of Sustainable Design and Architectural Engineering. He is a Chartered Architect (RIBA) and a Chartered Engineer (CIBSE) with more than 25 years of academic and practical experience in Architecture, Engineering, and Construction (AEC) within the Built Environment across the UK, Europe, the Middle East, and North Africa (MENA) regions.

Dr. Young Ki Kim is an Assistant Professor in the Department of Architectural Engineering at the United Arab Emirates University (UAEU). His research primarily focuses on sustainable building development, environmentally and climate-responsive building design, building operation and management, health and well-being, and indoor environmental quality (IEQ).

Dr. Wei Shi received his doctorate degree in Architecture and Sustainability at the University of East London in 2019. His research originated from a collaboration between the local authority and the university, aimed at enhancing the energy efficiency of various social housing projects. It focuses on the soft landing of the roll-out of the government's social housing refurbishment schemes while examining their drawbacks.

ENERGY EFFICIENCY APPLICATIONS IN BUILDINGS

Volume 1: Theories, Methods, and Tools

Bertug Ozarisoy, Hasim Altan, Young Ki Kim, and Wei Shi

Routledge
Taylor & Francis Group
LONDON AND NEW YORK

Designed cover image: Bridging the energy performance gap 2024 © Bertug Ozarisoy

First published 2025
by Routledge
4 Park Square, Milton Park, Abingdon, Oxon OX14 4RN

and by Routledge
605 Third Avenue, New York, NY 10158

Routledge is an imprint of the Taylor & Francis Group, an informa business

British Library Cataloguing-in-Publication Data
A catalogue record for this book is available from the British Library

ISBN: 9781032762296 (hbk)
ISBN: 9781032790725 (pbk)
ISBN: 9781003490364 (ebk)

DOI: 10.1201/9781003490364

Typeset in Times New Roman
by codeMantra

To my family, for their unconditional support throughout
my research journey and academic career.
Dr. Bertug Ozarisoy

To my family, my wife and two daughters, and
parents for their unconditional love and support.
Prof. Dr. Hasim Altan

To my family for their unconditional love and support.
Dr. Young Ki Kim

To my family, my wife and son, and
parents for their unconditional love and support.
Dr. Wei Shi

CONTENTS

PREFACE

This reference book focuses on improving home energy efficiency by considering occupants' behaviour on home energy performance and thermal comfort in dwellings in the UK. This study is aimed to develop viable intervention strategies to reduce home operational energy in the UK social housing sector. The study suggested that the provision of energy information and advice needs to strategically target occupants with different socio-demographic characteristics via varying approaches as an integral aspect of any retrofit programme. This study contributes to the studies on global energy crisis and proposes evidence-based energy policy design. It must be stressed that field study investigations were conducted by the previous scholars on optimising occupants' thermal comfort but most of the studies are neglected to consider both occupants' thermal comfort and energy use. This is because this reference book could be an effective roadmap to the policymakers for acting plans to tackle both with global energy crisis and warming.

This reference book presents the outcome and implications of current energy performance of base-case representative housing typologies by embedding building energy modelling and performance simulation methods across the UK. It investigated the significance of the patterns in energy use of the occupants to calibrate the energy performance of existing social housing stock and the occupants' thermal comfort during the overheating period under summer conditions. Furthermore, the study elaborated on the significance of dominant occupancy patterns embedded into the building simulation parameters to calibrate the actual energy end-use of case study buildings presented in Part 2, in order to provide representative energy consumption data to improve building energy efficiency using the questionnaire-based survey and performance simulation data.

This reference book lays out the guiding principles of holistic and inclusive practice, followed by a useful explanation for the adoption of building physics and the impact of the occupants' habitual adaptive behaviour on assessing domestic energy use in conjunction with investigating their thermal comfort. This reference book should be used alongside a range of other sources on building performance evaluation, design, and construction—and references are included. Furthermore, researchers, practitioners, architects, and engineers could also use this book as a guidance tool in their projects for building energy modelling and performance simulation. At the same time, it sets out to illustrate households' socio-demographic characteristics in energy use, thermal conductivity of building properties and environmental parameters for developing evidence-based holistic retrofitting design strategies to uptake delivery of long-term energy efficiency of the existing building stock. The layout of this reference book aims to provide a practical and balanced view of a current novel epistemological approach developed and a road map of the way forward in the construction sector that is rapidly changing in both its structures and processes as well as in the demands placed on its outputs and the solutions that are available.

London, UK	Dr. Bertug Ozarisoy
Kyrenia, Cyprus	Prof. Dr. Hasim Altan
Al Ain, United Arab Emirates	Dr. Young Ki Kim
London, UK	Dr. Wei Shi

GRAPHICAL ABSTRACT

Stages of development for the methodological framework of the study. Drawn by Ozarisoy, 2024

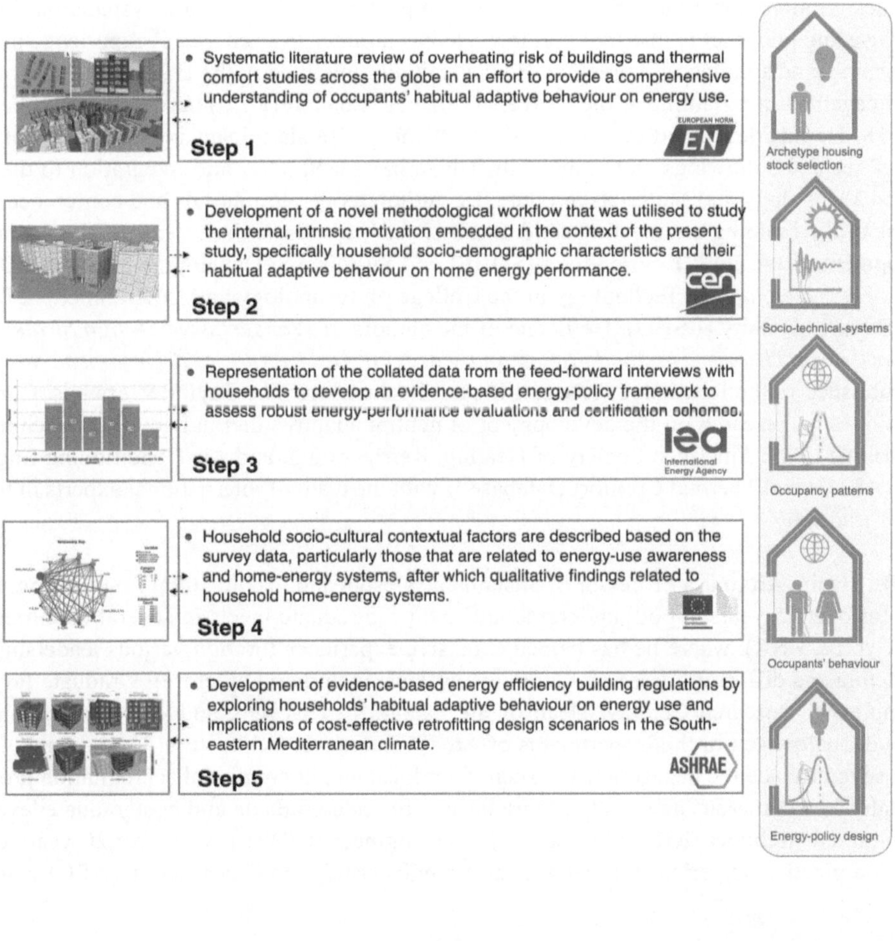

ABOUT THE AUTHORS

Dr. Bertug Ozarisoy is an architect and expert in parametric statistical analyst both in the UK and Cyprus. His research focuses on understanding the theory between architecture and energy policy design in conjunction with exploring the impact of passive cooling systems on domestic energy use and households' thermal comfort. He has an intensive ten years' experience in teaching, research, and architectural practice. He was initially involved in the BSc (Hons) in Construction Management programme at the University of East London (UEL) between September 2016 and 2018. He has developed teaching skills to demonstrate significant intellectual input to the students' technical drawings and monitor their learning adaptability and integration to the 'Construction Technology and Materials' module. He authored over 45 journals and conference papers relevant to the building engineering field and assisted with research projects in the Sustainable Development of the Built Environment (SDBE) and Sustainable Development Goals (SDGs) in Europe. He is a lecturer in Technology in the College of Technology and Environment at London South Bank University (LSBU). He is one of the authors of *Transgressive Design Strategies for Utopian Cities: Theories, Methodologies and Cases in Architecture and Urbanism*, which has been published in the Routledge Research in Architecture series by Taylor & Francis. Currently, he continues his research on the development of neutral adaptive thermal comfort benchmarking to contribute to the American Society of Heating, Refrigerating, and Air-Conditioning Engineers (ASHRAE) Global Thermal Comfort Database II with the team of international experts in thermal comfort studies.

Prof. Dr. Hasim Altan is a Professor of Sustainable Design and Architectural Engineering. He has taught for over 25 years at both undergraduate and postgraduate levels in several countries (UK, UAE, Cyprus, KSA), where he has gained extensive experience through various leadership roles in academia and different higher education institutions, including Director of Graduate Programs and University Research Centers; Dean of the Faculties of Design, and Architecture and Fine Arts; and Chairperson of the Departments of Architecture and Architectural Engineering. He has guided several academic departments through their local and international accreditation processes and established numerous new study programs at both undergraduate and postgraduate levels. He is a Chartered Architect (RIBA) and a Chartered Engineer (CIBSE) with over 20 years of academic and practical experience in Architecture, Engineering, and Construction (AEC) within the

Built Environment across the UK, Europe, the Middle East, and North Africa (MENA) regions. He serves on several editorial boards and evaluates project proposals for the European Commission, UK Research Councils, and the Qatar National Research Fund (QNRF) under the Qatar Foundation. Since 2004, he has individually or collaboratively secured and directed 32 research grants totaling over £21 million. He is a co-founding member of the International Network on Zero Energy Mass Custom Home (ZEMCH), which has organized ten international conferences, numerous design workshops, and various technical visits to date. Additionally, he has supervised 19 successful PhD students (11 as the primary supervisor) and over 80 Master's (MSc, MArch, and MEng) students. He has published more than 300 refereed international journal and conference papers, technical reports, authored or edited books and chapters, and other editorial works in related fields.

Dr. Young Ki Kim obtained a BEng degree in the School of Architecture in South Korea (1995–2002). He holds an MSc degree (2005) in Environmental Design and Engineering from the University College London (UCL). He finished his PhD degree at the University of Sheffield with the title of thesis *Adapting the UK Housing to Climate Change: Energy Consumption and Overheating*. He was a research associate, teaching fellow, and computer simulation expert in the Building Environments Analysis Unit (BEAU) in the School of Architecture at the University of Sheffield (2007–2014). He was a senior executive manager in the EAN Technology (2015–2017) in Seoul and he achieved a LEED AP BD+C certification and was involved in several LEED certification projects in South Korea. He also achieved the BREEAM Assessor and was involved in the IKEA Korea BREEAM certification projects. His main role in EAN Technology involved energy analysis, commissioning, sustainable building design, engineering consulting, and project management. In 2017, he joined the Department of Architectural Engineering, University of Sharjah, as a visiting assistant professor and was involved in several design studios and graduate senior projects. In 2018, he moved to the United Arab Emirates University (UAEU) as an Assistant Professor where he is currently working in the Department of Architectural Engineering at the College of Engineering.

Dr. Wei Shi received his doctorate degree in Architecture and Sustainability, at the University of East London in 2019. His research originated from the collaboration between the local authority and the university, aiming to increase the energy efficiency of numerous social housings. His research focuses on the soft landing of the roll-out of the government's social housing refurbishment schemes and investigates the downsides. With guidance and advice, and by adopting a mixed research methodology, future similar schemes were developed and presented to the local authority. Having completed his courses in Professional Practice Architecture (Part III), Professional Diploma in Architecture (Part II), and BA Architecture (Part I), Wei is now a RIBA registered architect working at Fereday Pollard Architects since 2019. In the residential work, Wei has been involved in design development, checking compliance with Building Regulations, and advancing the production of drawings for their submission for planning approval. In addition, Wei has been involved in several transport and infrastructure projects from their inception through detailed design. Wei is a very experienced user of AECOsim CAD software and the production of co-ordinated 3D models and drawings, Wei is familiar with the Network Rail Assurance process and relevant design standards associated with transport infrastructure. His experience and education to date have given him a strong understanding of the importance of the place and context, and the awareness of energy saving in a project.

ACKNOWLEDGEMENTS

First, we would like to express our deepest gratitude to the College of Technology and Environment at London South Bank University (LSBU) for providing a research environment for Dr. Bertug Ozarisoy to develop this reference book on energy efficiency guidelines in the UK. Special thanks are due to the Head of Division in Architecture – Prof. Igea Troiani, Associate Dean of Research and Enterprise – Prof. Issa Chaer, and Dean of the College of Technology and Environment – Prof. Chris Harty for their unwavering support throughout the development and writing of this reference book. Second, we wish to acknowledge the Graduate School at the University of East London (UEL) for offering a research environment for Dr. Wei Shi to pursue his PhD research in the School of Architecture, Computing, and Engineering. Special thanks go to Assoc. Prof. Mihaela Anca Ciupala for her steadfast support and for fostering a rigorous research culture at UEL. Third, we would like to extend our gratitude to the University of Sheffield – School of Architecture, and to Building Science professors – John Page, Peter Tregenza, Ian C. Ward, Steve Sharples, Jian Kang, and Adrian Pitts for their guidance and support for the remarkable research conducted in the 1990s and 2000s. Last but not least, we are grateful to Taylor & Francis for providing us the opportunity to publish this reference book.

Dr. Bertug Ozarisoy
Prof. Dr. Hasim Altan
Dr. Young Ki Kim
Dr. Wei Shi

PART 1

Energy Efficiency

INTRODUCTION

Background

As broadly accepted, global warming is now forever on the world's agenda. According to the UN Intergovernmental Panel on Climate Change, an immediate reduction in carbon dioxide (CO_2) emissions of at least 60% is required to stabilise concentrations in the atmosphere in the long term (Carry & Stephens, 2024). Moreover, the Royal Commission on Environmental Pollution Report on Climate Change has warned that increasing atmospheric concentrations of greenhouse gases, particularly CO_2, will have serious impacts on the world's climate (Srimath et al., 2016; Barker et al., 2022). The report further suggests that the evidence is mounting that climate change could have a significant effect on the UK and therefore the UK would have to reduce its emissions by 60% by 2050 (Kim et al., 2024).

On the other hand, the UK Government's Climate Change Programme set to deliver 20% reduction from the 1999 level by 2010 has suggested that this may be insufficient with current measures (Bassett et al., 2021). Given that the domestic sector is a major contributor to the UK's total CO_2 emissions (about 30%), this is an important area to tackle effectively. In the UK, while new buildings achieve a high level of energy efficiency through the standards enforced by Part L of the building regulations, existing housing stock experiences from one of the lowest energy efficiency levels in Europe, with an average SAP rating of 51 (Bobrova et al., 2023). Therefore, action in this area is evidently needed if both social and environmental sustainability is to be achieved.

Being aware of all this, the author has become more concerned about energy efficiency in housing and the possible impact it has caused on the environment (Morey et al., 2020). His main concern and focus areas have become 'energy conservation' and 'why it is vital in our environment'. During the past 25 years, energy for heating buildings has been relatively low-priced, mainly due to the effect of utility privatisation on prices. As a result, designers felt able to work without taking energy conservation into account. However, this situation has now started to change due to increasing prices and potential shortage of world oil, and therefore, designers will have to show careful consideration not only to building regulations, which prescribe a minimum standard but also to other factors in design, which will also reduce energy consumption (Gregory & Geels, 2023). Energy conservation is important for many reasons, such as economic, environmental, and social.

DOI: 10.1201/9781003490364-2

Before 1973, little action was taken towards energy conservation. With the rapid increase in energy prices in the 1970s, international attention has been given to an energy conservative way of life (Pata & Ertugrul, 2023). The problem of energy use and availability is common to a greater or lesser extent throughout the world. While the industrialised nations depend heavily upon fossil fuels for their industrial processes, the developing nations also desire to increase their technological capabilities and thus the use of energy in its various forms (Quinn et al., 2022). In the last 50 years, there has been a tremendous increase in world energy use, partly as a result of the availability of easily extractable fossil fuels (coal, gas, and oil) and awareness of the limited nature of these reserves that existed since the 1970s (Taylor et al., 2015). Although it is unlikely that the world will completely run out of fossil fuels in the next millennium, the majority of easily extractable reserves are located in a small part of the world, and the prices are expected to rise due to the increasing demand for energy.

Nowadays, fossil fuels are too costly to be burned; they should be used wisely, not wastefully, or/and again not for purposes of cooling, heating, etc. It must be stressed that the world's fossil fuel energy is limited by geological conditions (Bobrova et al., 2023). We are living in a period where energy demands must be reduced through any means available. These should also include conservation of energy, employment of renewable energy, energy-conscious planning, and design of buildings. In this stage, it could be said that renewable energy sources should be the primary energy sources of the 21st century, as being a free source of energy (derived from solar radiation, including direct use of solar energy for heating or electricity and indirect forms such as energy from the wind and waves) and thereby reducing the associated environmental impact on our very environment in worldwide (Levermore et al., 2018).

Solar energy is a renewable resource, which can easily be a plentiful contribution to heating and lighting in buildings and has an important role in sustainable development in our environment (Dhimish & Mather, 2019). There is plenty of sunshine available and if only properly controlled, it can help to reduce energy use in buildings. This energy is also a non-polluting source, and therefore, its effective utilisation helps to reduce emissions of carbon dioxide and other gases resulting from the use of fossil fuels (Kotthaus & Grimmond, 2014). Every building has a certain amount of its heating requirements met by solar energy. Sunlight passing through windows is a source of heat; however, most buildings are not specifically designed to utilise this solar energy. The value of passive solar heating is enhanced by proper building insulation. A well-insulated building requires less energy for heating; and thus, passive solar features can meet much of the heating load (Mandel & Pató, 2024).

Actions to improve housing energy efficiency could result in great energy savings, both for cooling and especially for heating (Ozarisoy, 2022). Among all efforts to improve energy efficiency, the application of thermal insulation in building envelope components (i.e., ceilings, walls, floors, and glazing) is the most effective and important factor. In addition to the reduction in space heating and cooling costs on a long-term basis, other benefits, such as occupant's comfort and smaller capacity requirements for heating and cooling systems, would be achieved. Due to considerable savings in operating costs and initial costs, insulation pays back its investment in a short period of time. It is a well-known fact that solar energy as an important and non-polluting renewable energy source can make a vast contribution to energy savings, which can again limit environmental damage, conserve fossil fuel reserves, and save money.

With the United Nations Framework Convention on Climate Change in 1992, the UK has committed to reducing the emission of the basket of six greenhouse gases with carbon dioxide being the most significant to 12.5% lower than the 1990 levels by 2030. Energy consumed by the building stock approaches 50% of the total energy consumption as opposed to transport being

28%. Furthermore, energy used only in the domestic building sector is responsible for about 30% of overall carbon dioxide emission and considerably is a major contributor to the global warming. Over the 25 years period (1973–1998), despite the increase mentioned above, the total energy consumed within a home by the average UK household fell by about a tenth (Skelhorn et al., 2018). Now this fall is mostly attributable to a decline in the average number of people per household, to an increase in insulation and draught proofing as new homes are built and existing dwellings are improved, and to the introduction of more energy-efficient heating systems.

Scope of the Problem

Today, if we were to analyse current conditions and energy standards of UK housing stock, we would easily conclude that there is potential for energy efficiency improvements, particularly in the private sector. According to the English House Condition Survey (EHCS) Reports published in 2021, there is considerable scope for improvements in the energy efficiency and thermal performance of the existing housing stock. Most of the existing housing has an average SAP rating of 51, which falls well below modern building regulations and energy standards (Ozarisoy, 2022).

In the UK, local authorities can only have a little influence on the private sector, which accounts for the majority of the stock, and this therefore remains a great challenge for the government, whilst there is no enforced legislation of energy standards in the private sector. Consequently, this research has focused on the privately rented properties within this sector that also accounts for about 15% of the total housing stock and demonstrates one of the worst conditions of housing standards in the country (Calderón et al., 2024).

Furthermore, the UK has a reputation for poorly heated homes, low indoor temperatures, and a general preference for sweaters over building insulation especially when compared to other European countries and therefore experiences higher levels of inequality and poverty (Taleghani et al., 2019). Although there has been rapid growth in central heating and greatly improved heating standards over the last two decades, low levels of insulation remain a problem.

Moreover, revised building regulations will only affect new housing and therefore legislation for energy efficiency improvements in existing housing should be properly enforced. Accordingly, standards need to be progressively raised and more effectively supervised (Zheng et al., 2023). Even so, it is clear that regulations on their own will not be sufficient to achieve a really substantial improvement in the overall environmental performance of dwellings. They will need to be supplemented and reinforced by stronger policy direction, and a more proactive programme to increase public and professional awareness of the objectives and the potential for improvement.

Actions to Tackle Existing Problems

The main mechanisms currently in place for addressing energy efficiency in homes are briefly explained below:

The Energy Efficiency Standards of Performance, the precursor of Energy Efficiency Commitment, was introduced as a mechanism to assist in the delivery of climate change policy, whereas Warm Front was initially more socially based to promote affordable warmth (Citadini De Oliveira et al., 2024).

The Home Energy Conservation Act requires local authorities to plan to improve the energy efficiency of housing in their area by 30% on a baseline of 1995 by 2030 (Colclough & Salaris, 2023).

Energy efficiency measures also form part of energy security; the performance and innovation unit report on energy policy has been published emphasising energy efficiency as a key strand of future energy policy. The report recommends an improvement target for home energy efficiency of 10% by 2010 and a further 20% by 2030 (Hurst & O'donovan, 2024).

The Government's Decent Homes standard is a four-part definition. Social housing providers are required to identify all homes that do not meet these standards and propose a plan to bring them up to standard by 2030.

Links between cold homes and ill-health are well established. The Health Act addressed the organisation of health care in the UK, such that the emphasis now falls on Primary Care Trusts to deliver local care working through the network of health professionals (Liu et al., 2024).

The UK Quality of Life headline indicators, include poverty, health, housing, and climate change and give a snapshot of the state of the country as a result of measures and events for sustainable development (Tomrukcu & Ashrafian, 2024). The broad question driving this research is: Can the UK meet its domestic sector goal to reduce carbon dioxide emissions to 20% below 1990 levels by 2030? (Pérez et al., 2024)

To do so, will improving energy efficiency and reducing carbon dioxide emissions within the privately rented sector help to deliver positive results and achieve long-term sustainability in the domestic sector? This research attempts to answer the primary question by taking into consideration the secondary question, which is investigating potentials for energy efficiency improvements and reduction of carbon dioxide emissions in the privately rented sector. It is important to emphasise that the aim of this study is to determine potentials for improving energy efficiency and reducing carbon dioxide emissions in existing housing, particularly in the private sector.

- To provide a database of the current conditions and energy standards of privately rented properties.
- To investigate potentials for energy efficiency improvements and reduction of carbon dioxide emissions in the privately rented sector.
- To evaluate and review energy and environmental assessment methods, and their applications in terms of energy efficiency and carbon index (CI).
- To identify a way of cost-effective modifications to dwelling base designs which can improve energy efficiency within privately rented properties.
- To develop case study results that can be used by landlords/landladies to encourage energy conservation.
- To promote general recommendations for immediate energy savings throughout privately rented properties analysed with supplementary documentation enclosed within feedback reports compiled.

This work for this research has investigated the potential for energy efficiency in privately rented properties within Sheffield and London. The research methodology is as follows: Current conditions of these properties including the efficiency of space and water heating systems in use were examined. SAP ratings of these dwellings with respect to CO_2 emissions were calculated. Reductions of CO_2 emissions due to modifications conducted to base designs were identified and compared to the current situation of the properties and therefore necessary energy efficiency improvements were suggested in order to optimise energy use and CO_2 emission consumed accordingly. In addition to SAP ratings and CO_2 emissions calculated, total costs of energy use for space and water heating, natural ventilation, and heat loss in these properties were also examined. A total of 245 properties have been studied. Existing conditions (with relevant SAP rating), total costs of energy use (both for space and water heating), CO_2 emissions, effective air change in the property (natural ventilation), and heat loss parameters have been determined. Furthermore, the research has also been followed up by the comparison between results obtained from Stelrad Software (Home Assessment Rating Procedure) and SAP Calculation Program (version 9.53).

References

Barker, P. A., Allen, G., Flynn, M., Riddick, S., & Pitt, J. R. (2022). Measurement of recreational N_2O emissions from an urban environment in Manchester, UK. *Urban Climate, 46*, 23–24. https://doi.org/10.1016/j.uclim.2022.101282

Bassett, R., Janes-Bassett, V., Phillipson, J., Young, P. J., & Blair, G. S. (2021). Climate driven trends in London's urban heat island intensity reconstructed over 70 years using a generalized additive model. *Urban Climate, 40,* 1–7. https://doi.org/10.1016/j.uclim.2021.100990

Bobrova, Y., Papachristos, G., Chiu, L. F., Tikhomirova, S., & Coon, T. M. (2023). Home for the Common Future (HCF): the use of home-meanings to promote domestic energy retrofit. *Energy Research & Social Science, 107*, 123–125. https://doi.org/10.1016/j.erss.2023.103358

Calderón, C., Cardenas, M. A., & Aoun, J. (2024). Quantifying peak heat demand in neighbourhoods: a UBEM approach and its implications for residential heating electrification in the UK—a case study of Newcastle upon Tyne. *Energy and Buildings 10*, 23–27. https://doi.org/10.1016/j.enbuild.2024.114609

Cary, M., & Stephens, H. M. (2024). Economic, environmental, and technical gains from the Kyoto protocol: evidence from cement manufacturing. *Resources Policy, 91*, 231–234. https://doi.org/10.1016/j.resourpol.2024.104926

Citadini De Oliveira, C., Catão Martins Vaz, I., & Ghisi, E. (2024). Retrofit strategies to improve energy efficiency in buildings: an integrative review. *Energy and Buildings, 321*, 32–33. https://doi.org/10.1016/j.enbuild.2024.114624

Colclough, S., & Salaris, C. (2023). Quantifying overheating in nZEB Irish residential buildings. An analysis of recorded interior temperatures of Irish newbuild and retrofit residential buildings against CIBSE, Passive House and WHO overheating criteria and recorded occupant satisfaction. *Energy and Buildings, 303*, 1–7. https://doi.org/10.1016/j.enbuild.2023.113571

Dhimish, M., & Mather, P. (2019). Exploratory evaluation of solar radiation and ambient temperature in twenty locations distributed in United Kingdom. *Urban Climate, 27*, 179, 1–3. https://doi.org/10.1016/j.uclim.2018.12.001

Gregory, J., & Geels, F. W. (2023). Unfolding low-carbon reorientation in a declining industry: a contextual analysis of changing company strategies in UK oil refining (1990–2023). *Energy Research & Social Science, 107*, 231–233. https://doi.org/10.1016/j.erss.2023.103345

Hurst, L. J., & O'donovan, T. S. (2024). Embodied impacts of key materials for UK decarbonised domestic retrofit: differences between sources of embodied carbon and embodied energy data. *Energy and Buildings, 319*, 91–97. https://doi.org/10.1016/j.enbuild.2024.114515

Kim, J., Sovacool, B. K., Bazilian, M., Griffiths, S., & Yang, M. (2024). Energy, material, and resource efficiency for industrial decarbonization: a systematic review of sociotechnical systems, technological innovations, and policy options. *Energy Research & Social Science, 112*, 23–27. https://doi.org/10.1016/j.erss.2024.103521

Kotthaus, S., & Grimmond, C. S. B. (2014). Energy exchange in a dense urban environment – part I: temporal variability of long-term observations in central London. *Urban Climate, 10*, 261, 21–27. https://doi.org/10.1016/j.uclim.2013.10.002

Levermore, G., Parkinson, J., Lee, K., Laycock, P., & Lindley, S. (2018). The increasing trend of the urban heat island intensity. *Urban Climate, 24*, 360, 1–7. https://doi.org/10.1016/j.uclim.2017.02.004

Liu, Y., Grimmond, S., Luo, Z., Hertwig, D., Mcgrory, M., Lo Piano, S., & Smith, S. T. (2024). Dwelling *conversion and energy retrofit modify building anthropogenic heat emission under past and future climates: a case study of London terraced houses*. Elsevier BV 112 (1–3). https://doi.org/10.1016/j.enbuild.2024.114668

Mandel, T., & Pató, Z. (2024). Towards effective implementation of the energy efficiency first principle: a theory-based classification and analysis of policy instruments. *Energy Research & Social Science, 115*, 431–435. https://doi.org/10.1016/j.erss.2024.103613

Morey, J., Beizaee, A., & Wright, A. (2020). An investigation into overheating in social housing dwellings in central England. *Building and Environment, 176*, 23–25. https://doi.org/10.1016/j.buildenv.2020.106814

Ozarisoy, B. (2022). Energy effectiveness of passive cooling design strategies to reduce the impact of long-term heatwaves on occupants' thermal comfort in Europe: climate change and mitigation. *Journal of Cleaner Production, 330*, 21–24. https://doi.org/10.1016/j.jclepro.2021.129675

Pata, U. K., & Ertugrul, H. M. (2023). Do the Kyoto Protocol, geopolitical risks, human capital and natural resources affect the sustainability limit? A new environmental approach based on the LCC hypothesis. *Resources Policy, 81*, 67–63. https://doi.org/10.1016/j.resourpol.2023.103352

Pérez, I. A., García, M. Á., Rasekhi, S., & Pazoki, F. (2024). The London pollution island under Lamb weather types. *Urban Climate, 53*, 121–123. https://doi.org/10.1016/j.uclim.2024.101834

Quinn, B., Gallagher, R., & Kuosmanen, T. (2022). Lurking in the shadows: the impact of CO_2 emissions target setting on carbon pricing in the Kyoto agreement period. *Energy Economics, 118*, 9–11. https://doi.org/10.1016/j.eneco.2022.106338

Skelhorn, C. P., Lindley, S. J., & Levermore, G. (2018). Urban greening and the UHI: seasonal trade-offs in heating and cooling energy consumption in Manchester, UK. *Urban Climate, 23*, 173, 1–3. https://doi.org/10.1016/j.uclim.2017.02.010

Srimath, S. T. G., Sokhi, R., Karppinen, A., Singh, V., & Kukkonen, J. (2016). Evaluation of an urban modelling system against three measurement campaigns in London and Birmingham. *Atmospheric Pollution Research, 8*(1), 38, 90–95. https://doi.org/10.1016/j.apr.2016.07.004

Taleghani, M., Marshall, A., Fitton, R., & Swan, W. (2019). Renaturing a microclimate: the impact of greening a neighbourhood on indoor thermal comfort during a heatwave in Manchester, UK. *Solar Energy, 182*, 245, 345–347. https://doi.org/10.1016/j.solener.2019.02.062

Taylor, J., Wilkinson, P., Davies, M., Armstrong, B., Chalabi, Z., Mavrogianni, A., Symonds, P., Oikonomou, E., & Bohnenstengel, S. I. (2015). *Mapping the effects of urban heat island, housing, and age on excess heat-related mortality in London*. Elsevier BV 181, (34–35). https://doi.org/10.1016/j.uclim.2015.08.001

Tomrukcu, G., & Ashrafian, T. (2024). Climate-resilient building energy efficiency retrofit: evaluating climate change impacts on residential buildings. *Energy and Buildings, 316*, 91–93. https://doi.org/10.1016/j.enbuild.2024.114315

Zheng, L., Mueller, M., Luo, C., Menneer, T., & Yan, X. (2023). Variations in whole-life carbon emissions of similar buildings in proximity: an analysis of 145 residential properties in Cornwall, UK. *Energy and Buildings, 296*, 11–13. https://doi.org/10.1016/j.enbuild.2023.113387

1

CLIMATE CHANGE AND UK FUTURE CLIMATE

Introduction

> The Earth's climate has changed many times in response to natural causes. The term climate change usually refers to man-made changes that have occurred since the early 1900s.
>
> —Met Office & Act on CO_2 (Anandarajah & Strachan, 2010)

> Even if global temperatures rise by only 2°C, 20–30% of species could face extinction.
>
> —Met Office & Act on CO_2 (Arnell et al., 2020)

Climate change, especially global warming, is the most important science issue of the 21st century, challenging the very structure of our global society in areas such as economics, sociology, geopolitics, local politics, and individuals' lifestyles (Arnell, 2022). Global warming is caused by the mass of greenhouse gases, such as carbon dioxide, in the atmosphere, resulting from the burning of fossil fuels and deforestation (Bevan et al., 2020). There is clear evidence that we already have elevated concentrations of atmospheric carbon dioxide at their highest level for the last half million years and perhaps even longer. The most recent report by the Intergovernmental Panel on Climate Change (IPCC) stated that there is clear evidence for a 0.75°C rise in global temperatures and 22 cm rise in sea level during the 20th century (Arnell et al., 2021). The report also predicts that global temperatures could rise further by between 1.1°C and 6.4°C by 2100 depending on the level of greenhouse gases (GHGs) in the atmosphere, and sea levels could rise by between 28 cm and 79 cm, and by more if the melting of Greenland and Antarctica accelerates. In addition, weather patterns will become less predictable and there will be an increase in frequency of extreme weather events, such as storms, floods, heatwaves, and droughts (Boardman, 2004). The Stern Review provided an economic perspective and stressed the importance of acting now to avert the most serious effects and economic costs of climate change in the future (Barker, 2008). Most recently, a major international review of the latest science has concluded that the speed and impacts of climate change are likely to be more severe than previously reported and that action to reduce CO_2 emissions and to adapt is most urgently needed (Jenkins et al., 2022). In deciding how to respond, we need evidence across all sectors of the economy on what actions might be most effective, the potential that they hold and how much they will cost. The Stern Review on

DOI: 10.1201/9781003490364-3

the Economics of Climate Change is released for the British Government on 30 October 2006 by Nicholas Stern, chair of the Grantham Research Institute on Climate Change and the Environment at the London School of Economics (LSE) and chair of the Centre for Climate Change Economics and Policy (CCEP) at Leeds University and LSE (Kay et al., 2019). The report discusses the effect of global warming on the world economy. The Review states that climate change is the greatest and widest-ranging market failure ever seen, presenting a unique challenge for economics. The Review's main conclusion is that the benefits of strong, early action on climate change far outweigh the costs of no action. According to the Review, without action, the overall costs of climate change will be equivalent to losing at least 5% of global gross domestic product (GDP) each year, now and forever, including a wider range of risks and an increase in the impact of around 20% of GDP or more (Wang et al., 2020).

The Review proposes that 1% of global GDP per annum needs to be invested, in order to avoid the worst effects of climate change. In June 2007, Stern increased the estimate for the annual cost of achieving stabilisation between 500 and 220 ppm CO_2 to 2% GDP, to account for faster than expected climate change (Yohe et al., 2007). This chapter studies global warming and climate change, it is followed by a review of how climate change has been observed. This chapter will cover the evidence showing that climate change is already occurring, its effect on human life and health, and future predictions at a global and national level. It also studies the impact of global warming in future. The GHG emissions' targets and policies are explained at a global and national level and are also set out in this chapter.

What Is Climate Change?

Climate change is a long-term change in the statistical distribution of weather patterns over period of time that ranges from decades to millions of years (Pielke, 2007). It may be a change in the average weather conditions or a change in the distribution of weather events with respect to an average, for example, after looking at rain gauge date, lake and reservoir levels, and satellite data, scientists can tell if during summer, an area was drier than average (Shipworth, 2007). If it continues to be drier than normal over the course of many summers, then it would likely indicate a change in the climate (Tol & Yohe, 2008). The difference between weather and climate is a measure of time. Weather is what conditions of the atmosphere are over a short period of time, and climate is how the atmosphere "behaves" over relatively long periods of time.

When scientists talk about climate, they are looking at an average of precipitation, temperature, humidity, sunshine, wind velocity, phenomena, such as fog, frost, and hailstorms, and other measures of the weather that occur over a long period in a particular place (Carrington & Stephenson, 2018). When we talk about climate change, we are talking about changes in long-term averages of daily weather. This study provides definitions of climate change used by the IPCC. Definitions of climate change in IPCC. Climate change in IPCC usage refers to a change in the state of the climate that can be identified (e.g., using statistical tests) by changes in the mean or the variability of its properties and that persists for an extended period, typically decades or longer. It refers to any change in climate over time, whether due to natural variability or as a result of human activity. This usage differs from that in the United Nations Framework Convention on Climate Change (UNFCCC), where climate change refers to a change of climate that is attributed directly or indirectly to human activity that alters the composition of the global atmosphere and that is in addition to natural climate variability observed over comparable time periods (Clarke et al., 2022). In the context of environmental policy, climate change usually refers to change in climate in recent decades.

It may be qualified as anthropogenic climate change, more generally known as global warming or anthropogenic global warming.

Historical Background

In 1896, a Swedish scientist published a new idea, namely, that as humanity burned fossil fuels, such as coal, which added carbon dioxide gas to the Earth's atmosphere, it would raise the planet's average temperature (Forster et al., 2020). This "greenhouse effect" was only one of many speculations about climate, and not the most plausible. Scientists found good reason to believe that our emissions could not change the climate. Major change seemed impossible except over tens of thousands of years due to a lack of knowledge about the Earth and climate. This study shows some of the very important early discoveries about climate change in the 18th and 19th centuries. Key discovery for early climate change 1753: Joseph Black discovered carbon dioxide, 1824: Joseph Fourier calculates that the Earth would be far colder if there is not an atmospheric effect which is the "greenhouse effect", 1896: Svante August Arrhenius publishes first calculation of global warming from human emissions of carbon dioxide.

The paragraphs below provide summaries from "The Discovery of Global Warming" by Cotterill et al. (2022) and these are general findings from the early 1930s to later 2000s. In the 1930s, people realised that the United States and North Atlantic region had warmed significantly during the previous half-century. Scientists supposed this was just a phase of some mild natural cycle, with unknown causes. In the late 1950s, carbon dioxide measurements were made on a mountaintop in Hawaii.

Over the next decade, these measurements confirmed that the level of CO_2 in the atmosphere was rising year after year (Kennedy-Aser et al., 2021). In 1967, an early computer simulation suggested that global temperatures might increase by more than 4°C, depending on the level of CO_2 (Kennedy-Aser et al., 2021). In the early 1970s, the rise of environmentalism raised public doubts about the benefits of human activity for the Earth. Alongside the greenhouse effect, some scientists pointed out that human activity was putting dust and smog particles into the atmosphere, where they could block sunlight and cool the world. Moreover, analysis of Northern Hemisphere weather statistics showed that a cooling trend had begun in the 1940s (Kennedy-Aser et al., 2021). One unexpected discovery was that the level of certain other gases was rising, which would add seriously to global warming. By the late 1970s global temperatures had evidently begun to rise again. The climate change first caught wide public attention in the summer of 1988, the hottest on record till then. By 2001, the IPCC managed to establish that our civilisation faced severe global warming. Since then, greatly improved computer models and the range and quantity data strengthened the conclusion that human emissions are very likely to cause serious climate change. This was reaffirmed by the IPCC in a report published in 2007 and it expected the planet's average temperature to rise anywhere between 1.4 and 6°C depending on the level of GHG emissions (Kennedy-Asser et al., 2022).

Cause of Climate Change – The Greenhouse Effect

The Sun is the primary source of energy for the Earth's climate. Satellite observations show that about 30% of the Sun's energy that reaches the Earth is reflected back to space by clouds, gases, and small particles in the atmosphere, and by the Earth's surface (Clarke et al., 2008). The remainder, about 240 W/m² when over the planet, is absorbed by the atmosphere and the surface. To

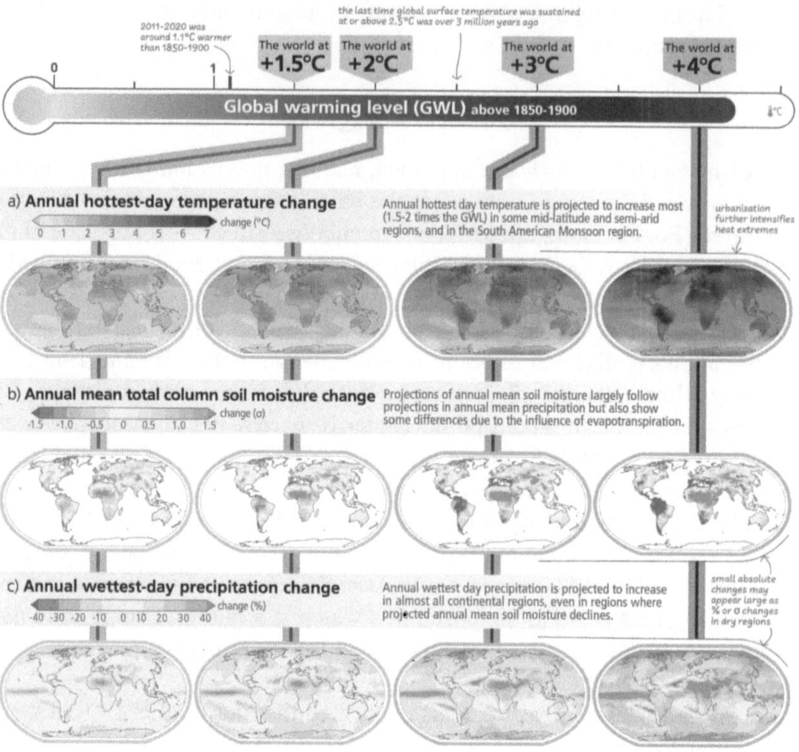

Figure 1.1 The Greenhouse gases effects mechanism.

balance the absorption of 240 W/m² from the Sun, the Earth's surface and atmosphere must emit the same amount of energy into space as infrared radiation. On average the surface emits significantly more than 240 W/m², but the net effect of absorption and emissions of infrared radiation by atmospheric gases and clouds is to reduce the amount reaching space until it approximately balances the incoming energy from the Sun (Clarke et al., 2008). The surface is thus kept warmer than it otherwise would be because, in addition to the energy it receives from the Sun, it also receives infrared energy emitted by the atmosphere. The warming that results from this infrared energy is known as the greenhouse effect (Goodchild et al., 2017).

Measurements from the surface, research aircraft and satellites, together with laboratory observations and calculations, show that, in addition to clouds, the two gases making the largest contribution to the greenhouse effect are water vapour followed by carbon dioxide. There are smaller contributions from many other gases including ozone, methane, nitrous oxide, and human-made gases such as CFCs (chlorofluorocarbons). Figure 1.1 provides a simplified, schematic representation of the flows of energy between space, the atmosphere, and the Earth's surface, and shows how these flows combine to trap heat near the surface and create the greenhouse effect.

The greenhouse effect is a process by which thermal radiation from a planetary surface is absorbed by atmospheric greenhouse gases and is re-radiated in all directions. Since part of this re-radiation is back towards the surface, energy is transferred to the surface and the lower atmosphere. As a result, the temperature there is higher than it would be if direct heating by solar radiation were the only warming mechanism (Klenert et al., 2020). Thus, the natural greenhouse effect makes the Earth habitable in the first place. Without the existence of the natural greenhouse effect, the average temperature on Earth would be about −19°C, that is, about 34°C colder than it is today (Klenert et al., 2020).

Greenhouse Gases

The Irish physicist John Tyndall was the first to realise, back in 1859, that without greenhouse gases in the atmosphere, it would quickly turn unpleasantly cold (Lacey-Barnacle, 2020). Tyndall correctly identified that the two most common gases in the air at his time, nitrogen and oxygen, are transparent to heat radiation and these gases are present in much smaller quantities. The most effective gases are water vapour, carbon dioxide, methane, and others, which have the heat-trapping effect (Lacey-Barnacle, 2020). Not all greenhouse gases are born equal. Carbon dioxide gets the most attention because it is the most abundant and lasts a long time in the atmosphere. Water vapour is also a greenhouse gas, but its emissions are irrelevant because it condenses out as rain in a matter of hours or days. The exception to this is water vapour emitted high in the atmosphere by aircraft, where it can have a longer-term warming effect. Table 1.1 shows the details of different gases and their ability to trap heat – their 'global warming potentials' – as well as their atmospheric lifetimes.

From the industrial revolution, humanity began exploiting fossil fuels, such as coal, then oil, and gas, for energy (Nieto et al., 2021). These fossil fuels burn to generate energy and have emitted greenhouse gases into the atmosphere over the decades and centuries. This study explains the general information of the GHGs. In 1750, there were 278 parts per million (ppm) of carbon dioxide in the Earth's atmosphere (Pettifor et al., 2015). Now that figure has risen to 380 ppm, and it is continuing to rise at 1.5–2 ppm per year. Methane levels have risen from 700 to 1745 parts per billion (ppb), while nitrous oxide has jumped from 270 to 314 ppb (Pettifor et al., 2015).

The other gases, such as CFCS, HFCS, and SF_6, are entirely products of the industrial age, so did not exist in 1750 (Pielke & Ritchie, 2020). Sources of greenhouse gases, *Carbon dioxide:* mainly produced by the combustion of coal, oil and gas from fossil fuels. Also results from deforestation and land clearance and from peat burning in the tropics. *Methane:* produced in places where oxygen-free decomposition occurs, such as in rice paddies and landfill sites. Can also be released by leaky pipelines and burping cows. *Nitrous oxides:* emitted from burning fossil fuels – in car exhausts, for example – and from agricultural fertilisers breaking down in the soil. *HFCs and other industrial gases:* as it says, these are produced by industrial processes. HFCs were introduced as replacements for CFCs, which are now banned to protect the ozone layer. They are used in refrigerants, in aerosols, and foam-blowing. SF_6 is used in high-voltage electrical switches but also in that squishy bit of some training shoes and inside tennis balls (Robinson et al., 2017).

Table 1.1 The Greenhouse gases and global warming potentials

Gas	*Chemical formula*	*Global warming potential (over a century)*	*Atmospheric lifetime (years)*
Carbon dioxide	CO_2	1	100–1,000
Methane	CH_4	23	12
Nitrous oxide	N_2O	296	114
Chlorofluorocarbons	CFCs	6,000–14,000	45–1,700
Hydrofluorocarbons	HFCs	12–12,000	0.3–260
Sulphur hexafluoride	SF_6	22,000	3,200

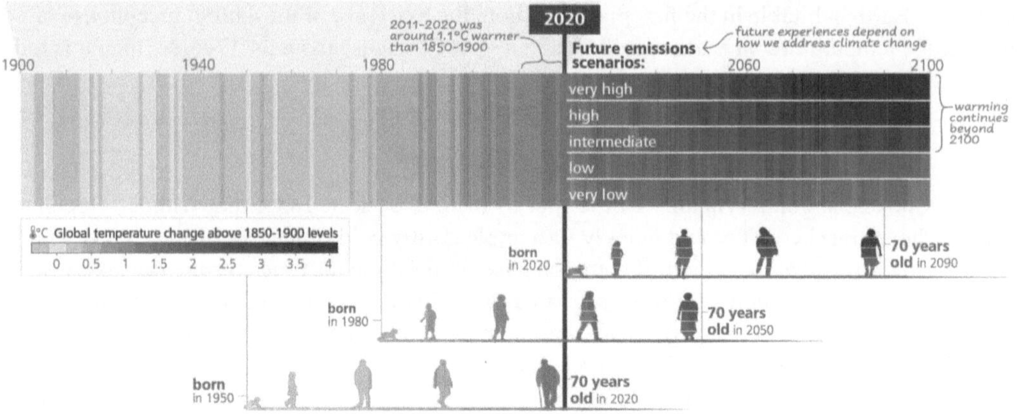

Figure 1.2 Changes in CO_2 and temperature in last 400,000 years.

Carbon dioxide levels have not been this high on the earth for millions of years. The temperature of the planet rose and fell with the cycles of the ice ages, but during the whole-time carbon dioxide levels were lower than they are now (Taneja & Mandys, 2022). Figure 1.2 shows that temperature and CO_2 shadow each other very closely over the long term, suggesting that the increase in CO_2 will indeed be followed by a temperature increase (the red line represents the level of the CO_2, which is currently around 380 ppm, and the blue line represents the changed temperature).

Man-made disaster shows that recent increase in temperature cannot be explained by natural causes. The report states that the main reason for recent global climate change is the emission of man-made greenhouse gases (Tingey & Webb, 2020). The black line represents the observed change in global average surface temperature, while the pink line represents the simulation result that accounts for both natural and anthropogenic forcings in changing the temperature. The blue line represents the simulation result that only considers natural forcing (Ekins & Lees, 2008).

Figure 1.3 shows that most climate change consisted of GHG emissions, which resulted from human activities. Emissions of CO_2 and the rest of the GHGs into the atmosphere from human activities have increased since the industrial revolution, particularly since about 1950. This study also shows changes in emissions from solid fuels (coal), liquid fuels (oil), and gas, as well as the total emissions from burning fossil fuels. It confirmed that CO_2 emissions from human activity, primarily from burning fossil fuel, are the main contributors to the concentration of CO_2 in the atmosphere.

The Evidence for Climate Change

> Warming of the global climate system is unequivocal, with global average temperatures having risen by nearly 0.8°C since the late 19th century, and rising at about 0.2°C decade over the past 25 years
>
> (Kelly, 2005)

The most often used indicator of climate change is global temperature, which is commonly known by terms such as global-mean temperature, annual-average temperature, and near-surface air temperature. The trends in global temperature increase over the past 25 years (1981–2005) have been

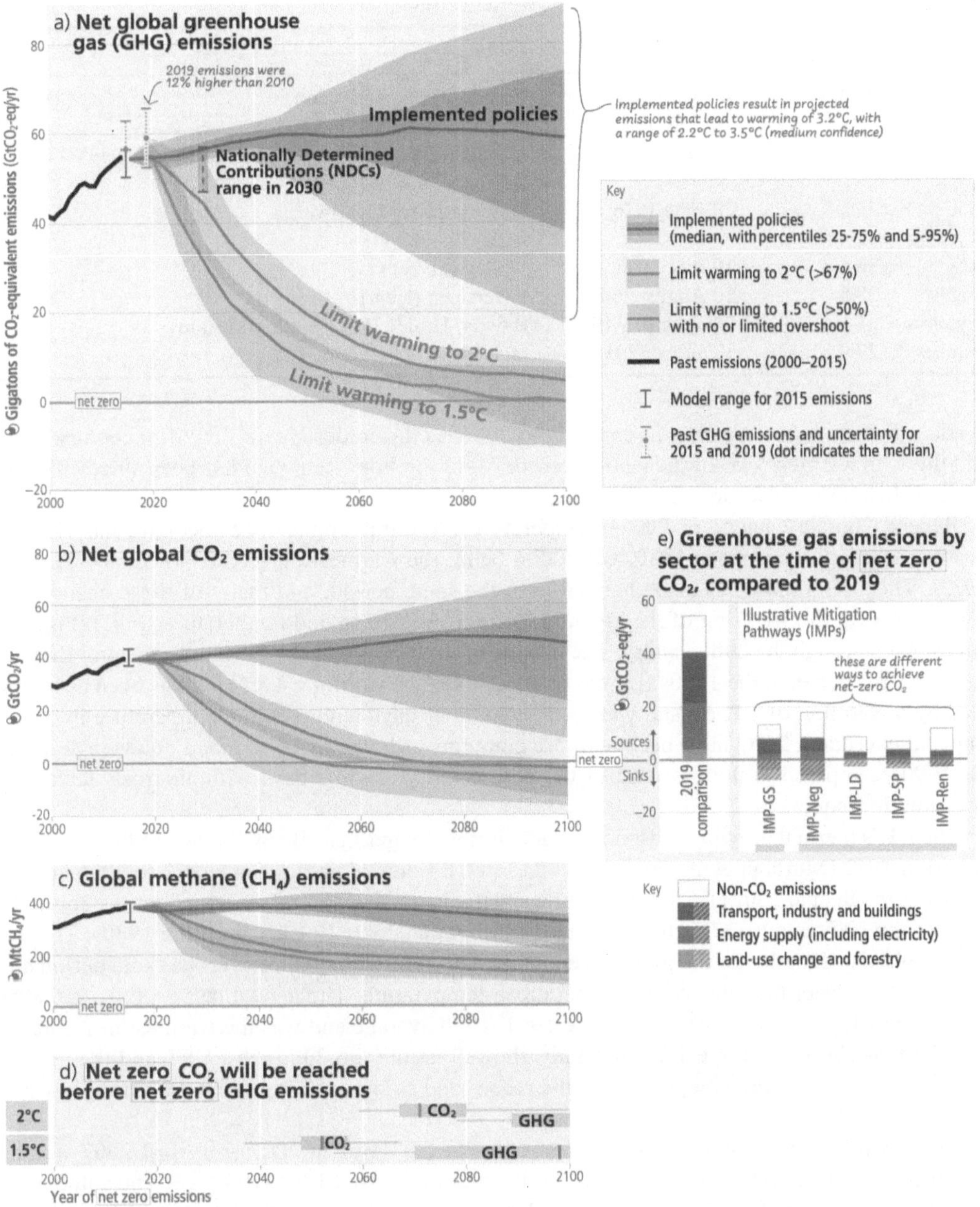

Figure 1.3 Emission of CO_2 from fossil fuel burning – rapid rise since 1950.

much greater than that over the whole period and have changed at about 0.2°C decade. The period 2001–2010 (0.44°C above 1961–1990 mean) was 0.20°C warmer than the 1991–2000 decade (0.24°C above 1961–1990 mean). The warmest year of the entire series has been 1998, with a temperature of 0.55°C above the 1961–1990 mean. After 1998, the next nine warmest years in the series are all in the decade 2001–2010 (Rosenow & Eyre, 2016). During this decade, only 2008

Table 1.2 The temperature recordings between 1949 and 2011

Month	*Temperature*	*Data*	*Location*
April	29.4	16.April 1949	Camden Square (London)
		22.May 1922	Camden Square (London)
May	32.8	29.May 1944	Horsham (West Sussex) Tunbridge Wells (Kent) Regent's Park (London)
June	35.6	28.June 1976	Southampton Mayflower Park
July	36.5	19.July 2006	Wisley (Surrey)
August	38.5	10.August 2003	Faversham (Kent)
September	35.6	2.September 1906	Bawtry, Hesley Hall (South Yorkshire)
October	29.9	1.October 2011	Gravesend (Kent)

is not in the ten warmest years. Even though 2008 was the coldest year of the 21st century it was still the 13th warmest year of the whole record. All of the 'top – ten' year has been since 1990 and most of them in last decade.

Recent Climate Change in UK – Temperature, "Central England Temperature has risen by about a degree Celsius since 1980, with 2006 being the warmest on record" *(Rosenow & Eyre, 2022).* The UK climate has also changed over the same period, and many of these changes are consistent with the warming of the global climate. Central England temperatures (CETs) rose by almost 1°C through the 20th century. The decade of the 1990s was the warmest in central England since records began in the 1660s (Lomas et al., 2024). The warming over land has been accompanied by a warming of UK coastal waters. Hot summer days with daytime temperature in central England exceeding 25°C have become more common – almost twice as many on average during the 1990s compared to the first half of the 20th century – while days with air frosts have been declining in frequency.

The UK's thermal growing season for plants is now longer than at any time since the start of the record in 1972 (Murtagh et al., 2021). Changes in CET annual values from 1877 to 2006 are also noted (Murtagh et al., 2021). This study shows that after a period of relative long-term stability for most of the 20th century, CET has increased by about a degree Celsius since the 1980s. This is a more rapid rise than that of the global average land-surface temperature over the same period and, considerably, faster than that of the global mean temperature. The annual mean CET of 10.82°C in 2006 was 1.35°C, or 0.18°C, above the 1961–1990 average and was the warmest in the period of record, as shown in Table 1.2. The study shows that the top 20 highest CET and the average temperature for 12 successive months in the record and coloured to represent the period in which the year lies.

As shown in Table 1.2, many of the top ranks of CET was in 2007 and April 2007 was the warmest of all. The average value of CET for the baseline period 1961–1990 is 9. The value of the top 20 highest ranks of the 12-month CET (Hamilton et al., 2013). In the UKCIP02 report, trends in several temperature quantities are presented for 14 regions of the UK. In terms of record warm individual days, the 10th of August 2003 was the hottest ever maximum temperature in the UK which was 38.5°C in Faversham, Kent and it exceeded the previous record in 1990 by 1.4°C. The study shows the highest daily temperature records in the UK from April to October since records began (Hanlon et al., 2021).

Table 1.2 shows the highest daily maximum temperature records in the UK (April to October) and highlights extreme weather in recent years. Few parts of the UK have been unaffected by extreme weather. For example, in 2003, around 2,000 excess deaths were recorded in England and Wales during the August heatwave (Khosravi et al., 2023). In London, the urban heat island effect meant that night-time temperatures were as much as 9°C higher than in the surrounding countryside. In 2007, summer floods affected England, Wales, and Northern Ireland, costing the economy more than *£3* billion in England alone (Tierney et al., 2020).

In 2009, the Cumbria floods resulted in £100s of millions of damages, including the loss of 20 road bridges and long-term disruption for local communities, in 2009 and 2010, prolonged cold caused wide-ranging problems for UK transport and water infrastructure (Fawcett & Hampton, 2020). For example, in December 2009, thousands of households in Northern Ireland experienced difficulties with water supplies; in December 2010, heavy snow across Scotland's Central Belt resulted in hundreds of motorists being stranded overnight. In 2011, parts of the UK experienced a much drier than average spring, resulting in insufficient rainfall for some crops. Over Easter, the higher temperatures, static weather pattern, and increased sunshine contributed to a pollution warning across England and Wales unusually early in the year. In Berkshire, forest fires led to the closure of businesses and schools and the evacuation of homes. Natural climatic variability means that, with or without climate change, events like these will continue to occur in future. However, events such as heatwaves, flooding, and drought are all projected to become more severe due to a changing climate, with vulnerable groups tending to experience disproportionate negative effects.

Future Global Changes

Continued GHG emissions at or above current levels would cause further warming and induce many changes in the global climate system during the 21st century that would be larger than those observed during the 20th century. The IPCC Special Report on Emissions Scenarios explained that for the next two decades, a warming of about 0.2°C per decade is projected. Even if the concentration of all GHGs and aerosols had been kept constant at year 2000 levels, a further warming of about 0.1°C per decade would be expected (Boomsma et al., 2016). Table 1.3 shows how the temperature and sea level projections increasingly depend on six different emissions scenarios. It also summarises the required emission levels for different groups of stabilisation concentrations and the resulting equilibrium global average temperature increases. From this emissions scenario, average temperature increases above the pre-industrial time would be from 2°C to more than 6°C and the sea level rise would be from 0.4 m to almost 4 m.

Table 1.3 CO_2 emissions and equilibrium temperature and sea level increase

Category	*CO_2 concentration (2005 = 379 ppm)**	*Change in global CO_2 emissions in 2050 (per cent of 2000 emissions)*	*Global average temperature increases above pre-industrial*	*Global average sea rise above pre-industrial*
	(ppm)	*(per cent)*	*(°C)*	*(meters)*
I	445–490	−85 to −50	2.0–2.4	0.4–1.4
II	490–535	−60 to −30	2.4–2.8	0.5–1.7
III	535–590	−30 to +5	2.8–3.2	0.6–1.9
IV	590–710	+10 to +60	3.2–4.0	0.6–2.4
V	710–855	+25 to +85	4.0–4.9	0.8–2.9
VI	855–1,130	+90 to +140	4.9–6.1	1.0–3.7

* CO_2 emissions including GHGS and aerosols.

Emissions Scenarios from UKCIP 02

The basis for the UKCIP 02 climate change scenarios is a set of four greenhouse gas emissions, which are taken from the IPCC SRES emissions scenarios. Each scenario represents a possible future, in terms of increasing temperature and carbon dioxide concentration levels in the atmosphere. The following subsections' findings were summarised from the UKCIP 02 report. Average annual temperatures across the UK may rise by between 2°C and 3.5°C by the 2080s, depending on the greenhouse gas emissions scenario. In general, there will be greater warming in the southeast than in the northwest of the UK, and there may be more warming in summer and autumn than in winter and spring. Under a High Emissions scenario, the southeast may be up to 5°C warmer in summer by the 2080s. The temperature of UK coastal waters will increase, although not as rapidly as over land, with again the greatest warming in the south. Offshore waters in the English Channel may warm in summer by between 2°C and 4°C by the 2080s. Annual average precipitation across the UK may decrease slightly, by between 0% and 15% by the 2080s, depending on the emissions scenario, although there are likely to be large regional and seasonal differences. Snowfall amounts will decrease significantly through the UK, perhaps by between 30% and 90% by the 2080s (Snell et al., 2015).

Seasonal Change

By the 2050s, typical spring temperatures may occur between one and three weeks earlier than at present and the onset of present winter temperatures may be delayed by between one and three weeks. This is likely to lead to a lengthening of the thermal growing season for plants. The amount of heating and cooling required in buildings will also change. The seasonal distribution of precipitation will change, with winters becoming wetter and summers perhaps drier across the UK and with the biggest relative changes in the south and east. Precipitation in the High Emissions scenario may decrease in summer by 50% by the 2080s in the southeast and increase in winter by up to 30%. Summer soil moisture by the 2050s may be reduced by about 30% over large parts of England for High Emissions scenario, and by 40% or more by the 2080s.

Weather File Generator and Simulation Weather File

Weather files which are generated from long-term measured data using statistical functions such as Test Meteorological Year 2 (TMY2) files are primarily suited for predicting a building's average energy performances for heating, cooling, and ventilation in order to dimension the required building services accordingly (Taylor et al., 2014). However, since these files represent 'typical' years and deliberately exclude peak year conditions, they are ill-suited for overheating assessment of naturally ventilated buildings. Therefore, in the UK, the Chartered Institution of Building Services Engineers (CIBSE) has taken an approach of producing two different simulation weather file sets, Test Reference Year (TRY) files for Heating Ventilation and Air Conditioning (HVAC) planning and Design Summer Year (DSY) files for overheating analysis (Yuan & Choudhary, 2023; Ozarisoy & Altan, 2024). Since 2006 files have been available for 14 sites throughout the UK utilising improved selection algorithms as discussed in detail by Zhang et al. in 2012. The current CIBSE, TRY, and DSY files are derived from measured UK Meteorological Office site data from 1983 to 2004. However, the solar radiation data has been modelled, as all 14 sites had little or no solar radiation data available. TRY files have been compiled in a similar way to TMY2 files using Finkelstein-Schafer statistics, whilst Design Summer Year (DSY0 files) represents a hot summer, which is defined as the third hottest summer of the 20-year data set in terms of daily mean dry bulb temperature from April to September.

Climate Change Impact in the Built Environment

> Warming of the global climate system is unequivocal, with global average temperatures having risen by nearly 0.8°C since the late 19th century, and rising at about 0.2°C decade over the past 25 years
>
> (Foulds & Powell, 2014)

The most often used indicator of climate change is global temperature which is commonly known by terms such as global-mean temperature, annual-average temperature, and near-surface air temperature.

This graph shows the change in global average temperature from 1850 to 2011. The bars represent the individual annual averages of temperature change in °C and the black curve shown a smoothed trend. The study indicates that global temperature has increased by about 0.8°C in mid-19th century and the period from 1910 to 2011. The trends in global temperature increase over the past 25 years (1981–2005) have been much greater than that over the whole period and have changed at about 0.2°C decade. The period 2001–2010 (0.44°C above 1961–1990 mean) was 0.20°C warmer than the 1991–2000 decade (0.24°C above 1961–1990 mean). The warmest year of the entire series has been 1998, with a temperature of 0.55°C above the 1961–1990 mean. After 1998, the next nine warmest years in the series are all in the decade 2001–2010. During this decade, only 2008 is not in the ten warmest years. Even though 2008 was the coldest year of the 21st century it was still the 13th warmest year of the whole record. All of the 'top-ten' years have occurred since 1990, with most of them in the last decade (Johnson et al., 2023).

Recent Climate Change in the UK

CET has risen by about a degree Celsius since 1980, with 2006 being the warmest on record. The UK climate has also changed over the same period, and many of these changes are consistent with the warming of global climate. CETs rose by almost 1°C through the 20th century (Keirstead, 2013). The decade of the 1990s was the warmest in central England since records began in the 1660s. The warming over land has been accompanied by a warming of UK coastal waters. Hot summer days with daytime temperature in central England exceeding 25°C have become more common – almost

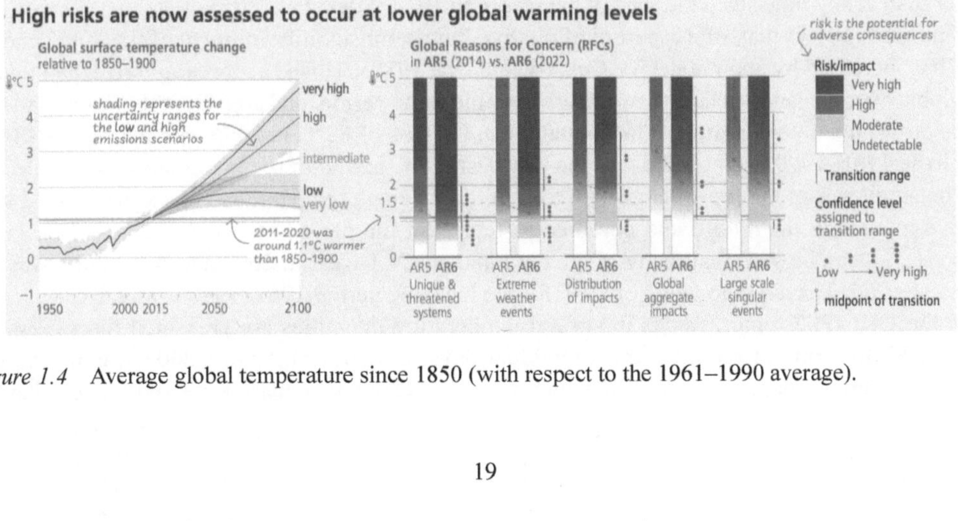

Figure 1.4 Average global temperature since 1850 (with respect to the 1961–1990 average).

Table 1.4 Highest daily maximum temperature records in the UK (April to October)

Month	*Temperature*	*Data*	*Location*
April	29.4	16.April 1949	Camden Square (London)
		22.May 1922	Camden Square (London)
May	32.8	29.May 1944	Horsham (West Sussex) Tunbridge Wells (Kent) Regent's Park (London)
June	35.6	28.June 1976	Southampton Mayflower Park
July	36.5	19.July 2006	Wisley (Surrey)
August	38.5	10.August 2003	Faversham (Kent)
September	35.6	2.September 1906	Bawtry, Hesley Hall (South Yorkshire)
October	29.9	1.October 2011	Gravesend (Kent)

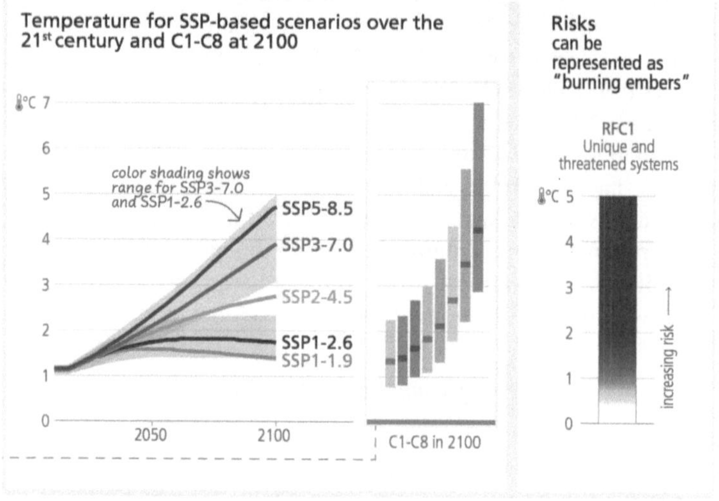

Figure 1.5 The value of the top 20 highest ranks of the 12-month CET (UKCIP 08e).

twice as many on average during the 1990s compared to the first half of the 20th century – while days with air frosts have been declining in frequency. The UK's thermal growing season for plants is now longer than at any time since the start of the record in 1972 (Sdei et al., 2015).

This study shows that after a period of relative long-term stability for most of the 20th century, CET has increased by about a degree Celsius since the 1980s. This is a more rapid rise than that of the global average land-surface temperature over the same period, and considerably faster than that of the global mean temperature. The annual mean CET of 10.82°C in 2006 was at 1.35°C–0.18°C above the 1961–1990 average and was the warmest in the period of record, as shown in Table 1.4.

The study shows the top 20 highest CET and the average temperature (black bar) for 12 successive months in the record and coloured to represent the period in which the year lies. From Figure 1.5, it can be seen that many of the top ranks of CET were in 2007 and April 2007 was the warmest of all. The average value of CET for the baseline period 1961–1990 is at 9.5°C.

In the UKCIP02 report, trends in several temperature quantities are presented for 14 regions of the UK. In terms of record warm individual days, the 10th of August 2003 was the hottest ever maximum temperature in the UK which was 38.5°C in Faversham, Kent and it exceeded the previous record in 1990 by 1.4°C (see Table 1.5 showing that the highest daily temperature records in the UK from April to October since records began.)

Extreme Weather in Recent Years

Few parts of the UK have been unaffected by extreme weather in recent years. For example, in 2003, around 2000 excess deaths were recorded in England and Wales during the August heatwave (CCRA, 2012). In London, the urban heat island effect meant that night-time temperatures were as much as 9°C higher than in the surrounding countryside. In 2007, summer floods affected England, Wales, and Northern Ireland, costing the economy more than £3 billion in England alone (Bonderup & Middlemiss, 2023). In 2009, the Cumbria floods resulted in £100s of millions of damages, including the loss of 20 road bridges and long-term disruption for local communities, in 2009 and 2010, prolonged cold caused wide-ranging problems for UK transport and water infrastructure. For example, in December 2009, thousands of households in Northern Ireland experienced difficulties with water supplies; in December 2010, heavy snow across Scotland's Central Belt resulted in hundreds of motorists being stranded overnight. In 2011, parts of the UK experienced a much drier than average spring, resulting in insufficient rainfall for some crops (Tovar, 2012). Over Easter, the higher temperatures, static weather pattern and increased sunshine contributed to a pollution warning across England and Wales unusually early in the year. In Berkshire, forest fires led to the closure of businesses and schools and the evacuation of homes. Natural climatic variability means that, with or without climate change, events like these will continue to occur in future. However, events such as heatwaves, flooding, and drought are all projected to become more severe due to a changing climate, with vulnerable groups tending to experience disproportionate negative effects.

Global Warming

Continued GHG emissions at or above current levels would cause further warming and induce many changes in the global climate system during the 21st century that would be larger than those observed during the 20th century. The IPCC Special Report on Emissions Scenarios explained that for the next two decades, a warming of about 0.2°C per decade is projected. Even if the concentration of all GHGs and aerosols had been kept constant at year 2000 levels, a further warming of about 0.1°C per decade would be expected (Kuzemko & Britton, 2019). Table 1.5 shows how the

Table 1.5 CO_2 emissions and equilibrium temperature and sea level increase

Category	*CO_2 concentration (2005 = 379 ppm)**	*Change in global CO_2 emissions in 2050 (per cent of 2000 emissions)*	*Global average temperature increases above pre-industrial*	*Global average sea rise above pre-industrial*
	(ppm)	*(Per cent)*	*(°C)*	*(meters)*
I	445–490	−85 to −50	2.0–2.4	0.4–1.4
II	490–535	−60 to −30	2.4–2.8	0.5–1.7
III	535–590	−30 to +5	2.8–3.2	0.6–1.9
IV	590–710	+10 to +60	3.2–4.0	0.6–2.4
V	710–855	+25 to +85	4.0–4.9	0.8–2.9
VI	855–1,130	+90 to +140	4.9–6.1	1.0–3.7

* CO_2 emissions including GHGS and aerosols.

temperature and sea level projections increasingly depend on six different emissions scenarios. It also summarises the required emission levels for different groups of stabilisation concentrations and the resulting equilibrium global average temperature increases. From this emissions scenario, average temperature increases above the pre-industrial time would be from 2°C to more than 6°C and the sea level rise would be from 0.4 m to almost 4 m.

Climate Change Projections

The basis for the UKCIP 02 climate change scenarios is a set of four greenhouse gas emissions scenarios, which are taken from the IPCC SRES emissions scenarios. Average annual temperatures across the UK may rise by between 2°C and 3.5°C by the 2080s, depending on the greenhouse gas emissions scenario. In general, there will be greater warming in the southeast than in the northwest of the UK, and there may be more warming in summer and autumn than in winter and spring. Under a High Emissions scenario, the southeast may be up to 5°C warmer in summer by the 2080s. The temperature of UK coastal waters will increase, although not as rapidly as over land, with again the greatest warming in the south. Offshore waters in the English Channel may warm in summer by between 2°C and 4°C by the 2080s. Annual average precipitation across the UK may decrease slightly, by between 0% and 15% by the 2080s, depending on the emissions scenario, although there are likely to be large regional and seasonal differences. Snowfall amounts will decrease significantly throughout the UK, perhaps by between 30% and 90% by the 2080s (Gooding & Gul, 2017).

Weather Variations

By the 2050s, typical spring temperatures may occur between one and three weeks earlier than at present and the onset of present winter temperatures may be delayed by between one and three weeks. This is likely to lead to a lengthening of the thermal growing season for plants. The amount of heating and cooling required in buildings will also change. The seasonal distribution of precipitation will change, with winters becoming wetter and summers perhaps drier across the UK and with the biggest relative changes in the south and east. Precipitation in the High Emissions scenario may decrease in summer by 50% by the 2080s in the southeast and increase in winter by up to 30%. Summer soil moisture by the 2050s may be reduced by about 30% over large parts of England for High Emissions scenario, and by 40% or more by the 2080s (Morton et al., 2020).

Weather Extremes

High summer temperatures will become more frequent and very cold winters will become increasingly rare. A very hot August, such as experienced in 1998 and 2003 with an average temperature 3.4°C above normal, may occur as often as one year in every five by the 2050s, and three years in every five by the 2080s. Even for the Low Emissions scenario, about two summers in three may be as hot as, or hotter than, the summer of 1995 by the 2080s. Extreme winter precipitation will become more frequent. By the 2080s, winter daily precipitation intensities that are experienced once every two years on average may become up to 20% heavier. Very dry summers, like 1995, may occur in half the years by the 2080s, while very wet winters like 1994/1995 may occur on average almost once a decade for the Medium-High Emissions scenario (Reid et al., 2015).

A combination of high temperatures and dry conditions in summer will also become more common. By the 2080s, virtually every summer over England and Wales – weather for the Low Emissions or High Emissions scenario may be warmer and drier than the typical summer. Weather File Generator and Simulation Weather File Weather files which are generated from long-term measured data using statistical functions, such as TMY2 files, are primarily suited for predicting a buildings' average energy performances for heating, cooling, and ventilation in order to dimension the required building services accordingly (Webborn et al., 2021). However, since these files represent 'typical' years and deliberately exclude peak year conditions, they are ill-suited for overheating assessment of naturally ventilated buildings. Therefore, in the UK, the CIBSE has taken an approach of producing two different simulation weather file sets, TRY files for HVAC planning and DSY files for overheating analysis (CIBSE, 2002). Since 2006 files have been available for 14 sites throughout the UK utilising improved selection algorithms as discussed in detail by Boomsma et al. in 2018.

The current CIBSE, TRY, and DSY files are derived from measured UK Met Office site data from 1983 to 2004. However, the solar radiation data has been modelled, as all 14 sites had little or no solar radiation data available. TRY files have been compiled in a similar way to TMY2 files using Finkelstein-Schafer statistics, whilst DSY files represent a hot summer, which is defined as the third hottest summer of the 20-year data set in terms of daily mean dry bulb temperature from April to September inclusive. CIBSE weather data is considered as the industry standard in the UK and is linked to building regulation compliance testing since TRY data has been used to determine monthly average values for the SBEM tool. Alternative compliance calculations such as those made by IES VE, for example, require full CIBSE weather files for compliance assessment (Gillich et al., 2019). TRY and DSY weather files are available from CIBSE in various file formats to facilitate integration into different simulation packages.

Climate Change Weather File Generator Tool

A climate change weather file generator tool that allows individual end user generation of such files has been developed using Microsoft Excel. This tool, which is shown as a screenshot in Figure 1.6, is listed as 'UKCIP02 extra'. 'Morphed' climate change weather data can be generated for any of the 14 CIBSE UK weather sites and saved in TMY2/EPW format, provided that the user possesses a licensed copy of a CIBSE TRY or DSY file and has obtained the appropriate license from UKCIP in order to download the required UKCIP02 climate change data.

The user can generate climate change weather files for all four UKCIP02 emissions scenarios for the 2020s, 2050s, and 2080s timeframes. Further details on obtaining the required data and a description of the tool's function are given in the help file provided on the tool's download page. It needs to mention here that UKCIP02 weather file generator could not cover the extreme weather events such as heat waves and heat island effect (Katris & Turner, 2021). Current climate models cannot model wind speed with sufficient accuracy to be able to give meaningful predictions of future wind speeds. Therefore, the wind speeds generated from the weather file generator essentially derived from the current wind speeds. Hence, the wind data for future climates is not reliable enough to simulate natural ventilation strategy. This is why this research uses a fixed air change rate for ventilation in order to simulate thermal comfort study under future climate scenarios, rather than using calculation methods. Hence, new methodologies were introduced to generate future climate weather data for simulating overheating and energy consumption, adapting to the Urban Heat Island effect in London (Ozarisoy, 2022).

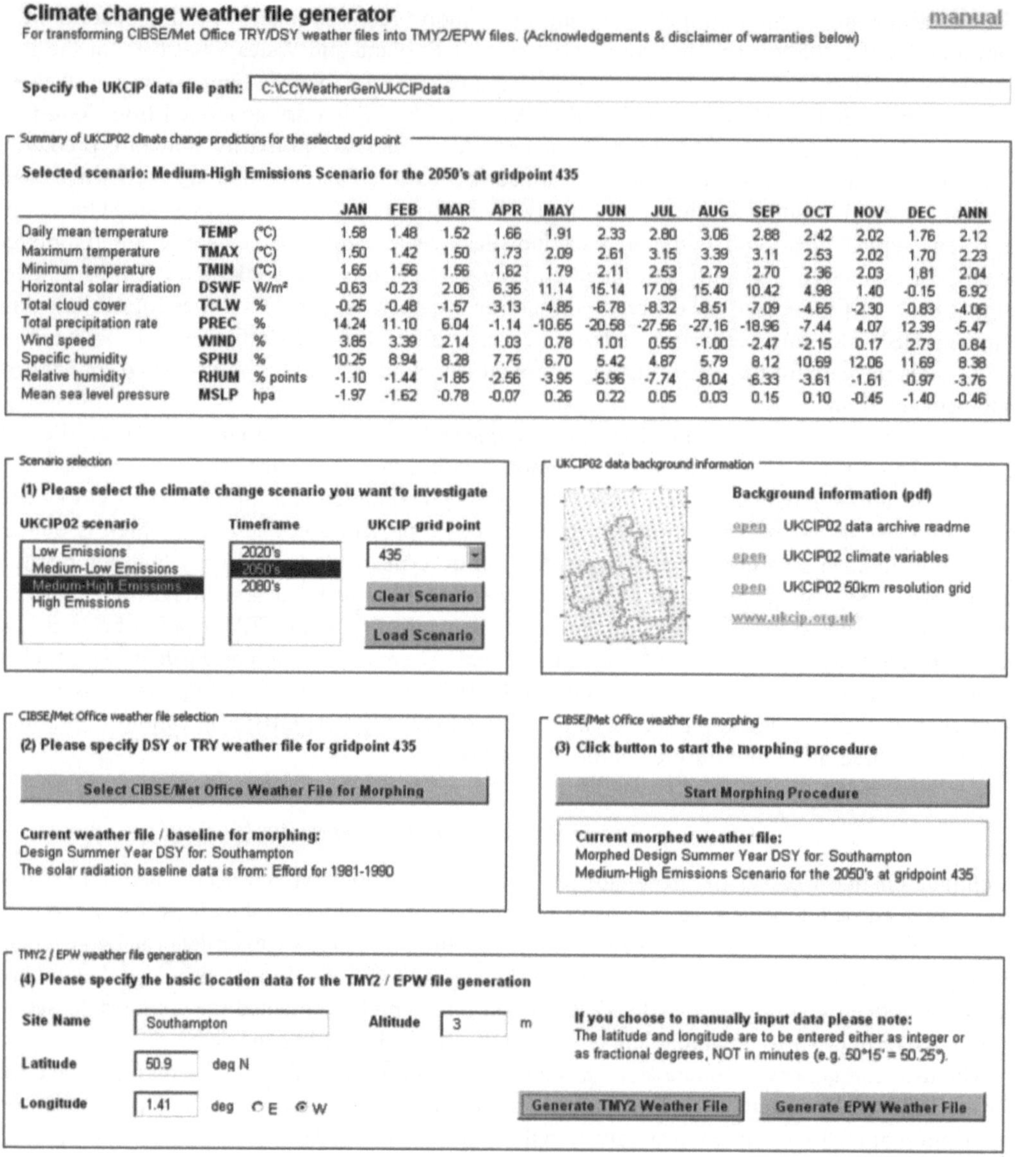

Figure 1.6 Screenshot of the climate change weather file generator tool.

Assessing the Difference – UKCIP02 and UKCP09

UKCP09 is the fifth generation of climate information for the United Kingdom and is the most comprehensive package produced to date. It has been produced in consultation with the user community and reflects advances in our understanding and modelling of the climate system, in computing power, and in the ability to analyse the modelled outputs. For the first time, it provides probabilistic projections of climate change based on quantification of the known sources of uncertainty. The climate

Table 1.6 The difference between UKCIP02 and UKCP09

Feature	*UKCIP02*	*UKCP09*
	2020s, 2050s, 2080s	Seven decadal overlapping 30-year time slices from 2010 to 2099
Spatial resolution	50 km grid	25 km grid
Temporal resolution	Daily, monthly seasonal, annual	Daily, monthly, seasonal, annual Weather generator for daily output
Aggregation	UK level	UK English administrative regions, devolved administrations, river-based regions
Modelling approach and quantification of uncertainty	Single UK model (HadCM3) for emissions scenarios	Ensemble approach: model parameter uncertainty explored in HadSM3, using perturbed physics experiments (PPE). PPE used to construct a statistical emulator from which a greater range of variability in the climate model parameters can be considered. Additional IPCC climate models included (MME) (model uncertainty). 3 emissions scenarios: low (B1), medium (A1B), and high (A1FI). Statistical framework used to produce probabilistic projections; uses will have a probability associated with each projection.
Outputs	Single mean value of projections of 15 climate variables	Probabilistic projections of same climate variables as UKCIP02 Observed trends in recent climate Marine projections
Delivery	UKCIP02 report Maps, underlying model output	3 science reports: land, marine, historical trends Web-based User Interface to provide customisable output Climate model output and associated probability Dedicated website

projections in UKCP09 should be seen as superseding the scenarios from UKCIP02. Although many of the projected changes are of a broadly similar nature, UKCP09 incorporates scientific advances that may have significant implications for the specifics of the projected climate change. It is strongly recommended that users no longer employ UKCIP02 in new projects. It is difficult to perform a comparison with UKCIP02 due to two reasons. First, the UKCP09 projections are probabilistic in nature, whereas UKCIP02 gave a (single) projection. The only way to get around this problem is to compare the 50th percentile of the UKCP09 projection with the UKCIP02 projection. But this is not strictly a correct comparison. Second, the UKCP09 and UKCIP02 projections utilise different emission scenarios. UKCIP02 had four emission scenarios, whereas UKCP09 had three emission scenarios. This difference results in it only being possible to compare the projections for each of the common high- and low-emission scenarios. The differences between the modelling techniques employed in UKCIP02 and UKCP09 are demonstrated in Table 1.6.

Possible Influence of Climate Change

While there is not yet a global concern on what levels of climate change might be defined to be 'dangerous', considerable support has developed for containing the rise in global temperature to a maximum of 2°C above pre-industrial levels. This is often referred to as 'the 2°C guardrail'.

Table 1.7 Degree change and CO_2 target

Degree change	*Action needed*	*CO_2 target*
One degree	Probably no action needed	350 ppm (today's level is 380 ppm)
Two degrees	Cut global emissions by 2015	400 ppm
Three degrees	Cut global emissions by 2030	450 ppm
Four degrees	Cut global emissions by 2050	550 ppm
Five degrees	Allow constantly rising emissions	650 ppm
Six degrees	Allow very high emissions	800 ppm

The IPCC, as well as more recent scientific research, indicates that even with temperature rises of less than 2°C, impacts can be significant, although some societies could cope with some of these impacts through pro-active adaptation strategies. Table 1.7 shows the required CO_2 emissions limit to meet global temperature degree change, in terms of peak global emissions year and target CO_2 level in the atmosphere.

The ultimate objective of the UNFCCC (which created the Kyoto Protocol) is to stabilise greenhouse gas concentrations to 'prevent dangerous anthropogenic (man-made) interference within the climate system' while allowing 'economic development to proceed in a sustainable manner'. In 2001, the IPCC produced a diagram to illustrate how the risks of adverse impacts from climate change increase with the magnitude of climate change (Katris & Turner, 2021).

Staying below two degrees with any level of certainty would require global greenhouse gas emissions to peak by 2015, especially given the reluctance of many people and nations to face up to the scale and urgency of the problem (Rosenow et al., 2018). However, even if we overshoot this target, any cuts we make will still help to slow the rate of change. At the very least, this gives human societies and natural species more time to adapt to rising sea levels, harsher weather and shifting climate zones. So, it's not too late to act, but procrastination is definitely over.

Heatwave and Mortality

In early August 2003, Europe experienced one of the hottest summer spells on record. The heat wave caused an estimated 35,000 additional deaths across Europe. For instance, in France, excess mortality between 1 to 20 August has been estimated at 14,800 and in the UK, the Office of National Statistics reported 2,045 deaths above the average for the previous 5 years between 4th and 13th August 2003. Scott et al.'s study found that heat-related deaths begin to increase when the average daily temperature rises above 19°C, with a linear relationship between deaths and temperature above this temperature (Scott et al., 2013). They found that there was a 3.34% increase in deaths for every 1°C increase in average temperature above 21.5°C. Figure 1.7 shows the June–August temperature anomalies (relative to 1961–1990 mean temperature) over parts of Europe. The black line represents as observed temperature, and the blue area represents as modelled temperatures from HadCM2 simulations. The figure shows that an event such as the summer 2003 heatwave in Europe will be common in the 2050s.

Global Policy

International policy on climate change is now a major influence on policy across the UK. The UNFCCC, established at the United Nations Conference on Environment and Development (UNCED) in 1992, is an international environmental treaty signed by the UK and over 180 countries and

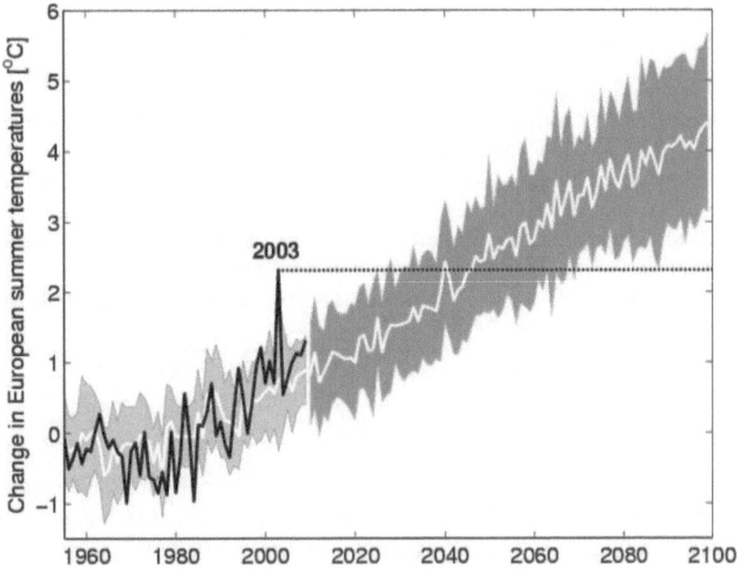

Figure 1.7 Comparison of the 2003 heat wave with future summer temperature.

provides the foundation for the subsequent development of international climate change policy. Its objective is: 'to achieve stabilisation of greenhouse gas concentrations in the atmosphere at a low enough level to prevent dangerous anthropogenic interference with the climate system'. The Kyoto Protocol, which came into force in 2005, was established under the UNFCCC. Industrialised countries (termed 'Annex 1' countries in the Protocol) that have ratified the Protocol, including the UK, are committed to reduce emissions of carbon dioxide and five other GHGs, or to engage in emissions trading (Rosenow et al., 2014).

The major distinction between the 'Protocol' and the 'Convention' is that the Convention encouraged industrialised countries to stabilise GHGs emissions, whereas the Protocol commits them to do so (Sovacool et al., 2023). Under the Treaty, countries must meet their targets primarily through national measures. However, the Protocol offers additional means of meeting targets through market-based mechanisms that operate across national boundaries. These are as follows: emissions trading (the European Union Emissions Trading Scheme is the largest in operation), the Clean Development Mechanism (CDM), and Joint Implementation (JI). These 'flexible mechanisms' are intended to allow countries with an emissions reduction of limitation commitment under the Kyoto Protocol (Annex B Party) to minimise their costs by buying or selling credits (trading scheme), or by implementing emissions-reduction projects in developing countries (CDM) or other Annex B countries (JI). Work is underway to develop a new regulatory framework for international action on climate change following the end of the first commitment period of the Kyoto Protocol in 2012. Negotiators aimed to secure a new deal in Copenhagen at the end of 2009. Complex accounting systems are required for the UNFCCC and the Kyoto Protocol to show whether parties to these agreements are meeting their commitments. The UNFCCC requires participating countries to submit national reports and annual inventories on implementation of the Convention. Annex I Parties that have ratified the Kyoto Protocol must include supplementary information on their emissions and removals of GHGs to demonstrate compliance with the Protocol's commitments, as shown in Table 1.8.

Table 1.8 Kyoto protocol

The Kyoto Protocol is an international agreement linked to the UNFCCC. The major feature of the Kyoto Protocol is that it sets binding targets for 37 industrialised countries and the European community for reducing greenhouse gas (GHG) emissions. These amount to an average of 5% against 1990 levels over the five-year period of 2008–2012. The major distinction between the Protocol and the Convention is that while the Convention encouraged industrialised countries to stabilise GHG emissions, the Protocol commits them to do so. Recognising that developed countries are principally responsible for the current high levels of GHG emissions in the atmosphere as a result of more than 150 years of industrial activity, the Protocol places a heavier burden on developed nations under the principle of "common but differentiated responsibilities". The Kyoto Protocol was adopted in Kyoto, Japan, on 11 December 1997, and entered into force on 16 February 2005.

Source: United Nations Framework Convention on Climate Change (UNFCCC).

UK Policy

The UK had a Climate Change Programme aimed to reduce carbon dioxide emissions by 20% on 1990 levels, by 2010. It is led by the Department of Energy and Climate Change (DECC) on behalf of the UK and devolved administrations, as shown in Table 1.9. The UK has taken a proactive approach in developing climate change policy by introducing a Climate Change Act in 2008. The first of its type in the world, it applies to England, Wales, and Northern Ireland and sets a target of at least 80% reduction by 2050, compared with 1990 levels (Wade et al., 2024). A Committee on Climate Change was established under the UK Act and is empowered to set successive five-year targets for the UK's GHGs account. In response to the Committee's advice (Committee on Climate Change, 2008), the Government has announced the budgets for 2008–2012, 2013–2017, and 2018–2022, leading to a 34% reduction in GHG emissions with respect to 1990 levels by 2020 (Committee on Climate Change, 2008).

A separate Climate Change Act is in force in Scotland. This sets a legally binding emissions reduction target of at least 80% by 2050, with an interim target in 2020 of a 34%, or possibly 42%, reduction from 1990 levels. The Act is supported by the Scottish Government Climate Change Delivery Plan, which describes how the Scottish Government would achieve the interim targets. Adoption of the 42% trajectory is dependent on the targets adopted by the EU and on advice from the Committee on Climate Change (Mcmichael &Shipworth, 2012). The Delivery Plan laid out woodland expansion targets associated with the GHGs reduction trajectories and also identified the growing role of biomass in renewable heat production (www.scotland.gov.uk). The UK Government has recently published a white paper, the UK Low Carbon Transition Plan which sets out how the Government will achieve the targets in the climate change budgets.

A UK Renewable Energy Strategy was published in 2009 and sets a target of renewable sources producing 15% of the UK's energy requirements by 2020. The Government has also created an Adapting to Climate Change (ACC) Programme to bring together existing adaptation work and co-ordinate future work. The ACC Programme has two phases. Phase 1, from 2008 to 2011, is laying the groundwork necessary to implement Phase 2, a statutory National Adaptation Programme, as required by the Climate Change Act. The objectives of Phase 1 of the Programme are to: develop a robust and comprehensive evidence base on the impacts and consequences of climate change in the UK; raise awareness of the need to take action new and help others to take action; measure success and take steps to ensure effective delivery; and work across Government at the national, regional and local level to embed adaptation into Government policies, programme, and systems (Goodchild et al., 2017).

Table 1.9 The Climate Change Act 2008

The Climate Change Act established a target to reduce the UK's Greenhouse Gase (GHG) emissions by at least 80% below base year levels by 2050, to be achieved through action on national and international levels. To achieve this target, the Act introduced a carbon budget system, which provide GHG emissions' limit in every five-year period, beginning in 2008. The first three carbon budgets were set into law in May 2009 and require emissions to be reduced by at least 34% below base year levels in 2020. The fourth carbon budget, for the period 2023–2027, was set in law in June 2011 and requires emissions to be reduced by 50% below base year, which is 1990 levels.

Carbon budget	1st Budgets	2nd Budgets	3rd Budgets	4th Budgets
level (Mt CO_2)	3,018	2,782	2,544	1,950
Percentage reduction below base year levels	23%	29%	35%	50%

The Programme is focused on England, although some elements will be UK-wide and will be developed in partnership with the other UK administrations. The Programme is essentially domestic in scope but the consequences of climate change overseas will have an impact on the UK because of the interconnected nature of our world. The Programme will therefore address those effects where there is a significant domestic risk. Furthermore, the UK Climate Change Act requires the Government to carry out, every five years, a national Climate Change Risk Assessment (CCRA), of which the first needs to be finished by 2012. An additional Adaptation Economic Analysis (AEA) will consider the costs and benefits of adaptation actions. The results of the CCRA and the AEA will inform a National Adaptation Programme, to be completed by the end of 2012, which will identify priority actions and allocate resources to meet them.

Conclusions

Climate change is caused by the massive increase of greenhouse gases, such as carbon dioxide, in the atmosphere, resulting from the burning of fossil fuels and deforestation. There is clear evidence that we have already elevated concentrations of atmospheric carbon dioxide to their highest level for the last half million years and perhaps even longer. IPCC (2007) states that there is clear evidence for a 0.75°C rise in global temperatures and 22 cm rise in sea level during the 20th century. The IPCC predicts that global temperatures could rise further by between 1.1°C and 6.4°C by 2100, and sea level could rise by between 28 cm and 79 cm (Webb & Van Der Host, 2021). To investigate the effect of climate change on energy consumption and thermal comfort/overheating of UK homes, this chapter studied an explanation of global warming and climate change, and this is followed by a review of how the global warming hypothesis was developed and the possible impact of various human societies. This chapter has not attempted to address all the issues within climate change; however, it has attempted to summarise what climate change is and to focus on the global warming effect on energy consumption, carbon dioxide emissions, and households' comfort. This chapter identifies three important facts: Recent temperature raises could not be explained by natural causes and the main reason of the recent global climate change is caused by man-made greenhouse gases. It was found that level of the CO_2 is currently around 380 ppm, and this level has not been this high on the Earth for millions of years. Temperature and carbon dioxide shadow each other very closely over the long term, suggesting that the increases in CO_2 will indeed be followed by a temperature increase. The dangerous temperature rise level is defined as 2°C above pre-industrial levels and the target CO_2 level is 400 ppm. Most recent research pointed out that keeping below 2°C could be impossible and our society needs to face beyond the dangerous situations (Webb & Van Der Host, 2021). In early August 2003, the heat wave hit over Europe, and it caused an estimated 35,000 additional deaths across Europe and 2,045 in the UK. The future climate simulation model from HadCM2 shows that an event such as the summer 2003 heat wave in Europe will be common in the 2050s. Findings point that we need to act in order to reduce greenhouse gases from our activity to avoid temperature rise above dangerous level and we need to prepare for very hot summers for the next 100 years, which could affect our environment and our health. It also gives some research questions which this research has tried to answer, such as how far the temperature could rise considering several different climate change scenarios and then how it could affect our life like energy consumption in heating and cooling, thermal comfort, overheating, and carbon dioxide emissions.

References

Anandarajah, G., & Strachan, N. (2010). Interactions and implications of renewable and climate change policy on UK energy scenarios. *Energy Policy, 38*(11), 6724. https://doi.org/10.1016/j.enpol.2010.06.042.

Arnell, N. W. (2022). The implications of climate change for emergency planning. *International Journal of Disaster Risk Reduction, 83*. https://doi.org/10.1016/j.ijdrr.2022.103425.

Arnell, N. W., Freeman, A., Kay, A. L., Rudd, A. C., & Lowe, J. A. (2021). *Indicators of climate risk in the UK at different levels of warming*. IOP Publishing. https://doi.org/10.1088/2515-7620/ac24c0.

Arnell, N. W., Kay, A. L., Freeman, A., Rudd, A. C., & Lowe, J. A. (2020). Changing climate risk in the UK: a multi-sectoral analysis using policy-relevant indicators. *Climate Risk Management, 31*. https://doi.org/10.1016/j.crm.2020.100265.

Barker, T. (2008). The economics of avoiding dangerous climate change. An editorial essay on The Stern Review. *Climatic Change, 89*(3–4), 173. https://doi.org/10.1007/s10584-008-9433-x.

Bevan, L. D., Colley, T. B., & Workman, M. (2020). Climate change strategic narratives in the United Kingdom: emergency, extinction, effectiveness. *Energy Research & Social Science, 69*. https://doi.org/10.1016/j.erss.2020.101580.

Boardman, B. (2004). New directions for household energy efficiency: evidence from the UK. *Energy Policy, 32*(17), 1921. https://doi.org/10.1016/j.enpol.2004.03.021.

Bonderup, S., & Middlemiss, L. (2023). Mould or cold? Contrasting representations of unhealthy housing in Denmark and England and the relation to energy poverty. *Energy Research & Social Science, 102*, 3–6. https://doi.org/10.1016/j.erss.2023.103176.

Boomsma, C., Goodhew, J., Goodhew, S., & Pahl, S. (2016). Improving the visibility of energy use in home heating in England: thermal images and the role of visual tailoring. *Energy Research & Social Science, 14*, 111. https://doi.org/10.1016/j.erss.2016.01.005.

Boomsma, C., Jones, R. V., Pahl, S., & Fuertes, A. (2018). Do psychological factors relate to energy saving behaviours in inefficient and damp homes? A study among English social housing residents. *Energy Research & Social Science, 47*, 146. https://doi.org/10.1016/j.erss.2018.09.007.

Carrington, G., & Stephenson, J. (2018). The politics of energy scenarios: are International Energy Agency and other conservative projections hampering the renewable energy transition? *Energy Research & Social Science, 46*, 103. https://doi.org/10.1016/j.erss.2018.07.011.

Clarke, B., Otto, F., Stuart-Smith, R., & Harrington, L. (2022). *Extreme weather impacts of climate change: an attribution perspective*. IOP Publishing. https://doi.org/10.1088/2752-5295/ac6e7d.

Clarke, J. A., Johnstone, C. M., Kelly, N. J., Strachan, P. A., & Tuohy, P. (2008). The role of built environment energy efficiency in a sustainable UK energy economy. *Energy Policy, 36*(12), 4605. https://doi.org/10.1016/j.enpol.2008.09.004.

Cotterill, D. F., Pope, J. O., & Stott, P. A. (2022). Future extension of the UK summer and its impact on autumn precipitation. *Climate Dynamics, 60*(5–6), 1801. https://doi.org/10.1007/s00382-022-06403-0.

Ekins, P., & Lees, E. (2008). The impact of EU policies on energy use in and the evolution of the UK built environment. *Energy Policy, 36*(12), 4580. https://doi.org/10.1016/j.enpol.2008.09.006.

Fawcett, T., & Hampton, S. (2020). Why & how energy efficiency policy should address SMEs. *Energy Policy, 140*, 1–6. https://doi.org/10.1016/j.enpol.2020.111337.

Forster, P. M., Forster, H. I., Evans, M. J., Gidden, M. J., Jones, C. D., Keller, C. A., Lamboll, R. D., Quéré, C. L., Rogelj, J., Rosen, D., Schleussner, C., Richardson, T. B., Smith, C. J., & Turnock, S. T. (2020). Current and future global climate impacts resulting from COVID-19. *Nature Climate Change, 10*(10), 913. https://doi.org/10.1038/s41558-020-0883-0.

Foulds, C., & Powell, J. (2014). Using the Homes Energy Efficiency Database as a research resource for residential insulation improvements. *Energy Policy, 69*, 57. https://doi.org/10.1016/j.enpol.2014.01.015.

Gillich, A., Saber, E. M., & Mohareb, E. (2019). Limits and uncertainty for energy efficiency in the UK housing stock. *Energy Policy, 133*. https://doi.org/10.1016/j.enpol.2019.110889.

Goodchild, B., Ambrose, A., & Maye-Banbury, A. (2017). Storytelling as oral history: revealing the changing experience of home heating in England. *Energy Research & Social Science, 31*, 137. https://doi.org/10.1016/j.erss.2017.06.009.

Gooding, L., & Gul, M. S. (2017). Achieving growth within the UK's Domestic Energy Efficiency Retrofitting Services sector, practitioner experiences and strategies moving forward. *Energy Policy, 105*, 173. https://doi.org/10.1016/j.enpol.2017.02.042.

Hamilton, I. G., Steadman, P. J., Bruhns, H., Summerfield, A. J., & Lowe, R. (2013). Energy efficiency in the British housing stock: energy demand and the Homes Energy Efficiency Database. *Energy Policy, 60*, 462. https://doi.org/10.1016/j.enpol.2013.04.004.

Hanlon, H. M., Bernie, D., Carigi, G., & Lowe, J. A. (2021). Future changes to high impact weather in the UK. *Climatic Change, 166*(3–4). https://doi.org/10.1007/s10584-021-03100-5.

Jenkins, K., Kennedy-Asser, A., Andrews, O., & Lo, Y. T. E. (2022). Updated *projections of UK heat-related mortality using policy-relevant global warming levels and socio-economic scenarios*. IOP Publishing. https://doi.org/10.1088/1748-9326/ac9cf3.

Johnson, E., Betts-Davies, S., & Barrett, J. (2023). Comparative analysis of UK net-zero scenarios: the role of energy demand reduction. *Energy Policy, 179*, 4–5. https://doi.org/10.1016/j.enpol.2023.113620.

Katris, A., & Turner, K. (2021). Can different approaches to funding household energy efficiency deliver on economic and social policy objectives? ECO and alternatives in the UK. *Energy Policy, 155*. https://doi.org/10.1016/j.enpol.2021.112375.

Kay, A. L., Watts, G., Wells, S. C., & Allen, S. (2019). The impact of climate change on U. K. River flows: a preliminary comparison of two generations of probabilistic climate projections. *Hydrological Processes, 34*(4), 1081. https://doi.org/10.1002/hyp.13644.

Keirstead, J. (2013). Benchmarking urban energy efficiency in the UK. *Energy Policy, 63*, 575. https://doi.org/10.1016/j.enpol.2013.08.063.

Kelly, N. (2005). The role of energy efficiency in reducing Scottish and UK CO_2 emissions. *Energy Policy, 34*(18), 3505. https://doi.org/10.1016/j.enpol.2005.07.019.

Kennedy-Asser, A. T., Andrews, O., Mitchell, D. M., & Warren, R. F. (2021). *Evaluating heat extremes in the UK Climate Projections (UKCP18)*. IOP Publishing. https://doi.org/10.1088/1748-9326/abc4ad.

Kennedy-Asser, A. T., Owen, G., Griffith, G. J., Andrews, O., Lo, Y. T. E., Mitchell, D. M., Jenkins, K., & Warren, R. F. (2022). *Projected risks associated with heat stress in the UK Climate Projections (UKCP18)*. IOP Publishing. https://doi.org/10.1088/1748-9326/ac541a.

Khosravi, F., Lowes, R., & Ugalde-Loo, C. E. (2023). Cooling is hotting up in the UK. *Energy Policy, 174*, 7–8. https://doi.org/10.1016/j.enpol.2023.113456.

Klenert, D., Funke, F., Mattauch, L., & O'callaghan, B. (2020). Five lessons from COVID-19 for advancing climate change mitigation. *Environmental and Resource Economics, 76*(4), 751. https://doi.org/10.1007/s10640-020-00453-w.

Kuzemko, C., & Britton, J. (2019). Policy, politics and materiality across scales: a framework for understanding local government sustainable energy capacity applied in England. *Energy Research & Social Science, 62*. https://doi.org/10.1016/j.erss.2019.101367.

Lacey-Barnacle, M. (2020). Proximities of energy justice: contesting community energy and austerity in England. *Energy Research & Social Science, 69*, 5–9. https://doi.org/10.1016/j.erss.2020.101713.

Lomas, K. J., Li, M., & Drury, P. (2024). How do energy efficiency measures affect the risk of summertime overheating and cold discomfort? Evidence from English homes. *Energy Policy, 188*, 3–7. https://doi.org/10.1016/j.enpol.2024.114108.

Morton, A., Reeves, A., Bull, R., & Preston, S. (2020). Empowering and Engaging European building users for energy efficiency. *Energy Research & Social Science, 70*. https://doi.org/10.1016/j.erss.2020.101772.

Mcmichael, M., & Shipworth, D. (2012). The value of social networks in the diffusion of energy-efficiency innovations in UK households. *Energy Policy, 53*, 159, 1–3. https://doi.org/10.1016/j.enpol.2012.10.039.

Murtagh, N., Owen, A. M., & Simpson, K. (2021). What motivates building repair-maintenance practitioners to include or avoid energy efficiency measures? Evidence from three studies in the United Kingdom. *Energy Research & Social Science, 73*, 12–15. https://doi.org/10.1016/j.erss.2021.101943.

Nieto, J., Pollitt, H., Brockway, P. E., Clements, L., Sakai, M., & Barrett, J. (2021). Socio-macroeconomic impacts of implementing different post-Brexit UK energy reduction targets to 2030. *Energy Policy, 158*. https://doi.org/10.1016/j.enpol.2021.112556.

Ozarisoy, B. (2022). Energy effectiveness of passive cooling design strategies to reduce the impact of long-term heatwaves on occupants' thermal comfort in Europe: climate change and mitigation. *Journal of Cleaner Production, 330*, 210. https://doi.org/10.1016/j.jclepro.2021.129675

Ozarisoy, B., & Altan, H. (2024). *Energy policy design in the Eastern Mediterranean Basin: a roadmap to energy efficiency*. Taylor & Francis Group, Routledge Explorations in Energy Studies, ISBN 9781032493046, https://www.routledge.com/Energy-Policy-Design-in-the-South-Eastern-Mediterranean-Basin-A-Roadmap/Ozarisoy-Altan/p/book/9781032493046.

Pettifor, H., Wilson, C., & Chryssochoidis, G. (2015). The appeal of the green deal: empirical evidence for the influence of energy efficiency policy on renovating homeowners. *Energy Policy, 79*, 161. https://doi.org/10.1016/j.enpol.2015.01.015.

Pielke, R. (2007). Mistreatment of the economic impacts of extreme events in the Stern Review Report on the Economics of Climate Change. *Global Environmental Change, 17*(3–4), 302. https://doi.org/10.1016/j.gloenvcha.2007.05.004.

Pielke, R., & Ritchie, J. (2020). Distorting the view of our climate future: the misuse and abuse of climate pathways and scenarios. *Energy Research & Social Science, 72*, 9. https://doi.org/10.1016/j.erss.2020.101890.

Reid, L., Mckee, K., & Crawford, J. (2015). Exploring the stigmatization of energy efficiency in the UK: an emerging research agenda. *Energy Research & Social Science, 10*, 141. https://doi.org/10.1016/j.erss.2015.07.010.

Robinson, C., Bouzarovski, S., & Lindley, S. (2017). 'Getting the measure of fuel poverty': the geography of fuel poverty indicators in England. *Energy Research & Social Science, 36*, 79. https://doi.org/10.1016/j.erss.2017.09.035.

Rosenow, J., & Eyre, N. (2016). A post-modern of the Green Deal: austerity, energy efficiency, and failure in British energy policy. *Energy Research & Social Science, 21*, 141. https://doi.org/10.1016/j.erss.2016.07.005.

Rosenow, J., & Eyre, N. (2022). Reinventing energy efficiency for net zero. *Energy Research & Social Science, 90*. https://doi.org/10.1016/j.erss.2022.102602.

Rosenow, J., Guertler, P., Sorrell, S., & Eyre, N. (2018). The remaining potential for energy savings in UK households. *Energy Policy, 121*, 542. https://doi.org/10.1016/j.enpol.2018.06.033.

Rosenow, J., Platt, R., & Demurtas, A. (2014). Fiscal impacts of energy efficiency programmes—The example of solid wall insulation investment in the UK. *Energy Policy, 74*, 610. https://doi.org/10.1016/j.enpol.2014.08.007.

Scott, F. L., Jones, C. R., & Webb, T. L. (2013). What do people living in deprived communities in the UK think about household energy efficiency interventions? *Energy Policy, 66*, 335. https://doi.org/10.1016/j.enpol.2013.10.084.

Sdei, A., Gloriant, F., Tittelein, P., Lassue, S., Hanna, P., Beslay, C., Gournet, R., & Mcevoy, M. (2015). Social housing retrofit strategies in England and France: a parametric and behavioural analysis. *Energy Research & Social Science, 10*, 62. https://doi.org/10.1016/j.erss.2015.07.001.

Shipworth, D. (2007). The Stern Review: implications for construction. *Building Research & Information, 35*(4), 478. https://doi.org/10.1080/09613210701330123.

Snell, C., Bevan, M., & Thomson, H. (2015). Justice, fuel poverty and disabled people in England. *Energy Research & Social Science, 10*, 123. https://doi.org/10.1016/j.erss.2015.07.012.

Sovacool, B. K., Bazilian, M. D., Kim, J., & Griffiths, S. (2023). Six bold steps towards net-zero industry. *Energy Research & Social Science, 99*, 3–8. https://doi.org/10.1016/j.erss.2023.103067.

Taneja, S., & Mandys, F. (2022). Drivers of UK household energy expenditure: promoting efficiency and curbing emissions. *Energy Policy, 167*. https://doi.org/10.1016/j.enpol.2022.113042.

Taylor, P. G., Upham, P., Mcdowall, W., & Christopherson, D. (2014). Energy model, boundary object and societal lens: 35 years of the MARKAL model in the UK. *Energy Research & Social Science, 4*, 32, 9–10. https://doi.org/10.1016/j.erss.2014.08.007.

Tierney, J. E., Poulsen, C. J., Montañez, I. P., Bhattacharya, T., Feng, R., Ford, H. L., Hönisch, B., Inglis, G. N., Petersen, S. V., Sagoo, N., Tabor, C. R., Thirumalai, K., Zhu, J., Burls, N. J., Foster, G. L., Goddéris, Y., Huber, B. T., Ivany, L. C., Kirtland Turner, S., … Zhang, Y. G. (2020). Past climates inform our future. *Science, 370*(6517). https://doi.org/10.1126/science.aay3701.

Tingey, M., & Webb, J. (2020). Governance institutions and prospects for local energy innovation: laggards and leaders among UK local authorities. *Energy Policy, 138*, 3–9. https://doi.org/10.1016/j.enpol.2019.111211.

Tol, R. S. J., & Yohe, G. W. (2008). The Stern review: a deconstruction. *Energy Policy, 37*(3), 1032. https://doi.org/10.1016/j.enpol.2008.11.008.

Tovar, M. A. (2012). The structure of energy efficiency investment in the UK households and its average monetary and environmental savings. *Energy Policy, 50*, 723. https://doi.org/10.1016/j.enpol.2012.08.019.

Wade, F., Britton, J., & Webb, J. (2024). Credible and comprehensive? Comparing policy mixes for Local Energy Systems in England, Scotland and Wales. *Energy Research & Social Science, 110*. https://doi.org/10.1016/j.erss.2024.103413.

Wang, T., Qu, Z., Yang, Z., Nichol, T., Clarke, G., & Ge, Y. (2020). Climate change research on transportation systems: climate risks, adaptation and planning. *Transportation Research Part D: Transport and Environment, 88* 2–9. https://doi.org/10.1016/j.trd.2020.102553.

Webb, J., & Van Der Horst, D. (2021). Understanding policy divergence after United Kingdom devolution: strategic action fields in Scottish energy efficiency policy. *Energy Research & Social Science, 78*. https://doi.org/10.1016/j.erss.2021.102121.

Webborn, E., Mckenna, E., Elam, S., Anderson, B., Cooper, A., & Oreszczyn, T. (2021). Increasing response rates and improving research design: learnings from the Smart Energy Research Lab in the United Kingdom. *Energy Research & Social Science, 83* 9–11. https://doi.org/10.1016/j.erss.2021.102312.

Yohe, G. W., & Tol, R. S. J. (2007). The Stern review: implications for climate change. *Environment: Science and Policy for Sustainable Development, 49*(2), 36–43. https://doi.org/10.3200/envt.49.2.36-43.

Yuan, M., & Choudhary, R. (2023). Energy Performance Certificate renewal — an analysis of reliability of simple non-domestic buildings' EPC ratings and pragmatic improving strategies in the UK. *Energy Policy, 178* 1–5. https://doi.org/10.1016/j.enpol.2023.113581.

Zhang, T., Siebers, P., & Aickelin, U. (2012). A three-dimensional model of residential energy consumer archetypes for local energy policy design in the UK. *Energy Policy, 47*, 102. https://doi.org/10.1016/j.enpol.2012.04.027.

2
ENERGY AND ENVIRONMENT

Introduction

This study is aimed at energy efficiency in housing and potential improvements in existing housing in the UK. This chapter outlines energy use and its impact on the environment and additionally examines the history of energy use and alternative energy sources in the UK. In this chapter, renewable energy is also reviewed to understand why it is essential to replace fossil fuels. Moreover, this chapter considers the action that is currently being taken both internationally and, in the UK, to tackle the effects, which our energy use impacts on our environment.

Energy plays a central role in human life and therefore its demand and use are increasing continuously as the world population increases (Bergman & Foxton, 2019). Energy sustains life and provides the facility that is certainly needed in people's lives. Today, the world is facing various environmental problems, which have come to the public consciousness. Many of these are largely the result of large-scale fuel use (Asadnabizadeh & Moe, 2024). One of the most significant problems appears to be that of 'global warming', a gradual increase in the global average air temperature at the earth's surface. The majority of scientists now believe that global warming is probably taking place, at a rate of around 0.3°C per decade, and that it is caused by increases in the concentration of so-called 'greenhouse gases' in the atmosphere (Akorede et al., 2012).

Despite the latest draft report from the UN Inter-Governmental Panel on Climate Change (IPCC) stating that the burning of fossil fuels has contributed substantially to the observed warming over the last 50 years, there is still a lobby that believes that there is no definitive answer to the question as to whether the current episode of global warming is being driven by human activity or is a fluctuation due to natural causes (Amin & Mourshed, 2023). However, there is widespread evidence that changes in the earth's climate are occurring at a rate almost unprecedented. It is the wide-scale evidence of anomalous climatic events coupled with the rate at which they are occurring that has convinced the IPCC scientists that the finger of blame points to human activities rather than natural causes (Averchenkova et al., 2020).

Global Warming and the Greenhouse Effect

There are four main gases that cause global warming: Carbon dioxide (CO_2), methane (CH_4), chlorofluorocarbons (CFCs) and nitrous oxide (N_2O). Carbon dioxide's warming contribution

DOI: 10.1201/9781003490364-4

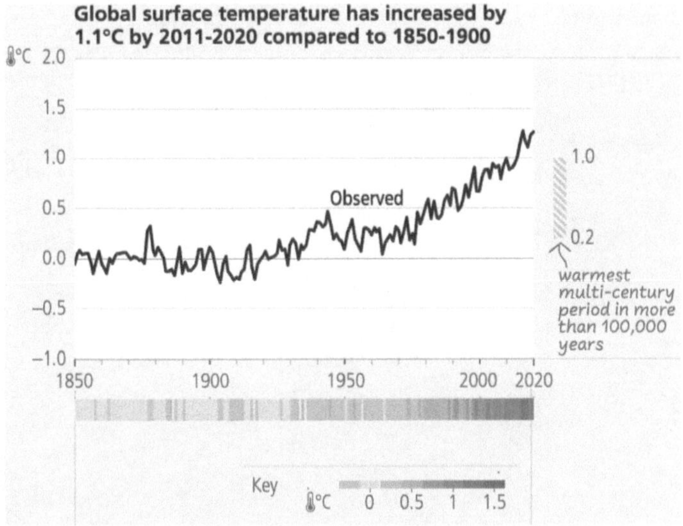

Figure 2.1 Relative warming effect of current emissions of greenhouse gases over the next 100 years.

is about 50%, whereas the contribution of methane, CFCs and nitrous oxide are approximately 18%, 14% and 6%, respectively (Arnell et al., 2015). The shockingly large contribution of these latter gases is due to the fact that methane is 25 times more effective than CO_2 as a greenhouse gas, CFCs up to 10,000 times and nitrous oxide up to 150 times. In contrast, carbon dioxide is more important due to the fact that it is integrally bound up with the carbon cycle and our use of fossil fuels.

The scientific evidence is growing that man-made greenhouse gas emissions are having a noticeable effect on the earth's climate. CO_2 has played a significant role in moderating the increase of the amount of the sun's energy falling on the earth since the beginning of life. Globally, seven of the ten warmest years on record were in the 1990s (Brand & Spencer, 2023). In the future, the earth's climate could warm by as much as 3°C over the next 100 years, as shown in Figure 2.1. We are in danger of destroying this delicate natural balance by exhausting and even increasing quantities into the atmosphere and the results could be catastrophic. During the energy crisis in the 1970s, the concern was with our diminishing reserves of gas, oil and coal. However today, the concern is to reduce CO_2 emission, which reaches the atmosphere. For more information on 'list of greenhouse gases' and 'how to compare the impact of climate change' the relative climate effects of greenhouse gasses in the built environment. (Fudge & Hopkin, 2023).

A balance between energy coming in from the sun in the form of visible radiation (sunlight) and energy constantly being emitted from the surface of the earth to space determines the temperature of the earth (Genus & Iskandarova, 2020). The energy coming in from the sun can go through the atmosphere almost unchanged and warm the earth; however, the infrared radiation emanating from the earth's surface is partly absorbed by some gases in the atmosphere and re-emitted downwards. This further warms the surface of the earth and the lower atmosphere (He et al., 2021). The gases that play a big role in this naturally are mainly water vapour and carbon dioxide. An analogy is made with the effect of a greenhouse, which allows sunshine to penetrate the glass that in turn keeps the heat in, hence the greenhouse effect, as shown in Figure 2.2.

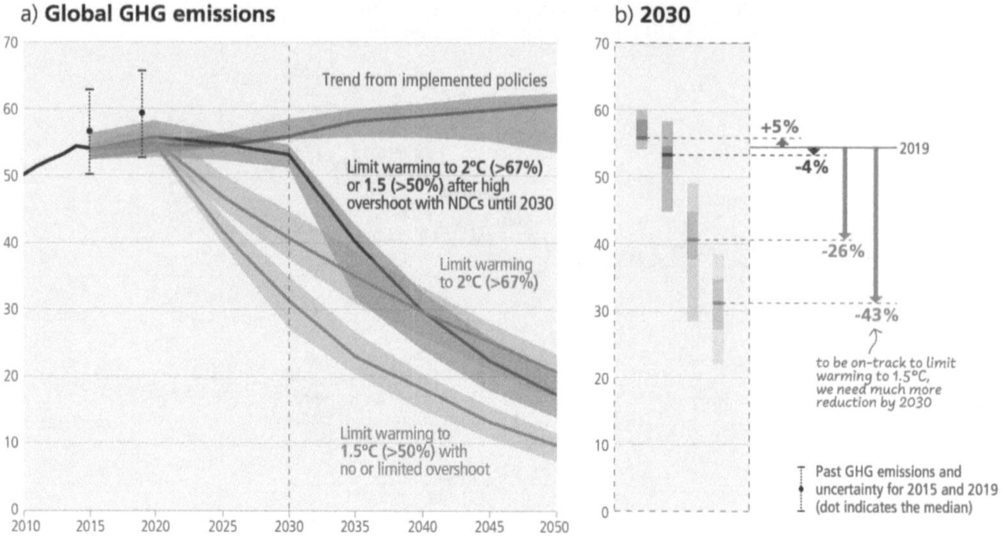

Figure 2.2 The greenhouse effect.

Without this natural greenhouse effect, the earth would be over 30°C cooler and would be too cold to be habitable. Nevertheless, as greenhouse gas concentrations increase well above their natural levels, the additional warming that will take place could threaten the future sustainability of the planet (Harkness et al., 2019).

World Energy Use

Today, modern societies and, particularly, industrial societies are entirely dependent upon the use of large quantities of energy, which are mostly in the form of fossil fuels (Giraudet, 2020). Energy is regarded as essential for the growing, distribution and preparation of foods, which are used for construction, manufacturing, communication and the organisation of other various activities (Chowdhury et al., 2018). World's energy use review is always useful in order to understand the modern use of energy in our time. Mankind has used wood fires to keep warm, to provide light and to cook food for over a million years (Dogan et al., 2022). Fire was also used to extract and work metals, fire clay pots and bricks. Yet again, animals were used for traction in agriculture and wind was used to power ships in the Mediterranean for over 6,000 years ago. The power of wind and water were the source of energy used in mills and, for many centuries, natural forces, such as wind and water, have been the main source of providing energy for transportation and production. These energy sources are still predominant in many parts of the less industrialised world (Carvalho et al., 2021).

It was the industrial revolution, which had crucial impact on changes in energy use and introduced the increase in dependence on fossil fuels (Xu et al., 2024). After the steam engine was invented, which was burning coal and coke fuels; water mills, which were running water as the power source, were replaced (Liu et al., 2022). Throughout the revolution, coal and iron ores were plentiful and, therefore, crude techniques and inefficient energy use were engaged in industrial processes, despite adverse environmental effects, which were not fully appreciated. Towards the end of the 19th and early 20th century, electricity and the internal combustion engine, oil and gas as additional fuels were developed. Energy was supplied directly from the combustion of oil or gas in engines or from electricity generated from burning coal, oil and gas or from hydroelectric plants (Croon et al., 2024).

The widespread distribution networks of electricity began in countries in the mid-20th century and progressed rapidly to the point that it was almost available universally (Kim et al., 2020). Industrial culture became entirely dependent on fossil fuels with the opening of major oil fields in the Middle East and North Africa (Ju et al., 2021). After the Second World War, nuclear sources of electricity were introduced as an additional energy source and fossil fuels were still seen as inexpensive and plentiful. The use of these fuels was mostly inefficient whilst their environmental effects were still largely ignored (Robinson & Mattioli, 2020). In the 20th century, manufacturing has continued to increase but is no longer the largest sector of the economy. Services, particularly, communications and information processing, have become dominant activities, accompanied by developments in support technologies (Shi et al., 2023).

Since the late 1960s there has been a growing recognition of the environmental impact of industrial societies, especially of the burning of fossil fuels. With the oil crisis in 1970s, there has been a significant growth in new techniques for making more efficient use of energy and providing energy from renewable sources (Ozarisoy, 2022). Major reductions in fuel use are now seen as technically possible, simply through care and attention to the energy aspects of the design of buildings, equipment and industrial processes, and many other methods. The understanding of energy efficiency was only beginning to be applied in the 1990s; however, economic constraints have been a major difficulty (Khanam & Reiner, 2022).

Interest in renewable energy has been growing steadily over the past 25 years and today only a few observers would perceive a future without renewables contributing to world energy demand (Nolden et al., 2020). Yet again, recent studies indicate that renewable energy will make a substantial contribution to global energy supplies in the longer term (Roddis et al., 2020). Increased concerns about the environment, particularly the impacts of conventional energy systems on global warming, have revitalised interest in renewable energy technologies, which have little or no net emissions of polluting gases and are widely seen as part of the solution (Mckenna et al., 2022). Furthermore, interest in non-conventional energy sources, new and renewable energy, also arises from the recognition that the commercial forms of energy in popular use today, such as liquid and gaseous hydrocarbons and solid forms of fossil fuels, represent limited and eventually exhaustible resources of energy (Cárdenas et al., 2021).

Energy Consumption

Modern societies, particularly industrial societies, are now largely dependent upon the use of large quantities of energy from fossil fuels (Kennedy et al., 2023). In 1992, the estimated total world consumption of primary energy (in all forms) was approximately 400EJ per year, equivalent to some 9,500 million tonnes of oil equivalent (mtoe) per year. Assuming, a world population of about 5,300 million, this gives an annual average world-fuel-use equivalent to about 1.8 tonnes of oil per person per year. This is equivalent to about 470 imperial gallons of oil per person per year (Rau et al., 2019).

The magnitude of the energy problem that may face future generations can be illustrated by the following calculation: the world's population stood at approximately 5 billion people, in 1990 (Paramati et al., 2021). The best UN estimates population trend show it continuing to increase to around 8 billion by 2025 but stabilising towards the end of the next century at somewhere between

Table 2.1 Increase in energy use expected as a result of population increases

Year	*Population*	*Total energy use*		*Energy use per person*	
	(Billion)	(EJ/y)	(TW)*	(GJ/y)	(KW)
1990 (dev)	1.2	284	9.0	237	7.5
1990 (ldc)	4.1	142	4.5	35	1.1
1990 (world)	5.3	426	13.5	80	2.5
2025 (dev)	1.4	167	5.3	120	3.8
2025 (ldc)	6.8	473	15.0	69	2.2
2025 (world)	8.2	640	20.3	78	2.5

Note: *Equivalent continuous power, dev: developed countries, ldc: less developed countries.

Table 2.2 Estimated annual energy consumption, 1992

Oil	*Coal*	*Gas*	*Biomass*	*Hydro*	*Nuclear*
32%	23%	19%	14%	6%	6%

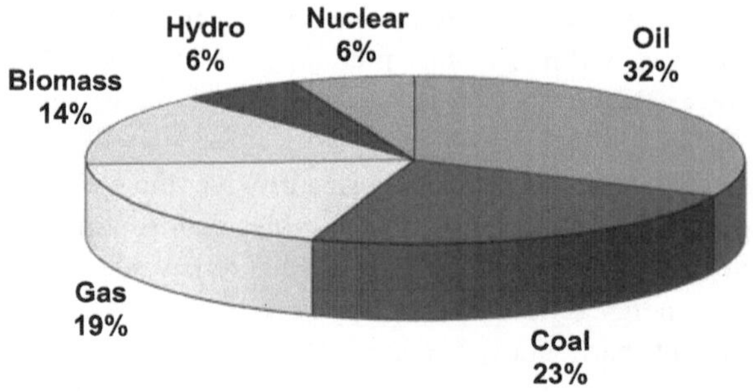

10 and 12 billion people. Most of that increase will be in the less developed countries (LDCs). Fuels are used at an average rate in the developed countries, which is more than six times that in the LDCs, as shown in Table 2.1.

It can be seen that the developed countries use nearly twice as much fuel as the LDCs, despite the fact that they have less than a third of their population. Given the expected population increase, there is still a rise of 50% in the overall level of global energy use (Liyanage et al., 2024). A breakdown of world primary energy consumption by source in 1992 is shown in Table 2.2. Oil is the dominant fuel, contributing some 32%, followed by coal at 23%. Coal was once the dominant world fuel but is now losing ground rapidly to oil and gas, which has a 19% share. Hydroelectricity and nuclear are much less used, at around 6% each, with biomass at about 14%.

Energy Use in the UK

The UK has substantial energy resources, including large reserves of coal, oil and gas, which make it self-sufficient. Figure 2.3 shows UK energy production and consumption between 1960 and 1992.

Gas and oil are known as primary sources and all made major contributions to UK energy consumption in 1992, as shown in Figure 2.4. Additionally, electricity from hydro and nuclear plants

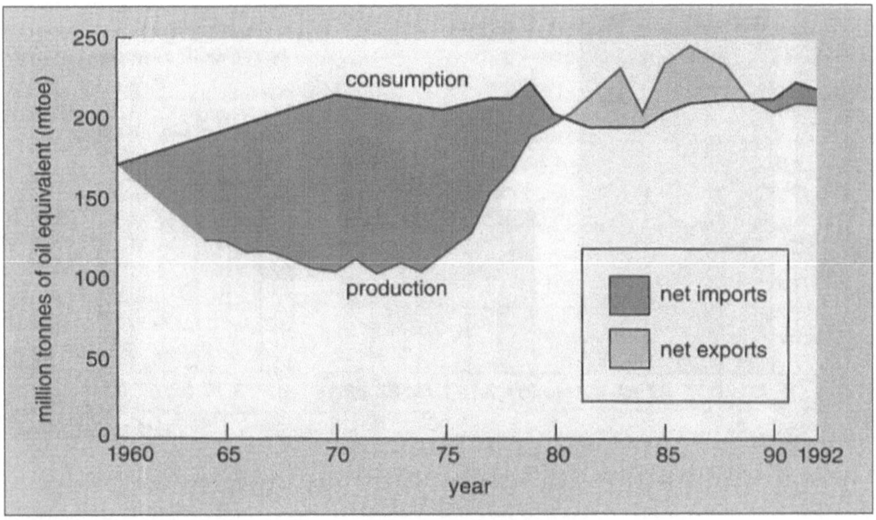

Figure 2.3 UK energy production and consumption, 1960–1992.

is known as primary electricity sources, while electricity generated by burning primary fuels is known as secondary electricity sources.

Potential Renewable Energy Use

To understand the importance of renewable energy use and the problem of present use of fuels, it is best to look at the energy that is currently used in industrial societies (Jahanbakhsh et al., 2023). If UK's energy use were to be divided into sectors, the following four main sectors can be identified: transport, domestic, commercial and institutional, and industrial (Miao et al., 2020).

Energy in the UK transport sector is dominated by road transport, and this accounts for about 80% of the energy used in this sector, which again is mainly consumed by private cars (Mcandrew et al., 2021). The main reason for using fossil fuels, particularly oil for transport purposes, is that they are inexpensive, easily stored in vehicles, widely available and have high energy density. The main energy use in the domestic sector is for space heating, water heating, cooking, lighting and electrical appliances (Tripathy et al., 2024). Most of the energy used, around 85%, is for low-grade heat, which is at a temperature below about 80°C, for space and water heating. This is generally provided directly by high-grade sources, by burning fuels directly, or much less efficiently, by electricity from thermal power plants (Alabid et al., 2022). There are now so many examples of housing with low heating requirements that the heat given off cooking, lights, appliances and the body heat of the occupants is sufficient to supply virtually all their space heating needs. This is again why energy-efficiency housing is particularly important, as we are facing global warming and other environmental problems in our environment.

The commercial and institutional sector energy use is quite similar to the domestic sector, which is currently dominated by space heating, followed by lighting, electrical appliances and equipment, air conditioning, water heating and cooking (Ozarisoy & Altan, 2024a). Then again, the energy use in this sector differentiates from building to building. The industrial sector varies both within and between countries. Generally, the largest category of end-use energy is process heat, often at high temperatures, as in the iron and steel or chemical industries. In this sector, the

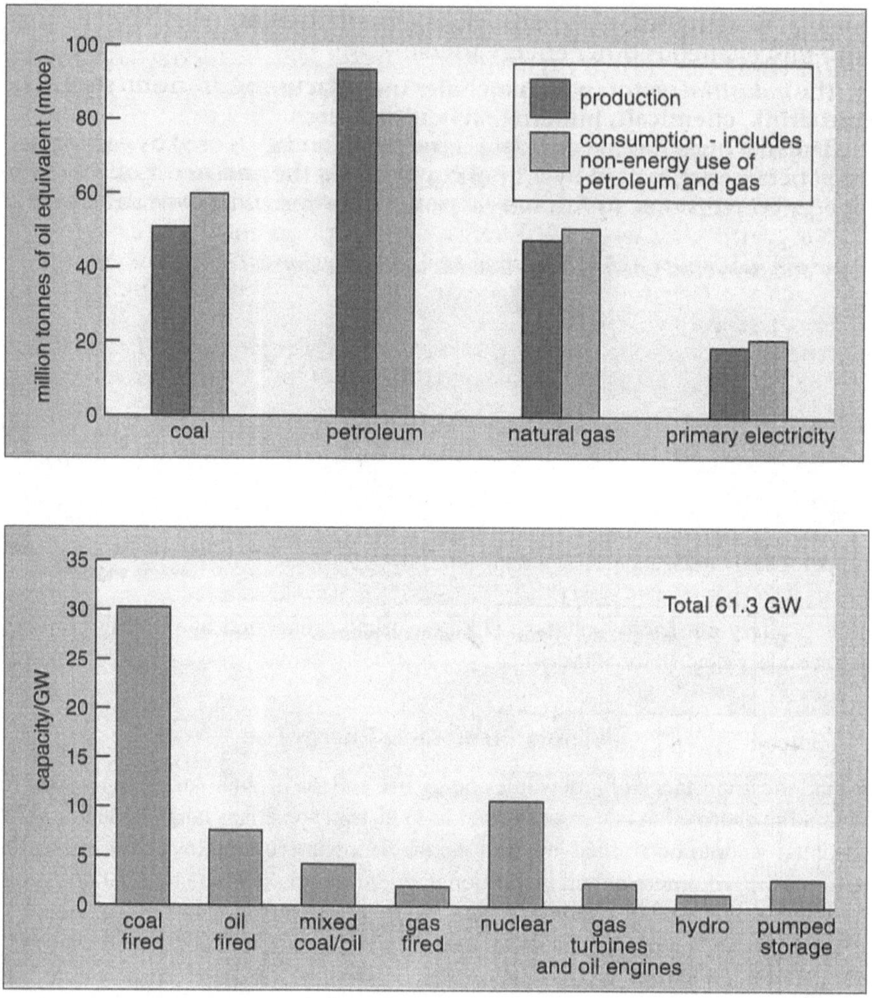

Figure 2.4 Production and consumption of primary fuels in the UK in 1992.

principle approaches to improved efficiency start with shifts from more to less energy-intensive industries; more efficient motors, drives and controls; more careful use of energy-intensive materials, such as aluminium; shortened process routes, which replace a larger number of energy consumption stages by a direct pathway to the end product; yield improvements on given process routes; technical changes to raise process efficiencies; and greater use of integrated energy technologies, such as Combined Heat and Power (CHP) and heat recovery schemes.

Environmental Problems

The main environmental problems currently loom in the public consciousness are briefly described below. This is another side effect of burning fossil fuels. Acid rain is a term, which is used to describe a variety of processes that might more accurately be referred to as acidic deposition. Natural rainfall is slightly acidic due to dissolved carbon dioxide, picked up in the atmosphere

Table 2.3 Direct fuel-use for the renewable-intensive global energy scenario

	1985	*2025*	*2050*
Coal	34.85	41.82	23.64
Oil	51.81	30.91	17.28
Gas	38.18	32.73	29.09
Biomass	0	43.64	56.36
PV/Wind	0	38.18	61.82

(Bastida et al., 2019). Organisms and ecosystems all over the planet have adapted to the slightly acidic nature of normal rain, and thus, it poses no environmental problems. It is an increase in the acidity of rain, caused by human activities, such as the combustion of fossil fuels, which again has turned acid rain into a problem (Huang et al., 2022).

Highly acidic rain can damage or destroy aquatic life, forests, crops and buildings, as well as pose a threat to human health. There are basically two ways of reducing acid rain: Emission control technologies can be attached to smokestacks at power plants and other industries, removing the acid gases before they are emitted into the atmosphere. Another alternative is to burn less high-sulphur fossil fuel. Switching to alternative sources of energy or improving the efficiency of our energy-consuming technologies can accomplish this.

Ultimately, the most effective methods of reducing acid rain are renewable energy and energy efficiency. Renewable energy technologies, such as solar and wind energy, can produce electricity without any emissions of sulphur dioxide or nitrous oxides and, additionally, both have the added benefit that also results in reduced emissions of carbon dioxide, the greenhouse gas most responsible for global warming (Jung & Schindler, 2022).

Power stations and oil refineries can cause significant visual intrusions and produce significant smells and other effluent locally. Although the transportation of oil is generally a safe industry, the scale of it and the size of the tankers mean that when accidents do occur, they have a large net effect on the environment. For example, in 1989 in Alaska, USA, about 40 million litres of crude oil spilled into the ocean from the supertanker 'Exxon Valdez', which caused massive environmental damage (clean-up cost over US$2 billion).

Environmental Sustainability and Climate Change

The continual man-made emissions of 'Greenhouse Gases' are resulting in global atmospheric warming, local climate changes and sea-level rise with the prospect of consequent serious environmental, social and economic impacts (Maby & Gwilliam, 2021). The most comprehensive scientific assessment of climate change was conducted by Working Group I of the IPCC, which is organized jointly by the World Meteorological Organization and the United Nations Environment Programme. For more information on Global Warming and Climate Change, the study examines the overheating risk of buildings and measure its impacts on the built environment.

Renewable energy is the term used to cover the energy flow, which occurs repeatedly in the environment and can be harnessed for human benefit. International Energy Agency statistics forecast that by the middle of the 21st century, renewable energy sources could account for three-fifths of the market for fuels used directly and two-fifths of the market for fuels used directly, as shown in Table 2.3. It is widely accepted within the international scientific community.

Table 2.4 World per capita emissions of CO_2 for renewable-intensive global energy scenario

	1985	*2025*	*2050*
Coal	4.16	2.94	1.62
Oil	6.02	2.06	1.03
Gas	2.06	1.03	1.18

Making a transition to a 'renewable-intensive' energy economy would provide environmental and other benefits not measured in standard economic accounts. Recent studies indicate that by 2025 global carbon dioxide emissions will be reduced to 75% of their 1985 levels, provided energy efficiency and renewable are both pursued aggressively, as shown in Table 2.4. Renewable energy is expected to be competitive with conventional energy, such benefits could be achieved at no additional cost.

The United Nations Conference on Environment and Development (UNCED) held in Rio de Janeiro, in June 1992, addressed the challenges of achieving worldwide sustainable development (Semple & Jenkins, 2019). The goal of sustainable development cannot be realised without major changes in the world's energy system. Accordingly, 'Agenda 21', which was adopted by UNCED, called for 'new policies or programs, as appropriate, to increase the contribution of environmentally safe and sound and cost-effective energy systems, particularly new and renewable ones, through less polluting and more efficient energy production, transmission, distribution and use'.

Passive Solar Energy

Many different techniques can be used to convert sunlight into useful forms of energy. Active and passive solar energy technologies are generally used for space conditioning (heating and cooling), while solar electric technologies, such as photovoltaic cells convert sunlight into electricity (Stevanović, 2013). Although the distinction between active and passive solar is blurred, the use of integral building components to capture the sun's energy is considered passive solar. Active solar technologies are generally add-on features, which utilise mechanical means to distribute captured solar energy. An example of active solar energy is a solar hot water heater, while passive solar features may be as simple as south-facing windows. Passive solar features can be used to heat and cool buildings, as well as provide light. The best time to incorporate passive solar technologies into a building is during the initial design (Ozarisoy & Altan, 2024b). These features can often be included in new buildings without significantly adding to construction costs while providing energy savings of up to 40%. Designing buildings to capture the ambient energy of the sun through passive solar features is one of the least expensive and most environmentally friendly methods of providing for our energy needs.

Each year, an enormous amount of solar energy reaches the earth's atmosphere. Much of this is reflected back into space by clouds before it reaches the planet's surface. Ninety-nine per cent of the sunlight that does reach the ground is converted into heat (the other 1% is captured by plants through photosynthesis) and radiated back into space. If only a small fraction of this energy could be captured, the world's energy demands could be met. The capture of solar energy by passive solar technologies has almost no negative impact on the environment. Passive solar energy gives off no air or water emissions and therefore does not contribute to any of the environmental problems, such as acid rain and global warming. The sun is an infinite and free source of energy, which is 'renewable' and will never become depleted like fossil fuels. There is nothing new about using the sun's energy to heat our living spaces; humankind has used passive solar techniques for thousands of years. In many countries, inexpensive and plentiful fossil fuels have led to the abandonment of passive solar energy. Rediscovering passive solar energy and incorporating technological advances can go a long way towards creating a more sustainable energy future.

Direct solar gain increased thermal mass and attached sunspaces are the most common features of passive solar heating. Direct solar gain, the main source of passive solar heat, is accomplished by capturing the sun's energy through large areas of south facing windows. Window glass is virtually transparent to incoming solar radiation. When sunlight strikes the interior of a building, it is converted into heat, which is not as readily transmitted back through the glass, thus resulting in a heat gain inside the house. Window glass, however, is generally not a good insulator, and increased solar heat gain during the day can be offset by loss of heat through windows at night. New high-efficiency, triple-glazed windows with special coatings have recently been developed that have such a high insulation value that they are net producers of heat even when facing north in the winter.

Solar cooling on the other hand may sound impractical, but there are a variety of ways in which the sun's energy, although indirectly, can be used to cool the interior of buildings. The two most common methods of passive solar cooling are the use of vegetation (trees and vines) and natural ventilation (wind). An example of painting buildings a light colour to reflect sunlight and keep them cool is also often considered to be a passive solar construction technique.

Daylighting is the use of sunlight to replace electric lighting in a building. There is no such technology at the current time capable of storing sunlight for release at a later time. Daylighting is therefore most valuable in applications, such as office buildings where most of the lighting demand occurs during the day. Windows provide light for the perimeter of buildings while atria, light-shelves and light-pipes, can transmit daylight into the interior of buildings. In combination with electronic 'photo-sensor' controls, which adjust electric lights according to light levels, daylighting features can drastically reduce the amount of electricity required to light a building. The use of daylighting has often been seen as contradictory to the need for keeping a building cool in the summer. Sunshine streaming through a window provides daylight but is also a source of heat. While this heat is valuable in the winter, it can make buildings unbearably hot in the summer. An example of new window technologies, such as films, which let in light but not heat, and 'smart windows' whose transparency can be adjusted by an electric current, have helped to reconcile the need for both light and heat in buildings.

Passive solar energy has the potential to supply a large proportion of energy needs for a properly designed building such as a house. The best opportunity for using passive solar features is in new construction. Before the proliferation of fossil fuels, architects usually designed buildings to utilise available solar energy for heating, cooling and lighting. Recent advances in technology and building materials have greatly expanded the available tools for architects to work with. Passive solar energy, while often seen as 'low technology', represents in many cases, the cleanest and least expensive possible source of useful energy for buildings.

The International Framework on Climate Change

In response to increasing concerns about climate change, the United Nations Framework Convention on Climate Change was agreed at the Earth Summit in Rio de Janeiro in 1992. Under the Convention, all developed countries agreed to aim to return their greenhouse gas emissions to 1990 levels by 2000.

It was then quickly recognised that the Convention commitments could only be a first step in the international response to climate change. Climate prediction models showed that deeper cuts in emissions would be needed to prevent serious interference with the climate. The Kyoto Protocol was designed to address this issue and was agreed in December 1997 (He et al., 2021). Accordingly, developed countries agreed to targets, which will reduce their overall emissions of a basket of six greenhouse gases (carbon dioxide, methane, nitrous oxide, hydrofluorocarbons, perfluorocarbons and sulphur hexafluoride) by 5.2% below 1990 levels over the period 2008–2012. This was the first time these targets would be legally binding and differentiated between Parties to the Convention (He et al., 2021).

Under the Kyoto Protocol, the European Union (EU) agreed to an 8% cut in greenhouse gas emissions and under a burden-sharing agreement amongst its member states, the United States (US) agreed to a 7% cut, Japan agreed to a 6%, Russia and Ukraine to return to 1990 levels and Australia was allowed an 8% increase (Dogan et al., 2022).

The UK's Climate Change Programme

Following the Kyoto Protocol, the UK agreed to a 12.5% cut in greenhouse gas emissions below 1990 levels by 2008–2012. The Government and the devolved administrations also have a domestic goal to reduce the UK's emissions of carbon dioxide, the main greenhouse gas, by 20% below 1990 levels by 2010 (Dogan et al., 2022). The UK will continue to be committed to working at the international level to encourage global responsibility and action, not only in relation to climate change but also more widely in terms of the environmental impact of energy generation and use.

In November 2000, the government with the devolved administrations published the UK Climate Change Programme aimed at meeting the UK's Kyoto target and moving towards its domestic goal. The programme sets out a far-reaching strategy for tackling climate change in the UK. It aims to ensure that the UK moves towards a more sustainable, low-carbon economy. It puts in place policies that indicate clear signals about the changes that need to be made in the longer term and outlines a variety of measures to reduce emissions across all sectors of the economy for the remainder of this decade and beyond.

The UK's programme is a significant contribution to the global response to climate change and the government estimates the package of measures included in the programme could cut the UK's greenhouse gas emissions to 23% below 1990 levels in 2010, or 19% in carbon dioxide alone (Kim et al., 2020). The government also believes that there will be clear benefits for the UK by moving beyond its Kyoto target towards its domestic goal. Consultation on the programme has revealed that many businesses agree that taking early action and starting the transition to a low-carbon economy will bring financial and economic gains. The programme includes a flexible, cost-effective set of policies and measures that will help to achieve these objectives, safeguard the UK's competitiveness and also deliver wider benefits for the environment and society.

On the other hand, the Utilities Act 2000 has contributed to the government's environmental objectives in a number of ways. It has introduced a power to place an obligation on companies to increase the proportion of electricity generated from renewable sources. It has transferred from the Office of Gas and Electricity Markets (OFGEM) to the government the power to set energy-saving targets by encouraging consumers to take up energy efficiency improvements. It has also set the framework for the government to issue guidance to the Regulatory Authority on its environmental and social policies to which the Authority must have regard in making its decisions.

Conclusions

Today, global warming is one of the most serious environmental problems the world faces. Floods, storms and droughts in the UK and across the world show clearly how vulnerable the countries are to climate extremes and how high the human, environmental and economic costs can be. As most scientists believe today, climate change is now inevitable, but the worst effects can be avoided if the world acts now to reduce greenhouse gas emissions. Although it is now generally understood that extreme weather patterns are probably due to global warming caused by increased carbon dioxide emissions to the atmosphere, there is no real consensus on how to tackle the problem and reduce emissions. With the media's attention to environmental concerns, people are becoming increasingly conscious of the fact that their lifestyle is affecting the earth's very natural resources and environment. This approach also creates a considerable challenge for the energy policy and should be based on integrated economic, environmental and social policies to ensure that energy resources and use of these resources are sustainable now and for generations to come.

This chapter has looked at energy use and the environment on the whole and environmental concerns rose due to global warming and carbon dioxide emissions. The recent IPCC report has stated that to stabilise the concentrations of CO_2 emissions in the atmosphere at the necessary level, a reduction in emissions due to human activity in the order of over 60% will be required. Accordingly, buildings have been identified as a major source of carbon dioxide emissions and therefore a major contributor to global warming. In the UK, energy consumed by the building stock approaches 50% of the total while transport is responsible for 28% and predicted to rise significantly. Domestic energy use represents a large proportion of total national energy use and has risen from 25% of the total in 1970 to 30% in 2001 (Xu et al., 2024). Clearly, housing energy consumption is a major contributor to global warming and has become one of the key elements of improving energy efficiency and reducing CO_2 emissions in the country.

References

Akorede, M. F., Hizam, H., Ab Kadir, M. Z. A., Aris, I., & Buba, S. D. (2012). Mitigating the anthropogenic global warming in the electric power industry. *Renewable and Sustainable Energy Reviews, 16*(5), 2747. https://doi.org/10.1016/j.rser.2012.02.037.

Alabid, J., Bennadji, A., & Seddiki, M. (2022). A review on the energy retrofit policies and improvements of the UK existing buildings, challenges and benefits. *Renewable and Sustainable Energy Reviews, 159*, 1–5. https://doi.org/10.1016/j.rser.2022.112161.

Amin, A., & Mourshed, M. (2023). Weather and climate data for energy applications. *Renewable and Sustainable Energy Reviews, 192*, 4–5. https://doi.org/10.1016/j.rser.2023.114247.

Arnell, N. W., Halliday, S. J., Battarbee, R. W., Skeffington, R. A., & Wade, A. J. (2015). The implications of climate change for the water environment in England. *Progress in Physical Geography: Earth and Environment, 39*(1), 93. https://doi.org/10.1177/0309133314560369.

Asadnabizadeh, M., & Moe, E. (2024). A review of global carbon markets from Kyoto to Paris and beyond: the persistent failure of implementation. *Frontiers in Environmental Science, 12*. https://doi.org/10.3389/fenvs.2024.1368105.

Averchenkova, A., Fankhauser, S., & Finnegan, J. J. (2020). *The impact of strategic climate legislation: evidence from expert interviews on the UK Climate Change Act (21–23)*. Informa UK Limited. https://doi.org/10.1080/14693062.2020.1819190.

Bastida, L., Cohen, J. J., Kollmann, A., Moya, A., & Reichl, J. (2019). Exploring the role of ICT on household behavioural energy efficiency to mitigate global warming. *Renewable and Sustainable Energy Reviews, 103*, 455. https://doi.org/10.1016/j.rser.2019.01.004.

Bergman, N., & Foxton, T. J. (2019). Reframing policy for the energy efficiency challenge: insights from housing retrofits in the United Kingdom. *Energy Research & Social Science, 63*, 1–3. https://doi.org/10.1016/j.erss.2019.101386.

Brand, J. H., & Spencer, K. L. (2023). Potential pollution risks of historic landfills in England: further analysis of climate change impacts. *WIREs Water, 11*(3), 4–5. https://doi.org/10.1002/wat2.1706.

Cárdenas, B., Swinfen-Styles, L., Rouse, J., Hoskin, A., Xu, W., & Garvey, S. D. (2021). Energy storage capacity vs. renewable penetration: a study for the UK. *Renewable Energy, 171*, 849. https://doi.org/10.1016/j.renene.2021.02.149.

Carvalho, D., Rocha, A., Costoya, X., Decastro, M., & Gómez-Gesteira, M. (2021). Wind energy resource over Europe under CMIP6 future climate projections: what changes from CMIP5 to CMIP6. *Renewable and Sustainable Energy Reviews, 151*, 35–37. https://doi.org/10.1016/j.rser.2021.111594.

Chowdhury, J. I., Hu, Y., Haltas, I., Balta-Ozkan, N., Matthew, G. J., & Varga, L. (2018). Reducing industrial energy demand in the UK: a review of energy efficiency technologies and energy saving potential in selected sectors. *Renewable and Sustainable Energy Reviews, 94*, 1153. https://doi.org/10.1016/j.rser.2018.06.040.

Croon, T. M., Hoekstra, J. S. C. M., & Dubois, U. (2024). Energy poverty alleviation by social housing providers: a qualitative investigation of targeted interventions in France, England, and the Netherlands. *Energy Policy*, 192, 9–11. https://doi.org/10.1016/j.enpol.2024.114247.

Dogan, E., Chishti, M. Z., Karimi Alavijeh, N., & Tzeremes, P. (2022). The roles of technology and Kyoto Protocol in energy transition towards COP26 targets: evidence from the novel GMM-PVAR approach for G-7 countries. *Technological Forecasting and Social Change, 181*, 5–6. https://doi.org/10.1016/j.techfore.2022.121756.

Fudge, C., & Hopkin, T. (2023). Achieving the 2025 future homes energy and carbon standard for England using AAC masonry. *Ce/Papers, 6*(2), 15. https://doi.org/10.1002/cepa.1980.

Genus, A., & Iskandarova, M. (2020). Transforming the energy system? Technology and organisational legitimacy and the institutionalisation of community renewable energy. *Renewable and Sustainable Energy Reviews, 125*, 5–6. https://doi.org/10.1016/j.rser.2020.109795.

Giraudet, L. (2020). Energy efficiency as a credence good: a review of informational barriers to energy savings in the building sector. *Energy Economics, 87*, 23–27. https://doi.org/10.1016/j.eneco.2020.104698.

Harkness, C., Semenov, M. A., Areal, F., Senapati, N., Trnka, M., Balek, J., & Bishop, J. (2019). Adverse weather conditions for UK wheat production under climate change. *Agricultural and Forest Meteorology, 282–283*. https://doi.org/10.1016/j.agrformet.2019.107862.

He, R., Luo, L., Shamsuddin, A., & Tang, Q. (2021). Corporate carbon accounting: a literature review of carbon accounting research from the Kyoto Protocol to the Paris Agreement. *Accounting & Finance, 62*(1), 261. https://doi.org/10.1111/acfi.12789.

Huang, W. T. K., Braithwaite, I., Charlton-Perez, A., Sarran, C., & Sun, T. (2022). *Non-linear response of temperature-related mortality risk to global warming in England and Wales*. IOP Publishing (41–43). https://doi.org/10.1088/1748-9326/ac50d5.

Jahanbakhsh, A., Louis Potapov-Crighton, A., Mosallanezhad, A., Tohidi Kaloorazi, N., & Maroto-Valer, M. M. (2023). Underground hydrogen storage: a UK perspective. *Renewable and Sustainable Energy Reviews, 189*. https://doi.org/10.1016/j.rser.2023.114001.

Ju, B., Kim, M., Kim, S., & Moreno-Ternero, J. D. (2021). Fair international protocols for the abatement of GHG emissions. *Energy Economics, 94*, 65–67. https://doi.org/10.1016/j.eneco.2020.105091.

Jung, C., & Schindler, D. (2022). A review of recent studies on wind resource projections under climate change. *Renewable and Sustainable Energy Reviews, 165*, 93–97. https://doi.org/10.1016/j.rser.2022.112596.

Kennedy, C., Bertram, D., & White, C. J. (2023). Reviewing the UK's exploited hydropower resource (onshore and offshore). *Renewable and Sustainable Energy Reviews, 189,* 78–84. https://doi.org/10.1016/j.rser.2023.113966.

Khanam, T., & Reiner, D. M. (2022). Evaluating gaps in knowledge, willingness and heating performance in individual preferences on household energy and climate policy: evidence from the UK. *Renewable and Sustainable Energy Reviews, 160*, 89–91. https://doi.org/10.1016/j.rser.2022.112229.

Kim, Y., Tanaka, K., & Matsuoka, S. (2020). Environmental and economic effectiveness of the Kyoto Protocol. *PLoS One, 15*(7). https://doi.org/10.1371/journal.pone.0236299.

Liu, J., Wang, J., & Cardinal, J. (2022). Evolution and reform of UK electricity market. *Renewable and Sustainable Energy Reviews, 161*, 7–11. https://doi.org/10.1016/j.rser.2022.112317.

Liyanage, D. R., Hewage, K., Hussain, S. A., Razi, F., & Sadiq, R. (2024). Climate adaptation of existing buildings: a critical review on planning energy retrofit strategies for future climate. *Renewable and Sustainable Energy Reviews, 199*, 13–14. https://doi.org/10.1016/j.rser.2024.114476.

Maby, C., & Gwilliam, J. (2021). Integrating energy efficiency into private home repair, maintenance and improvement practice in England and Wales. *Building Research & Information, 50*(4), 424. https://doi.org/10.1080/09613218.2021.1963652.

Mcandrew, R., Mulcahy, R., Gordon, R., & Russell-Bennett, R. (2021). Household energy efficiency interventions: a systematic literature review. *Energy Policy, 150*, 89–91. 10.1016/j.enpol.2021.112136.

Mckenna, R., Mulalic, I., Soutar, I., Weinand, J. M., Price, J., Petrović, S., & Mainzer, K. (2022). Exploring trade-offs between landscape impact, land use and resource quality for onshore variable renewable energy: an application to Great Britain. *Energy, 250*. https://doi.org/10.1016/j.energy.2022.123754.

Miao, C., Teng, K., Wang, Y., & Jiang, L. (2020). *Technoeconomic analysis on a hybrid power system for the UK household using renewable energy: a case study*. MDPI AG (23–27). https://doi.org/10.3390/en13123231.

Nolden, C., Barnes, J., & Nicholls, J. (2020). Community energy business model evolution: a review of solar photovoltaic developments in England. *Renewable and Sustainable Energy Reviews, 122*, 35–37. https://doi.org/10.1016/j.rser.2020.109722.

Ozarisoy, B. (2022). *Energy effectiveness of passive cooling design strategies to reduce the impact of long-term heatwaves on occupants'* thermal comfort in Europe: Climate change and mitigation. https://doi.org/10.1016/j.jclepro.2021.129675.

Ozarisoy, B., & Altan, H. (2024a). *Energy policy design in the Eastern mediterranean basin: a roadmap to energy efficiency*. Taylor & Francis Group, Routledge Explorations in Energy Studies, ISBN 9781032493046, https://www.routledge.com/Energy-Policy-Design-in-the-South-Eastern-Mediterranean-Basin-A-Roadmap/Ozarisoy-Altan/p/book/9781032493046.

Ozarisoy, B., & Altan, H. (2024b). *A novel methodological framework in energy policy*. Taylor & Francis Group, Routledge Explorations in Energy Studies, https://www.taylorfrancis.com/chapters/mono/10.4324/9781003393092-3/novel-methodological-framework-energy-policy-bertug-ozarisoy-hasim-altan.

Paramati, S. R., Shahzad, U., & Doğan, B. (2021). The role of environmental technology for energy demand and energy efficiency: evidence from OECD countries. *Renewable and Sustainable Energy Reviews, 153*, 91–93. https://doi.org/10.1016/j.rser.2021.111735.

Rau, H., Moran, P., Manton, R., & Goggins, J. (2019). Changing energy cultures? Household energy use before and after a building energy efficiency retrofit. *Sustainable Cities and Society, 54*. https://doi.org/10.1016/j.scs.2019.101983.

Robinson, C., & Mattioli, G. (2020). Double energy vulnerability: Spatial intersections of domestic and transport energy poverty in England. *Energy Research & Social Science, 70* (3–7). https://doi.org/10.1016/j.erss.2020.101699.

Roddis, P., Roelich, K., Tran, K., Carver, S., Dallimer, M., & Ziv, G. (2020). What shapes community acceptance of large-scale solar farms? A case study of the UK's first 'nationally significant' solar farm. *Solar Energy, 209*, 235. https://doi.org/10.1016/j.solener.2020.08.065.

Semple, S., & Jenkins, D. (2019). Variation of energy performance certificate assessments in the European Union. *Energy Policy, 137*. https://doi.org/10.1016/j.enpol.2019.111127.

Shi, Y., Sorrell, S., & Foxon, T. (2023). The impact of teleworking on domestic energy use and carbon emissions: an assessment for England. *Energy and Buildings, 287*, 110–112. https://doi.org/10.1016/j.enbuild.2023.112996.

Stevanović, S. (2013). Optimization of passive solar design strategies: a review. *Renewable and Sustainable Energy Reviews, 25*, 177. https://doi.org/10.1016/j.rser.2013.04.028.

Tripathy, P., Jena, P. K., & Mishra, B. R. (2024). Systematic literature review and bibliometric analysis of energy efficiency. *Renewable and Sustainable Energy Reviews, 200*, 12–13. https://doi.org/10.1016/j.rser.2024.114583.

Xu, C., Qin, Z., Chen, J., & Zhang, J. (2024). Heterogeneous technology-induced global CO_2 emission reduction and emission forecasting since the Kyoto era. *Applied Energy, 371*, 23–27. https://doi.org/10.1016/j.apenergy.2024.123678.

3

OCCUPANTS' ENERGY CONSUMPTION BEHAVIOUR AND BEHAVIOURAL INTERVENTIONS

Introduction

This chapter undertakes further exploration of their successes and failures by focusing on the impact of occupants' behavioural issues. The focus is primarily within the UK's social housing sector but also extended to owner-occupied and private rent housing sectors for more comprehensive and meaningful discussion. It is believed that although the UK's retrofit programmes are relatively successfully delivered, energy performance may still not meet expectations. The reason for this is the way occupants operate their homes will also lead to the building energy performance gap. The methods of regulating energy consumption behaviours and consequently improving energy performance are another focus of the current study. The approaches adopted in order to improve occupants' energy consumption behaviours are widely discussed in the following sections. Additionally, in the background of the transition of the UK's energy network, a great number of components are adopted to help improve occupants' energy-saving awareness such as smart meters. Their impact on occupants' behaviours is also discussed. Through examination of energy management applications, there is an aim to explore the potential opportunities to contribute to home energy conservation in a new and innovative direction. The review of applications helps to identify gaps and inform the research questions on how to provide more effective behavioural interventions through energy management applications.

It is noted that behavioural interventions for occupants need to be designed by considering both, policymakers (top-down approach) and end users (bottom-up approach). The ways that occupants operate their homes could be influenced by tailored policies supported by strategic interactions with energy management applications or smart meters. Recent studies suggested that policymaking needs to consider occupants' varying socio-demographic backgrounds, such as age and family size, to promote behavioural change. Due to insufficient energy knowledge, user manuals for home operations need to be simple and clear. Besides, 'face-to-face' and other format of interactive approaches could be implemented by policymakers to suit different groups of people. This chapter

DOI: 10.1201/9781003490364-5

aims to examine the importance of occupants' energy consumption behaviour on home energy performance and existing approaches to promote home operation.

Impact of Energy Consumption Behaviour

As previously stated, the significant impact of occupants' energy consumption behaviour has been increasingly realised as energy performance may still not meet the set targets albeit retrofit measures are successfully installed. In general, there are two types of energy performance deficits identified: the 'rebound effect' and the 'prebound effect'. The former is identified as when real energy performance is higher than expectations and the latter is when actual energy performance is lower than expectations. As many retrofit projects could not achieve expected levels of energy savings, the research focuses on the issues that lead to 'rebound effect', which mainly happens following project handover. This is mostly related to occupants' lifestyle and occupancy patterns and the way in which they operate their retrofitted properties.

The importance of addressing 'rebound effect' is emphasised by Galvin (2014) as being "more useful for performance evaluation of thermal retrofits of existing homes: defining the 'energy savings deficit' and the building energy performance gap". The 'rebound effect' has been widely discussed by experts and scholars. Ozarisoy and Altan (2024a) argued that the definition of rebound is not clear as it has various interpretations. It is stated that the primary issue that leads to rebound in the residential sector is the space heating and space cooling behaviour of energy end users. A range of 0–50% rebound was identified for a 100% increase in energy efficiency. The impact of 'rebound effect' was also advised by, who stated that serious and extensive rebound may destroy policy measures for energy efficiency in domestic and public sectors. The computable general equilibrium model is generally used to predict how economy reacts to the changes of policy by employing actual economic data. They suggest that energy subsidies reformation, which includes replacing the current fossil fuel subsidies and providing clean energy subsidies could fundamentally offset it.

The mitigation of energy savings due to occupants' behavioural issues was also identified by the researchers based on the investigation of social housing estates. Through the investigation of social housing tower block in Newham, London. The pilot study indicated that home energy consumption patterns and indoor thermal comfort are significantly influenced by occupants' energy behaviours and their socio-demographic characteristics through a variety of approaches such as simulation modelling, site monitoring and structured interviews. The study has found that an improper method of operating homes may lead to energy over-consumption. Through the evaluation of 10 recently retrofitted multi-family residential buildings that cover 1,100 flats in Geneva, Jad et al. (2016) asserted that only 42% of the predicted energy savings could be achieved in reality. The failure of the retrofit is due to building operation and occupants' behaviours. Other scholars supporting this include Gupta and Gregg (2016), Jad et al. (2016) and Mills and Schleich (2012).

By monitoring the electricity consumption of 72 UK dwellings from four social housing estates and one owner-occupied estate for a two-year period, research by Firth et al. (2008) identified that the built form does not play an important role in dwellings' energy performance. Instead, various social and behavioural issues determine how much energy will be consumed, such as number of occupants, types of electronic appliances used and occupants' behavioural patterns. In another study by Guerra-Santin and Itard in 2010, households with a programmatic thermostat were found to be consuming more energy than households with a manual thermostat because users tended to turn the heating on for longer. The same study also indicates that elderly occupants use heating and ventilation systems more than younger occupants. Ürge-Vorsatz et al. (2007) pointed out that reducing the life-cycle cost of materials will significantly reduce energy consumption and greenhouse gas emissions. Moreover, non-technological approaches can also contribute to energy savings, such as behavioural and cultural elements and energy-use patterns. Lee et al. (2013) argued that in order to achieve 60% CO_2 reduction by 2050, it is crucial to emphasise behavioural changes and increase carbon tax. Through investigation of social housing estates in the pilot project, Gupta and Gregg (2016) also stated that a low level of communication between professionals and occupants regarding behavioural issues may lead to the failure of a project. During the retrofit delivery of large-scale social housing estates, contractors are encouraged to communicate with occupants and support them in understanding appropriate methods of operating their homes (if they are not moving out during the energy upgrade).

From a different perspective, Galvin (2014) highlighted the 'rebound effect' as a metric that is not precise. Three types of 'rebound effect' are identified in relation to different metrics and employed in empirical studies of three 30-apartment buildings in Germany. They include classic 'rebound effect', 'energy saving deficit' and the 'energy performance gap'. Each metric is used for a specific condition in order to be precise. As there were concerns around the 'rebound effect', energy performance was monitored in the project in terms of energy consumption and CO_2 emissions. Sunikka-Blank and Galvin (2012) also argued that behaviour changes (non-technical measures) play a much more important role in energy saving than policies tend to assume. Additionally, it has been reported that the Retrofit Delivery Programme finds total investment in the programme is £17 million, of which £20,000 was used to award the 194 best retrofit strategies to stimulate household engagement in the social housing sector and promote energy consumption behaviour.

The 'rebound effect' has been proved by Ben and Steemers (2014) in a project based in the Brunswick Centre in London. The researchers examined the implications of people's behaviour in relation to listed buildings being retrofitted. To explore the effects of householders' behaviour, a physical retrofit model framework was developed at a domestic level, based on a validated energy simulation tool using the Integrated Environmental Solutions (IES) software suite. Factors of occupants' behaviour are converted to different parameters and taken into account in this model framework. It was found that a lower level of behaviour change effect is associated with a higher retrofit level. As there were certain degrees of retrofit restrictions on the listed buildings, the effect of behaviour change was more obvious. Heating and indoor temperature has the highest saving potential among other sectors in relation to behaviour change.

Wei et al. (2016) then extended their research to the public building sector. By employing the same approach, they suggest that certain behaviour, such as heating-related behaviour, has a large impact on low-carbon retrofit outcomes.

Elsharkawy and Rutherford (2015) conducted a survey questionnaire to investigate householders' awareness of the relationship between behaviour changes and household energy efficiency. They did this in one pilot area of social housing estates in the Community Energy Saving Programme (CESP) in Aspley, Nottingham. It was concluded that the majority of participants had a basic awareness of energy-saving actions such as 'turn off unwanted lights', 'boil only water needed', 'unplug unused equipment', etc. However, the big challenge is still addressing habitual energy consumption. The researchers believe that policies and schemes need to incorporate awareness-raising strategies. It was stated that tenants may consider using electrical appliances and heating systems to suit their lifestyle and gain reasonable value more than utilising them at optimum efficiency (Elsharkawy & Rutherford, 2018).

Other studies also support the findings (Smith & Pett, 2005; Pretlove & Kade, 2015). Through the investigation of identical low-energy-standard houses in the southwest of the UK, Jones et al. (2016) indicated that buildings with an extremely high degree of insulation may be considered uncomfortably warm. A sample of houses was identified as being susceptible to overheating in the summer period, which was caused by higher insulation and air tightness of the building fabric. However, occupants' energy consumption behaviour such as the use of windows, blinds and ventilation systems may mitigate that influence as it plays an important role in reducing and upsurging indoor air temperature in order to achieve a higher level of thermal comfort.

At first, a few data collection technologies need to be employed, for example, sensors, to gather all necessary behavioural data. Then, the data need to be quantified through the implementation of 'behavioural initiated programmes' such as the International Energy Agency – Annex 66: Definition and Simulation of Occupant Behavior in Buildings. It is indicated that due to a lack of general scientific standards, the model validation technologies are not substantial. In the end, the developed behavioural model can be incorporated into one of the existing Building Energy Simulation (BES) tools such as EnergyPlus, DOE-2, TRNSYS, etc. Several challenges have been identified, such as the requirement of advancements of behavioural data collection and behaviour quantifications processes. With the same purpose, an innovative simulation approach has been developed by Sun and Hong (2017) to take account of occupants' behavioural issues in energy performance. By tackling five important measures, lighting, plug-loads, Heating Ventilation Air Conditioning and Refrigeration systems, openings and thermostats, the potential energy savings affected by behavioural issues is 41% of total energy consumption.

In addition, occupants' behaviour can be determined by many factors. Through a survey of occupants' behaviour in New Zealand, Isaacs et al. (2010) find New Zealanders are comfortable living at a lower temperature than people in the rest of the world. The survey also identifies that New Zealand homes tend to have less heating appliances because central heating systems are not commonly implemented in homes. New Zealanders also seem comfortable heating rooms individually. It is believed that there are certain normative standards or 'pride' factors that drive their energy use patterns. The importance of occupants' energy consumption behaviour on home energy performance are explored in this section. The factors that potentially impact on the way occupants operate their homes are broadly discussed by researchers and scholars. However, to incorporate behavioural factors into the design of retrofit requests further investigation. The approaches applied for doing that are also discussed in the following section.

Mechanisms for Investigating Energy Consumption Behaviour

Home energy performance is believed to be significantly determined by energy-related behaviour, which is influenced by household profiles (Santangelo & Tondelli, 2017). Therefore, exploring the interrelationship between home energy performance, occupants' energy behaviour and their socio-demographic characteristics is very important (Steemers & Yun, 2009). The exploration methods of data collection are mainly from questionnaire surveys, interviews, observations and reading statistical data from smart meters or other types of In-House Displays (IHDs).

A number of instruments are discussed in this section to identify the influential factors of energy consumption. There is an indication that occupants' selection of retrofit measure is determined by a number of socio-demographic characteristics, such as age, housing typology, income level and education level (Poortinga et al., 2003). Using a questionnaire survey of 2,000 randomly selected households in the Netherlands by post, respondents' preferences for a number of retrofit measures were explored in the format of '1-5' Likert Scale questions. The measures range from energy conservation actions to retrofit installations such as 'appliances not on stand-by', 'switch off lights in unused rooms', 'shorter showers', 'house insulation', 'applying radiator insulation', etc. With 455 returned questionnaires, Poortinga et al. (2003) argued retrofit measures are more welcomed than behavioural measures. The latter are applied to reduce indirect energy use. Therefore, they did not have the full awareness of the respondents. Retrofit design needs to be strategic by addressing the identified socio-demographic groups.

Another pilot study identified that as well as the conventional 'top-down' approach the success of low carbon retrofits needs to incorporate a 'bottom-up' approach that focuses on occupants' understanding of energy consumption behaviour. In order to measure this for the CESP scheme in Nottingham, a 'pragmatic' approach was employed with a mixed quantitative and qualitative research methodology. Data was collected by carrying out questionnaires and accessing documents and databases in two phases, before and after technical intervention. This was in order to investigate the implications of occupants' behaviour and its influence on the uptake of schemes and policies. The data was analysed by employing SPSS (Statistical Package for the Social Service) to identify the correlation between energy consumption behaviour and change in energy performance. The same approach was applied by Watts et al. (2011) in order to explore the real impact of energy performance certificates (EPCs). They focused on 2,000 new homebuyers in Southampton. The questionnaire survey was carried out in the first year the EPCs were released. It found that although the EPCs had been implemented, this did not impact much on occupants' decision-making processes. This shows energy efficiency is not the most significant issue considered by householders. It is suggested that the certification schemes need to be further refined in order to be widespread and meet occupants' interests.

Guerra-Santin (2011) also tried to examine the impact of household profiles and their energy consumption behaviour on home energy performance through a household questionnaire survey that was carried out with 6,000 households by the OTB Research Institute in two districts of the Netherlands. There were 313 usable questionnaires completed which comprised 5% of the data sample. The reason for the low response rate was found to be due to lengthy questionnaires, which required a lot of time to be completed. Additionally, questions relating to respondents' lifestyles and possessions might be too sensitive to answer. The study concluded that energy use in space heating is significantly associated with household profile and behavioural patterns. The efficiency of retrofit programmes is different depending on household groups as their profiles significantly impact energy use patterns.

Apart from the research methods encompassing questionnaire surveys to collect quantitative data, interviews and other approaches have also been adopted by scholars for the same purpose. Lowery (2012) conducted research using a practical low-carbon retrofit case study of a social housing estate in collaboration with Gentoo Group, which was responsible for the project's construction and ownership. Applying a qualitative approach to data collection, interviews were carried out with local householders, covering aspects of their lifestyle patterns and the impact of these on energy performance. This was divided into two phases: before technical intervention and after technical intervention. The recorded data was processed and organised by employing template analysis. MAXqda which is used for analysing qualitative and mixed method data was applied to analyse the frequency of occupants' behaviour. Ultimately, an energy-related lifestyle pattern was identified, and key energy consumption behaviour was pinpointed.

It is noted that mixed methods are widely adopted in this domain to capture data through questionnaire surveys supplemented with feedback from interviewees. Lindén et al. (2005) conducted a household questionnaire survey with 600 respondents followed by a focus group interview with 12 households. Similar to Guerra-Santin (2011), the questionnaire was constructed with '1-9' Likert Scale questions in order to explore occupants' environmental attitudes and energy consumption behaviour. The addresses of households and birth years were accessed through the energy supplier's database. The selection of a data sample for questionnaire surveys is based on occupants' age groups and housing typologies. As a result, several poor behavioural measures were identified along with suggestions for future policymaking in the aspects of economic measures, administrative measures and user-friendly technologies.

With mixed methods adopted, Pelenur (2013) also explored the impact of occupants' behaviour towards energy performance such as occupants' opinions and motivations. In order to explore the implications of those factors on energy performance, a thematic analysis was employed. The researchers carried out semi-structured interviews on householders' attitudes to retrofits (the motivations and barriers) and home energy consumption. Since the interviews were semi-structured, with multiple responses from a single participant, the single-by-multiple response test was employed, instead of Pearson's chi-square test of association.

Differing from conducting a questionnaire survey or interview, empirical tests of several types of short-term and long-term occupant feedback techniques were conducted for two different housing typologies. Significant gaps between actual and predicted performance were identified in the aspects of indoor CO_2 level, daylight, indoor temperature and noise transmission. It was shown that the selection of user-centred retrofit measures is significantly determined by those issues which need to be considered by designers of the retrofit programme to optimise time, cost and process.

On the other hand, a conceptual framework to identify the most influential factors of energy consumption behaviour before and after occupancy-focused technical interventions was developed by Karatas et al. (2015). Consequently, suggestions on technical interventions that are targeted to different energy use patterns and household profiles could be provided through the framework. The criteria examined include occupants perceived retrofit interventions with their levels of involvement, occupants' knowledge in interpreting retrofit measures and energy conservation, and the opportunities for uptake of the programme in terms of availability, accessibility and time.

The determinants of occupants' energy behaviour's influence on home energy performance have been identified by looking at a variety of approaches. Understanding of those determinants will help promote occupants' energy consumption behaviour for home energy efficiency. Strategies adopted for effective behavioural intervention are demonstrated in the following section.

Strategies to Improve Energy Consumption Behaviour

The impact of occupants' energy behaviour on home energy performance has been broadly discussed in different aspects. It is believed that providing effective strategies to rationalise energy use behaviour is crucial for bridging the BPG. Steg (2008) stated that behavioural strategies can be provided in two ways: informational strategies to directly improve occupants' knowledge, awareness, perceptions and motivation; and structural strategies which aim to create a more attractive context in which decisions are made. The former is conducted through the provision of relevant information or education while the latter is conducted through the improvement of infrastructure or services. Both strategies are initiated by designers of a retrofit programme.

To promote occupants' energy consumption behaviour with long-term and sustained change is not easy as ingrained habits may have been adopted for decades. Therefore, the method of regulating occupants' energy consumption behaviour is crucial. It is believed that habitual actions are not easily changed, as a habitual action 'undermines attention to information of other possible courses of action'. It is suggested use should be made of disrupting methods to change people's behaviour such as making specific plans or changes in the living environment (Verplanken et al., 1997). The challenge of changing occupants' energy consumption behaviour is noted by Barthelmes et al. (2018), who stated change of behaviour requires long periods to internalise and adopt new behaviour. Earlier research also highlights the importance of informational interventions when regulating occupants' energy consumption behaviour. An experiment on providing informational interventions for building users was conducted by focusing on a large office building across two consecutive winters. The impact of the interventions was obvious within a short period of time and remained effective after a year.

To cope with the 'rebound effect' and increase domestic energy efficiency, Walker et al. (2014) discussed the possibilities of regulating occupants' behaviour in order to allow retrofitted social housing properties to reach expected energy performance levels. A number of factors have a significant impact on occupants' energy consumption behaviour, such as occupants' relevant knowledge, their long-term habits, the adopted retrofit measures, quality of installations, conveniences of the measures, thermal comfort and external environments. In order to improve energy consumption behaviour, knowledge and skills need to be delivered along with the implementation of physical installations. In the case under study, providing guidance of necessary knowledge and skills to occupants only worked for heating control and showers but failed to affect the use of thermostats and radiator valves. The approach only seemed effective on existing technologies but not on newly installed technologies. It is recommended (Walker et al., 2014) that an integral approach needs to be adopted by policymakers to help social renters balance various control devices and understand how all parts of the system work. This could be achieved through the interactions provided by end-user devices, such as energy management applications.

Further, the importance of inhabitants' energy consumption behaviour was emphasised by Baborska-Narozny et al. (2017). Based on a case study of 18 overheated flats in a tower block, they suggested that although overheating in retrofitted dwellings is caused by inappropriate approaches, proper inhabitant behaviour will significantly reduce this issue. Baborska-Narozny et al. (2017) also indicated that the conventional way of providing user guidance and home demonstration tours is proven to have failed. More widespread behavioural practices are those that can be easily seen by neighbours, such as opening the windows to mitigate overheating. Additionally, occupants may change their behaviour effectively through monitoring and sensory systems. The monitoring and sensory systems could provide instant feedback of energy use to occupants and remind them to improve their energy behaviour. As a result, a collective approach needs to be carefully designed and implemented.

Failures to provide appropriate and sufficient information and education to end-users have also been argued by researchers. Through analysis of more than 100 retrofit programmes, Moloney et al. (2009) argued that the rational choice model for decision-making which is based on technology and economic measures failed to achieve expected energy efficiency. The reason is that the model does not consider the impact of social context, such as systems, standards and norms. Additionally, improvement of energy consumption behaviour through information and education may not be stable as it requires long-term interactions. Owens and Driffill (2008) suggested that mixed messages could generate confusion for occupants. For example, occupants are urged to turn on the ventilation system to improve indoor climate, but it is suggested not to do this in terms of utility price at peak hours. Some approaches are also difficult to follow as they conflict with everyday life. Additionally, Owens and Driffill (2008) argued that the system adopted to promote occupants' energy consumption behaviour may not reflect reality. Therefore, a systematic approach needs to be adopted to incorporate complex socio-economic and socio-technical systems.

Moreover, through interviews of middle-class homeowners and social housing tenants, Gram-Hanssen and Georg (2018) stated that the change of policymaking is significant as it needs to be diverted from focusing on building energy efficiency only to the considerations of how new technologies and measures affect indoor comfort and occupants' everyday life. It is argued that the implementation of new energy efficiency measures needs to facilitate occupants' behavioural changes and actually reduce energy consumption rather than increase it by adding additional appliances. Gram-Hanssen and Georg (2018) indicated that building characteristics and householders' backgrounds need to be taken into consideration to give insights to policymakers for tailored designs.

Apart from promoting occupants' energy consumption behaviour through information and feedback, education is also used as a powerful instrument for behavioural intervention. However, all of these approaches would not work effectively on their own. As previously stated, Moloney et al. (2009) argued that the transition to a low-carbon community is significant as it provides a broad sense of system change and shifts the behavioural intervention from simple behaviour changes to a set of sustainable social practices. It is stated that a systematic practice integrating community approaches and their associated technologies, infrastructures and institutions needs to be adopted. The relationship between components of the system needs to be understood. Gram-Hanssen (2010) explored the impact of user-centred approaches on heating consumption by adopting a practice-theory approach. By analysing heating and occupancy patterns of five households with different socio-demographic characteristics living in similar buildings, the indication is that a systematic approach needs to be applied to provide comfort practices for different household typologies. This includes technologies, embodied habits, knowledge and meanings. The advantage of community-based programmes is also highlighted by Heffner and Campbell (2011) as that occupants' awareness of retrofit barriers has highly increased due to multiple benefits possibly achieved in the same environment thus bringing a stronger impact. The home energy performance which only represents a single issue in the complex framework needs to be considered together with non-energy measures.

Another possible way of regulating occupants' energy consumption behaviour is to provide them with regular feedback on energy usage. Energy consumption indicators were installed in 44 UK households for two months. Among these properties, energy savings increased by 10% in 14 properties and by 20% in 6 properties. Barthelmes et al. (2018) also highlighted the importance of energy feedback that changes need to be attractive, easy to understand and adaptable on a daily basis. In order to incentivise occupants to change their energy consumption behaviour, real-time energy consumption and indoor environmental quality (IEQ) feedback need to be provided in a user-friendly way and supplemented with tailored information campaigns. Further, suggestions on how to optimise energy performance based on feedback ought to be provided.

Based on the review of strategies to improve occupants' energy use behaviour, energy information and education provided by the government or local authorities may be difficult to popularise due to differences in social and technical contexts. However, fully relying on occupants to take the initiative will also prove ineffective. As a result, behavioural intervention through smart technology could be deployed to mediate between the two different methods. The impact of smart technology on occupants' energy consumption behaviour is discussed in the next section.

Behavioural Interventions for Home Energy Conservation

The role of IHDs and energy management applications in behavioural change has been recognised in many studies (Schultz et al., 2015; Zhang et al., 2016; Pritoni et al., 2017; Shetty et al., 2015). The important role of smart technology in behavioural intervention relies on the transition of the UK's smart grid energy network and the fast development of the Advanced Metering Infrastructure (AMI). This section starts with a brief account of the development of smart technologies and follows with a discussion of the impact of smart meters and energy management applications.

Development of Smart Meters and Energy Management Applications

During the transition of the UK's energy network in the last few decades, the importance of feedback mechanisms from energy end users has been noted. The UK's new bi-directional energy network requires more installations of smart devices at end users' level to effectively manage energy. Other than BES tools, which are mainly used by energy experts, energy monitoring and management applications have been widely employed in households. The most distinct motivation for this trend was the mandatory requirement of implementation of smart metering devices by the UK government. The government stated that in order to secure affordable, secure and sustainable energy supply, smart metering devices would play an important role in low-carbon energy transitions. The government set up a target that smart metering devices need to be installed in each UK home across the country by 2030. Energy companies have been required to lead the roll-out of smart meters.

The benefits of smart meters include a series of intelligent functions, such as providing near real-time energy use monitoring to occupants, effectively organising energy consumption behaviour, paying exact bills instead of estimated ones and more flexibly switching energy suppliers. The smart metering device may also be connected to an in-home display for further details on energy consumption and credit balance. Energy companies are responsible for installing smart meters taking into account energy consumers' interests in the aspects of privacy, security, product quality and the needs of vulnerable consumers. The smart meter captures near real-time energy consumption of each household and transmits data back to energy companies for monitoring purposes. Traditional manual meter readings and estimated billings are replaced by accurate bills, which help energy users avoid energy and financial losses.

The development of the UK's energy network strengthens the relationship between energy management level and energy end user's level. It helps energy management diagnose current conditions of energy performance in the domestic sector and more effectively manage energy distribution. Additionally, occupants have much more opportunity to clearly understand their energy usage patterns from a number of end-use devices, such as smart metering, IHDs, desktops and smart phone applications. A variety of tools and applications has been evolving for the purpose of improved energy efficiency of new built and retrofit of existing buildings. Amongst those, BES tools and energy management applications play imperative roles. The former is widely used by policymakers and designers during the design stage of retrofit programmes in order to understand the conditions of targeted properties and identify problems. It has been asserted in the previous chapter that imprecise BES is one of the acknowledged barriers for efficient retrofit programmes.

In regard to energy management applications, those are used by occupants to understand their energy usages and improve energy conservation awareness and behaviour. As previously discussed, the manner occupants operate their homes significantly influences home energy performance. Therefore, implementation of energy management applications as a tool to help achieve the efficiency of retrofit programmes may be considered a viable solution.

Regulating Occupants' Energy Consumption Behaviour by Smart Meters

The implementation of smart meters indicates positive results in Northern Ireland. Through an experimental large-scale case study, Gans et al. (2013) monitored residential electricity consumption since April 2002 when pre-payment meters were introduced. Data collected between two different periods (with pre-payment meters and with advanced metering systems) show a 11–17% decrease of energy consumption.

Zhang et al. (2016) indicated promising energy savings by implementing smart meters and in-house displays in Shanghai, China. They demonstrated 9.1% energy savings and 11% financial savings in targeted homes with IHDs. However, further work can be conducted to show how to successfully affect energy performance with IHD feedback in different national and cultural contexts. The importance of smart meters and the great contribution brought by in-house displays are also recognised by Schultz et al. (2015). To progress further, Schultz et al. (2015) investigated what types of information and feedback framework provided through the IHD will make the most significant impact on the occupants. The research was carried out to monitor energy consumption through an experimental case study in which IHDs with different frameworks were installed. Energy consumption was compared between two periods: just after the installation of IHD and after three months. This showed that although occupants prefer cost-framed feedback, significant energy reduction was demonstrated in the examples of households with electricity-framed feedback. The electricity-framed feedback displays energy usage by kilowatt hour, but cost-framed feedback displays energy usage by the amount of energy bills. Further, educational information provided by IHDs may increase knowledge but may not, in effect, help reduce energy consumption. The aspects shown on IHDs are also examined by Stromback et al. (2011), who showed the most effective aspects for energy reduction are up-to-date reporting on consumption, historical feedback and recent bills. The comparison of consumption which was believed effective did not help with energy reduction in the pilot study of this report.

Although the implementation of smart metering is believed to be an effective method for energy conservation, a number of studies do not advocate applying smart meters. Research in the Netherlands and US find that the negative aspects smart meters bring are more obvious than their advantages due to varying reasons. General reasons include the invasion of privacy, increased energy consumption due to smart technology and general unwillingness of investment in this new device.

McDaniel (2009) claimed that occupants' privacies have been unintendedly publicised due to the implementation of smart meters, especially information about their habits and energy-using signatures. The reveal of private behavioural data is always aligned with financial or political incentives and from the utility companies. He proposed that in order to solve this problem, the government needs to set up strict rules for the smart grid on how the data will be collected and with whom data can be shared. By raising the same concern of retaining customers' privacy, Raj Rajagopalan et al. (2011) urged that existing solutions may enhance privacy protection but in doing so sacrifice the benefit of the utility. In order to balance both, a novel privacy-utility trade-off model has been developed and tested on a stationary Gaussian Markov model in order to balance privacy and utility requirements. In detail, the model is able to quantify the utility-privacy trade-off in smart meter data. The guarantee of privacy is defined based on the calculated least information leakage while the utility is preserved.

As mentioned, it has been proved that real-time feedback from IHDs may not help to reduce energy consumption effectively in particular circumstances. Petkov et al. (2011) also supported the view that although an increased adoption rate of energy monitors and displays is demonstrated, they are not able to increase energy users' motivations by addressing their particular needs. In addition, the interface of current energy monitors lends itself unengaging and mundane. The technical feedback provided by IHDs is dry numbers and intangible units, which are not suitable for most occupants, this point of view is supported by Carroll et al. (2014).

Through an experimental case study in Ireland, it was found that although real-time energy suggestions and educational information can help increase occupants' knowledge and awareness, this fails to reduce short-term energy demand. Carroll et al. (2014) argued that the reason for this is the double-side impact of the IHDs. Real-time feedback can be seen as either a reminder or motivator for occupants. Focusing on a larger scope, Hargreaves et al. (2017) examined the impact of smart home technologies based on the AMI. Through the implementation of smart home technologies in 10 UK homes for nine months, it has been shown that those technologies have not substantially reduced energy consumption. Besides, the same study highlights that training occupants and making them familiar with new technologies is important but time-consuming.

Improving Energy Management by Smart Meters

The advantages that smart meters and AMI bring are widely recognised by a number of researchers and scholars (Gans et al., 2013; Stromback et al., 2011; Schultz et al., 2015). However, efforts are not only made towards effectively reducing energy consumption by using smart meters but also towards improving the effectiveness of the energy management system. The challenges of energy load are variable demand and distinct peak loads compared with other periods. The introduction of a smart grid in the UK has improved monitoring and management of electricity load, which is commonly benefited by smart meter devices. Other grid-connected devices also have big potential for improving the electricity load (Cetin & O'Neill, 2017).

A new intraday energy load forecasting system was introduced by Quilumba et al. (2015) to replace the conventional aggregated forecasting system at the management level. It enhances energy management and operations by identifying groups of occupants with similar energy use patterns before performing load forecasting. This is supported by others including Haben et al. (2016). The same approach is employed by Hayes et al. (2015) but with an adverse result. Their research demonstrates that there are a few limitations on short-term load forecasting using smart meter data. They also indicate similar prediction accuracies between using real-time data and historical data from smart meters.

It was found that the current electricity load for residential use is highly variable and the fluctuations of the electricity load may lead to financial loss. As the diversified electricity load is influenced a lot by occupants' behaviour, a new modelling and optimisation framework has been developed based on smart meter data to forecast residential electricity demand and improve the efficiency of the energy network. This first applies a nonintrusive load monitoring algorithm to monitor, for example, an air conditioning unit only instead of monitoring a whole building. Further, a smart meter and thermostat are applied to form a simplified model to predicate energy consumption of the air conditioning unit. The end result is electricity load of a whole community, which will be reduced by shifting a set point of thermostats in each home via a centralised control scheme. One year after their initial research, Perez et al. (2017) explored the correlations between temperature and energy use patterns among different housing typologies. They show that the energy-consuming behaviour of air conditioning units will keep steady and low until a certain changing point of the temperature. Subsequently, a linear increase and an 'energy slope' gradually occur. As a result, this changing-point model can act as a screening tool to compare energy use patterns between different houses and target the ones with the most significant energy slopes. This helps to effectively identify peak load candidates for energy reductions. The impact of smart meters is broadly discussed in terms of regulating occupants' energy consumption behaviour and improving energy management in this section. With the raised-up awareness of smart meters and IHDs, the strategies for engaging them in behavioural intervention are discussed in the following section.

Behavioural Interventions for Behavioural Change

Along with the development of IHDs and energy management applications, interactions between them and humans were predominantly covered in the field of Human-Computer Interactions (HCI).

The approach that is applied in order to address climate change and environmental sustainability through HCI is called 'sustainable HCI' (DiSalvo et al., 2010). The aim of sustainable HCI is to motivate people to save energy and live in more sustainable environment. Energy management applications which are computerised could potentially motivate people through facilitating the adoption of pro-environmental behaviour. In order to understand people's motivation for behavioural change, psychological knowledge has also been embedded. It must be stressed that change of behaviour is based on two psychological models: rational-choice models and norm-activation models, where the former indicates that people tend to change their behaviour to avoid punishment or seek rewards and the latter indicates people do so to remain moral or due to personal norms. Specific elements of how energy management applications influence users' motivations need to be discussed.

Energy management applications may act as an effective media to encourage behavioural interventions to occupants. Several studies discussed the impact of different types of interventions on stimulating occupants' energy consumption behaviour in terms of curtailment and investment. There are a variety of intervention strategies that can influence occupants' behavioural decisions. They are categorised by Han et al. (2012) as 'antecedent interventions', 'consequence interventions' and 'structural interventions'. 'Antecedent strategies' influence occupants' energy consumption behaviour at an early stage without the condition of each home's energy performance being known. These strategies strengthen occupants' sustainable awareness and help them to be aware of potential energy problems. 'Consequence strategies' involve knowing the conditions of energy performance by providing feedback reports or rewards. 'Structural strategies' focus on contextual changes, such as financial legislation, in order to facilitate the improvement of energy consumption behaviour.

According to Figure 3.1, both 'structural strategies' and 'antecedent strategies' only need to take into consideration a series of background information to influence occupants' motivations. This includes, for example, occupants' socio-demographic characteristics, their financial abilities and housing characteristics. However, 'consequence strategies' impact occupants' motivations by tackling specific energy problems and providing energy suggestions based on occupants'

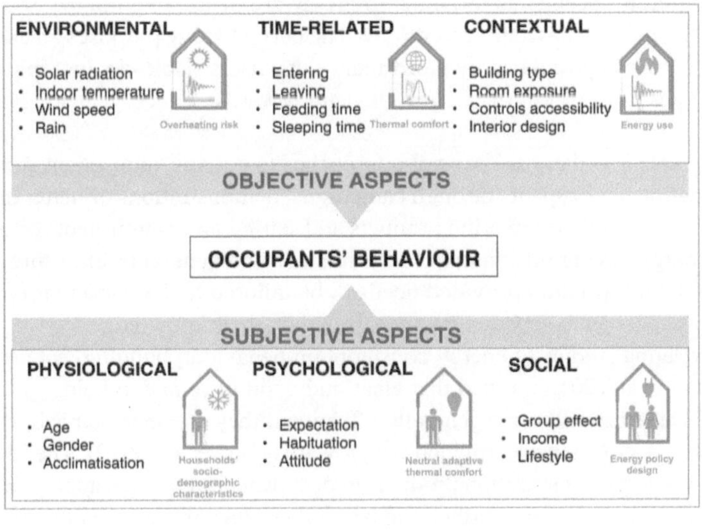

Figure 3.1 Behavioural intervention strategies model. Drawn by author.

Table 3.1 Possible intervention approaches

Antecedent strategies			
Information			
Mass media	Website	Brochure	TV/Radio
Newspaper	Posters	Mobile advertisement	
Demonstration			
Neighbours' performance	Nearby companies	Association ambassadors	Display measures in model
Free products			
Commitment/goal-setting			
With municipality	With homeowner association	With own household	With neighbourhood
Consequence strategies			
Feedback			
Energy use of electronic appliances	Benchmark comparable households	Current energy usage	Behavioural suggestions
Rewards			
Awards/prizes	Tax deduction from land value tax		
Structural strategies			
Financial legislation			
Price policies	Product tax	Energy tax	Removal tax
Subsidies	Costs measures	Loans	Low rents
Green loan	Legislation	Building regulation	Building performance certificate

backgrounds. Therefore, occupants are much more aware of the importance of change and alter their energy consumption behaviour accordingly. The current study proposes an energy application that requires the input of current energy performance and users' socio-demographic characteristics and produces output in the form of suggestions and rewards. Hence, it belongs to consequence strategies.

As shown in Table 3.1, there are several approaches to deliver 'antecedent strategies', such as providing information to occupants through campaigns, demonstrations of better energy consumption behaviour, free energy conservation products and setting up commitments or goals. It is stated that to provide energy-saving information in the manner of a general brief or through mass media is not effective. The information provided needs to be tailored to the particular type of household to actually increase occupants' awareness and strengthen their knowledge.

Based on correlating studies of energy consumption behaviour, building technology and energy performance, Zhao et al. (2016) argued that clear and valid information being given to homeowners is significant. Demonstrations are generally effective if they are understandable and meaningful. They will be more effective when combined with rewards (Han et al., 2012). The same study asserts that providing free energy conservation products to occupants is a passive approach as it only opens an opportunity for them to try out. Further, the larger the scope of the commitment made, the more possible it is that occupants will change their behaviour. It was found by Lucas et al. (2008) and

Cialdini (2005) that if commitments to energy savings are made in public, participants tend to more effectively regulate their energy consumption behaviour due to expectations of the society.

The aim of consequence strategies is to stimulate a change in occupants' energy consumption behaviour by showing the relationship between them and energy performance. Energy use feedback is usually provided based on the assumptions of energy performance. Abrahamse (2007) argued that the approach would prove effective if the feedback was frequently provided and especially connected with energy-saving commitments. The effectiveness of rewards was proven by Geller (2002) but was thought to be short-lived. In addition, financial legislation is an effective approach to stimulate energy consumption behaviour changes by changing the energy use context. Legislation on dwellings in terms of energy performance certificates and more specific building regulations has been undertaken by the government. It is also most effective to combine financial factors such as subsidies or fines/taxes (Han et al., 2013). Xu et al. (2012) and Zhao et al. (2016) also indicated that the integration of a monitoring system into tax incentive programmes may as well make the approach more robust and impactful.

In addition, energy management applications are developed based on the smart energy management system with metering devices to help occupants better understand their energy consumption and effectively save energy. The energy management applications are mainly developed by major energy companies and have become one of the most important components of today's smart energy management. Additionally, scholars and researchers have been working on the development of different types of energy management applications which are driven by occupants' behaviour. The reason for that is because generalised behavioural suggestions based on statistical thermal comfort will not be suitable for each individual case. Thus, energy performance will be improved by compromising an individual's living comfort. The successes and failures of energy management applications are discussed and presented in the below sections.

Raising awareness is needed to make occupants familiar with using energy applications and, in a further step, with the smart energy management system as applications are mainly developed based on it. Further, occupants' energy conservation awareness also needs to be motivated. The first stage of that could be the roll-out of smart meters as this increases interactions between occupants and the energy management system and generates behavioural change. A review of energy management applications developed by energy suppliers and other application developers is presented in the next section.

Applications from Major Energy Companies

It is asserted that many applications in the current market have employed relatively effective behavioural intervention approaches (Ozarisoy, 2022; Ozarisoy & Altan, 2024a). By focusing on application features and impact, a comparison of the existing applications was undertaken in order to inform more effective methods of behavioural interventions. Some of the remarkable features will be identified and quickly demonstrated to participants in the focus group to critically appraise benefits, practicality and feasibility.

Several significant energy management applications developed by major energy companies such as British Gas, EDF Energy and E.ON were selected for comparison. The majority of the selected tools focus on applications developed by independent application developers as they normally integrate more innovative application features, for example, real-time monitoring, remote control, and energy consumption comparison, as shown in Figure 3.2.

As shown in Figure 3.2, energy management applications which are developed by energy companies mostly tend to provide easy and convenient customer experiences in order to increase

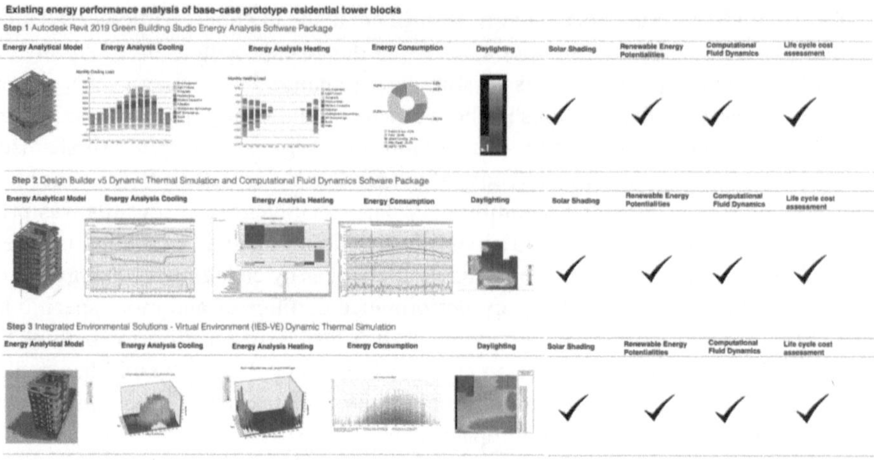

Figure 3.2 Comparison of energy management applications in the current market. Drawn by author.

their customers in the market. Thus, there are many similarities in terms of features and functions. Although some of the energy conservation approaches have been gradually integrated into these applications, they focus more on better user experience and account management for better customer service. Besides, most energy management applications developed by energy suppliers, strive to facilitate convenient communication between customers and their technical teams to solve problems and schedule home visits. However, a potential conflict has been identified by Hannon et al. (2013) that Energy Utility Companies (EUCo) make a profit on selling energy units. The models that are developed by EUCo are coupled with their revenue. It may not help towards profits to thoroughly reduce unnecessary energy consumption of clients. On the other hand, more efforts need to be made to facilitate the development of Energy Service Companies (ESCo) because they do not sell energy units.

By contrast, applications developed by independent application developers are more stimulating, such as Efergy Engage, OVO, Voltaware, Wink, RetrofitLab, HIVE Active Heating, Joulebug, Carbon Calculator, HomeBeat and Homeselfe. A comprehensive comparison of these applications is carried out, which indicated those tools are more effective than the ones developed by energy suppliers in terms of energy savings by motivating behavioural change. This is also supported by Hannon et al. (2013), who stated EUCo do not thoroughly help energy reduction as their nature is to make a profit. More interesting aspects are found in the more innovative applications, such as 'recommended retrofit scenarios', 'tailored behavioural suggestions', 'energy saving comparisons' and 'energy performance mock-ups'. These applications are compared and hereby discussed.

Although the more innovative and advanced aspects of applications have been recognised, they have not been widely implemented and incorporated into existing energy management systems. Several applications are able to automatically read domestic energy consumption and assign consumption to individual appliances by connecting to a smart meter and associated sensors, for example, Voltaware, Wink, HomeBeat and Eferge Engage. Users of those applications are able to know real-time home energy consumption in different categories of energy consumed and control the usage of each category remotely. Furthermore, behavioural suggestions are constantly given based on thresholds set up by the users.

In addition, gamification design is believed to attract more attention and increase motivation for occupants to rationalise their energy use. This has been incorporated into the design of applications, such as JouleBug and HomeBeat. In considering disabled people, voice recognition features are incorporated in some applications as well, including Wink, OVO and HIVE Active Heating. Additionally, Wink not only provides remote control for appliances and openings but also ensures a more secure home by providing surveillance and detection services when users are away (Wink, 2018). Moreover, Home Selfe puts the function of energy simulation for energy users first. Energy performance diagnosis is conducted based on entering a number of simple parameters, such as building type, heating type and number of occupants. Then, suggestions for potential improvements are provided to users through energy performance optimisation options. Occupants' knowledge of their home energy performance may also improve during this process.

Many innovative aspects and successes of these applications are identified, but some shortcomings are also recognisable. Several applications focus on real-time energy consumption and remote control of electricity appliances and heating systems. This facilitates energy management but may not be adequate to achieve tangible savings as there is a lack of interaction with users to stimulate incentives. In some of the applications, users are allowed to set up their preferred energy alert boundaries while they may not have substantial knowledge of good energy performance practices. Thus, the parameters set up by occupants may not be effective. Additionally, gamification design, which potentially increases occupants' motivation, could be utilised better in the future. For instance, gamifying energy-saving actions into tasks with points earned and rankings may motivate occupants to adopt more energy-saving actions and compete with neighbours or friends.

Due to the rising concern about occupants' energy consumption behaviour, a number of occupant-driven applications are developed to undertake in-depth exploration of occupants' thermal comfort and provide tailored behavioural suggestions accordingly (Konis & Annavaram, 2017). In order to achieve that, occupants' energy use patterns and indoor environment are monitored by sensors for analysis. A discussion of this is presented in the next section.

Occupant-Driven Energy Management Applications

The behavioural suggestions provided by energy management applications are normally general guidelines, which are based on static comfort criteria. It is hard to provide tailored behavioural suggestions to each individual with different thermal comfort requirements. D'Oca et al. (2014) argued that standardised parameters and fixed schedules are not reliable to promote occupants' behaviour. The data needs to be obtained from field/on-site measurements. To overcome this hurdle, occupant-driven energy-saving strategies have been considered. One of the differences between typical energy management applications and occupant-driven applications is that the latter not only requires smart metering devices but also involves a series of sensors fitted into occupied spaces.

An innovative mobile-based application, Occupant Mobile Gateway, was developed to first acquire occupants' thermal comfort and consequently provide behavioural suggestions accordingly. By connecting to embedded sensors, the application captures and models personal thermal comfort preferences. Potential savings caused by occupants' behaviour are recommended through a data-driven thermal management programme without sacrificing occupants' personal comfort. Studies of occupant-driven energy efficiency have been considered in recent years (Pritoni et al., 2017; Shetty et al., 2015; Gupta et al., 2014).

With the same purpose, a programme for providing energy and water advice to social housing occupants based on their energy- and water-related behaviour was developed by Laskari et al. (2015). In contrast to smartphone-based applications, this programme is based on the energy use display interface from a TV screen and designed in the framework of Information and Communication Technology. Among energy-saving measures with different levels of investment, it has been found that the most suitable approaches for low-income social housing households are cost-effective approaches without major interventions in homes. Advice is given on a TV screen with different categories, such as electricity, heating, gas, hot water, etc. According to the user experience survey, the most followed energy-saving advice is related to electricity use, followed by IEQ settings, and general user behaviour. Advice about heating and gas was found to be difficult to follow.

Through the review of a number of data-driven models in the world. Hong et al. (2015a) developed a framework to represent the influence of occupant energy consumption behaviour with four key elements: drivers, needs, actions and systems (DNAs). They argued that relying technology alone could fail to achieve energy efficiency as a number of aspects involved in energy performance are influenced by occupants' energy consumption behaviours, such as building management, operation, design and retrofit. The framework was also expected to be used to standardise the impact of occupant energy consumption behaviour by the international research community. The DNA framework could be also adopted for the future development of occupant-driven models by researchers and scholars.

Based on the abovementioned examples of occupant-driven energy tools, it can be seen that in order to achieve greater energy savings, the combined approach of utilising both BES and energy management applications has been considered the latest trend. The most advanced strategy is to first capture and measure occupants' personal preferences through mobile apps and then set up tailored approaches accordingly through BES and energy management systems.

Other Energy Management Applications

Apart from typical energy management applications developed by energy companies and occupant-driven applications, scholars and specialists also focus on developing more innovative application suites for various purposes. Due to the unengaging formats of current IHDs, a novel and motivation-specific energy management application, EnergyWiz, was proposed by Petkov et al. (2011) in order to provide more effective behavioural interventions to users. EnergyWiz was tested by 17 interviewees through semi-structured interviews. Occupants were actively motivated in different approaches such as energy self-comparisons, energy comparisons with neighbours in the same conditions and the challenge tasks of the competition features. However, the ranking system was only preferred by part of the interviewees. Further, competition with friends was seen as preferable by the interviewees than with other anonymous users.

The PowerPedia application aimed to increase users' energy consumption behaviour through action-guiding and visualised feedback rather than traditional feedback with values only. By connecting to smart meters, users are given the opportunity to understand the energy consumption of each individual appliance with a variety of analyses available, such as historic energy patterns and accumulated consumption. Petkov et al. (2011) introduced a ranking system that was also developed in PowerPedia in order to show application users how they performed compared with their neighbours. Energy comparison can be conducted at the household level or device level. In addition, energy conservation measures may be uploaded onto the application and shared with other users. As a result, application users could clearly understand the energy efficiency levels of their devices and methods to improve them. PowerPedia also introduced a socialised platform for energy feedback which has also improved the overall outcome.

According to McKechnie (2015), as a complement to the physical energy monitoring device, EnergyCloud has been developed in Canada to control energy performance in the domestic sector on smartphones. A comparison of energy consumption can be provided on this application, which allows occupants to upload and compare their energy consumption with their neighbours. The same aspect can be also seen in Chai Energy, which provides not only neighbour energy comparisons but also energy tariff comparisons. Additionally, the application separates the energy performance of weekdays and weekends into different scenarios in its data analysis section. What can be noted is that the Chai Energy application analyses the feasibility of installing solar panels on the roofs of occupants. Through analysis of local weather, roof area, cost and current tariffs, occupants will be informed if it is a good choice to install solar panels. According to reviews of the abovementioned energy management application, more functions have been developed for occupants' interests.

Recent research also indicated an innovative mobile-based application for occupants to be informed of anomalous energy consumption (Ozarisoy & Altan, 2024b). It is argued that some of the existing approaches failed to rapidly obtain and analyse present data, so do not effectively promote occupants' behavioural changes. However, this framework is able to provide a real-time early warning interface to occupants who lack in-depth technical knowledge. The energy data is then collected by sensors and analysed through an enhanced module to understand users' normal energy use patterns. Then, predictions will be displayed through a smartphone application to forecast unusual energy uses and alert users. The records of anomalous data will also be shared with building managers for further improvement and retrofit strategies. Some of the newly developed application features which used to be only considered for professionals have now been applied to normal energy users, for example, feasibility analysis of solar panel installations. However, this trend also indicates that more efforts need to be made to optimise this function and cover as many retrofit approaches as possible. A few of the most effective energy applications which were developed by application-developing companies have been selected for in-depth discussion in the following paragraphs. Their successes and failures are criticised in order to form the design of an energy management application for the research.

Conclusions

This chapter focused on the impact of occupants' behaviour on home energy performance. It first continues the discussion of the barriers of retrofit programmes in the UK. As discussed in Chapter 2, there are several issues identified during the design and delivery stage of a programme. In addition, the way that occupants operate their homes after the handover of the programme also significantly impacts energy performance. One of the major issues is also the 'rebound effect' and 'building energy performance gap'. Although there has been an awareness of this for several decades, it has not been fully addressed. In a number of case studies, the results of energy conservation following retrofit may have not met the expectations that were set up in the planning stage. The reason for this is that home energy performance is subject to a wider range of variables, such as technical, social and behavioural. However, occupants' behaviour, energy use patterns and socio-demographic characteristics have not been thoroughly explored. It is crucial to consider and incorporate these factors in the design stage of projects as the way that occupants operate their homes will significantly impact on home energy performance.

In addition, approaches adopted to address occupants' behavioural issues and improve retrofit efficiency are also discussed. One of the effective approaches is through the implementation of smart meters and energy management applications. As the importance of smart meters has been gradually realised in the UK and Europe in the last decade, a series of policy and regulatory documents have been published to facilitate their implementation. The impact of smart metering devices on occupants' energy consumption behaviours and home energy performance is also explained through a number of experimental case studies. However, the invasion of occupants' privacy has been flagged as the main concern. Researchers and scholars are making efforts on improving the artificial intelligence with thorough protections on occupants' privacy. Further, a few innovative approaches based on smart meters are introduced to further increase the effectiveness of energy performance.

A review and analysis of energy management applications is presented. The tools that are mostly applied have been developed by the major UK energy companies such as E.ON, British Gas, EDF Energy, Scottish Power and NPower. In addition, there are a number of energy management applications developed by specialising companies. These tools will normally have some innovative aspects in contrast to those developed by the major companies. In general, as the tools are designed for energy users, their aspects show huge variation in comparison to BES tools. Energy management applications have relatively simple tool structures, user-friendly interfaces and visualised simulation results to attract energy users' attention. There are some innovative aspects found in the tools, but they have not been widely spread, for example, comparison scenarios, behavioural suggestions and voice recognition. Due to the nature of application audiences, energy management applications focus primarily on being concise, straightforward, convenient and interactive.

Through examination of energy management applications, this chapter has explored the potential opportunities to contribute to domestic energy conservation in a new and innovative direction. A review of the applications helps to identify gaps and inform the research questions on how to provide more effective behavioural interventions through energy management applications as a potential solution.

References

Baborska-Narożny, M., Stevenson, F., & Grudzińska, M. (2016). Overheating in retrofitted flats: occupant practices, learning and interventions. *Building Research & Information, 45*(1–2), 40. https://doi.org/10.1080/09613218.2016.1226671.

Ben, H., & Steemers, K. (2014). Energy retrofit and occupant behaviour in protected housing: a case study of the Brunswick Centre in London. *Energy and Buildings, 80*, 120. https://doi.org/10.1016/j.enbuild.2014.05.019.

Carroll, J., Lyons, S., & Denny, E. (2014). Reducing household electricity demand through smart metering: the role of improved information about energy saving. *Energy Economics, 45*, 234. https://doi.org/10.1016/j.eneco.2014.07.007.

Cetin, K. S., & O'neill, Z. (2017). Smart meters and smart devices in buildings: a review of recent progress and influence on electricity use and peak demand. *Current Sustainable/Renewable Energy Reports, 4*(1), 1. https://doi.org/10.1007/s40518-017-0063-7.

Cialdini, R. B. (2005). TARGET ARTICLE: basic social influence is underestimated. *Psychological Inquiry, 16*(4), 158. https://doi.org/10.1207/s15327965pli1604_03.

Elsharkawy, H., & Rutherford, P. (2014). Retrofitting social housing in the UK: home energy use and performance in a pre-Community Energy Saving Programme (CESP). *Energy and Buildings, 88*, 25. https://doi.org/10.1016/j.enbuild.2014.11.045.

Elsharkawy, H., & Rutherford, P. (2018). Energy-efficient retrofit of social housing in the UK: lessons learned from a Community Energy Saving Programme (CESP) in Nottingham. *Energy and Buildings, 172*, 295. https://doi.org/10.1016/j.enbuild.2018.04.067.

Firth, S., Lomas, K., Wright, A., & Wall, R. (2008). Identifying trends in the use of domestic appliances from household electricity consumption measurements. *Energy and Buildings, 40*(5), 926. https://doi.org/10.1016/j.enbuild.2007.07.005.

Galvin, R. (2014). Making the 'rebound effect' more useful for performance evaluation of thermal retrofits of existing homes: defining the 'energy savings deficit' and the 'energy performance gap'. *Energy and Buildings, 69*, 515. https://doi.org/10.1016/j.enbuild.2013.11.004.

Gans, W., Alberini, A., & Longo, A. (2012). Smart meter devices and the effect of feedback on residential electricity consumption: evidence from a natural experiment in Northern Ireland. *Energy Economics, 36*, 729. https://doi.org/10.1016/j.eneco.2012.11.022.

Gram-Hanssen, K. (2010). Residential heat comfort practices: understanding users. *Building Research & Information, 38*(2), 175. https://doi.org/10.1080/09613210903541527.

Gram-Hanssen, K. (2013). New needs for better understanding of household's energy consumption – behaviour, lifestyle or practices? *Architectural Engineering and Design Management, 10*(1-2), 91. https://doi.org/10.1080/17452007.2013.837251.

Gram-Hanssen, K., & Georg, S. (2017). Energy performance gaps: promises, people, practices. *Building Research & Information, 46*(1), 1. https://doi.org/10.1080/09613218.2017.1356127.

Guerra-Santin, O. (2011). Behavioural patterns and user profiles related to energy consumption for heating. *Energy and Buildings, 43*(10), 2662. https://doi.org/10.1016/j.enbuild.2011.06.024.

Guerra-Santin, O., & Itard, L. (2010). Occupants' behaviour: determinants and effects on residential heating consumption. *Building Research & Information, 38*(3), 318. https://doi.org/10.1080/09613211003661074.

Gupta, R., & Gregg, M. (2016). Do deep low carbon domestic retrofits actually work? *Energy and Buildings, 129*, 330. https://doi.org/10.1016/j.enbuild.2016.08.010.

Gupta, S. K., Kar, K., Mishra, S., & Wen, J. T. (2014). *Building temperature control with active occupant feedback*. Procceedings of the 19th World Congress, The International Federation of Automatic Control, Cape Town, South Africa, August 24th-29th, 2014. https://www.sciencedirect.com/science/article/pii/S1474667016417210.

Han, Q., Nieuwenhijsen, I., De Vries, B., Blokhuis, E., & Schaefer, W. (2012). Intervention strategy to stimulate energy-saving behavior of local residents. *Energy Policy, 52*, 706. https://doi.org/10.1016/j.enpol.2012.10.031.

Hannon, M. J., Foxon, T. J., & Gale, W. F. (2013). The co-evolutionary relationship between Energy Service Companies and the UK energy system: implications for a low-carbon transition. *Energy Policy, 61*, 1031. https://doi.org/10.1016/j.enpol.2013.06.009.

Hargreaves, T., Wilson, C., & Hauxwell-Baldwin, R. (2017). Learning to live in a smart home. *Building Research & Information, 46*(1), 127. https://doi.org/10.1080/09613218.2017.1286882.

Hong, T., Doca, S., Turner, W. J. N., & Taylor-Lange, S. C. (2015a). An ontology to represent energy-related occupant behavior in buildings. Part I: introduction to the DNAs framework. *Building and Environment, 92*, 764. https://doi.org/10.1016/j.buildenv.2015.02.019.

Hong, T., Taylor-Lange, S. C., D'oca, S., Yan, D., & Corgnati, S. P. (2015b). Advances in research and applications of energy-related occupant behavior in buildings. *Energy and Buildings, 116*, 694. https://doi.org/10.1016/j.enbuild.2015.11.052.

Isaacs, N., Saville-Smith, K., Camilleri, M., & Burrough, L. (2010). Energy in New Zealand houses: comfort, physics and consumption. *Building Research & Information, 38*(5), 470. https://doi.org/10.1080/09613218.2010.494383.

Karatas, A., Menassa, C. C., & Stoiko, A. (2015). A framework for delivering targeted occupancy interventions to reduce energy usage in buildings. *Procedia Engineering, 118*, 752. https://doi.org/10.1016/j.proeng.2015.08.510.

Konis, K., & Annavaram, M. (2017). The occupant mobile gateway: a participatory sensing and machine-learning approach for occupant-aware energy management. *Building and Environment, 118*, 1. https://doi.org/10.1016/j.buildenv.2017.03.025.

Laskari, M., Karatasou, S., & Santamouris, M. (2015). The design of an energy and water advice programme for low-income households. *Energy and Buildings, 110*, 426. https://doi.org/10.1016/j.enbuild.2015.11.008.

Lee, W. V., & Steemers, K. (2016). Exposure duration in overheating assessments: a retrofit modelling study. *Building Research & Information, 45*, 60–82. https://doi.org/10.1080/09613218.2017.1252614.

Lindén, A., Carlsson-Kanyama, A., & Eriksson, B. (2005). Efficient and inefficient aspects of residential energy behaviour: what are the policy instruments for change? *Energy Policy, 34*(14), 1918. https://doi.org/10.1016/j.enpol.2005.01.015.

Lucas, K., Brooks, M., Darnton, A., & Jones, J. E. (2008). Promoting pro-environmental behaviour: existing evidence and policy implications. *Environmental Science & Policy, 11*(5), 456. https://doi.org/10.1016/j.envsci.2008.03.001.

Mills, B., & Schleich, J. (2012). Residential energy-efficient technology adoption, energy conservation, knowledge, and attitudes: an analysis of European countries. *Energy Policy, 49*, 616. https://doi.org/10.1016/j.enpol.2012.07.008.

Moloney, S., Horne, R. E., & Fien, J. (2009). Transitioning to low carbon communities—from behaviour change to systemic change: lessons from Australia. *Energy Policy, 38*(12), 7614. https://doi.org/10.1016/j.enpol.2009.06.058.

Owens, S., & Driffill, L. (2008). How to change attitudes and behaviours in the context of energy. *Energy Policy, 36*(12), 4412. https://doi.org/10.1016/j.enpol.2008.09.031.

Ozarisoy, B. (2022). Energy effectiveness of passive cooling design strategies to reduce the impact of long-term heatwaves on occupants' thermal comfort in Europe: climate change and mitigation. *Journal of Cleaner Production, 330*. https://doi.org/10.1016/j.jclepro.2021.129675.

Ozarisoy, B., & Altan, H. (2023). State-of-The-Art I: energy efficiency directives and policy aspirations in retrofit interventions. In: *Handbook of retrofitting high density residential buildings*. Springer. https://doi.org/10.1007/978-3-031-11854-8_2.

Ozarisoy, B., & Altan, H. (2024a). *Energy policy in the eastern mediterranean climate*, Taylor & Francis Group, Routledge Explorations in Energy Studies, https://www.taylorfrancis.com/chapters/mono/10.4324/9781003393092-2/energy-policy-eastern-mediterranean-climate-bertug-ozarisoy-hasim-altan.

Ozarisoy, B., & Altan, H. (2024b). *A novel methodological framework in energy policy*. Taylor & Francis Group, Routledge Explorations in Energy Studies, https://www.taylorfrancis.com/chapters/mono/10.4324/9781003393092-3/novel-methodological-framework-energy-policy-bertug-ozarisoy-hasim-altan.

Perez, K. X., Cetin, K., Baldea, M., & Edgar, T. F. (2017). Development and analysis of residential change-point models from smart meter data. *Energy and Buildings, 139*, 351. https://doi.org/10.1016/j.enbuild.2016.12.084.

Pretlove, S., & Kade, S. (2015). Post occupancy evaluation of social housing designed and built to code for sustainable homes levels 3, 4 and 5. *Energy and Buildings, 110*, 120. https://doi.org/10.1016/j.enbuild.2015.10.014.

Pritoni, M., Salmon, K., Sanguinetti, A., Morejohn, J., & Modera, M. (2017). Occupant thermal feedback for improved efficiency in university buildings. *Energy and Buildings, 144*, 241. https://doi.org/10.1016/j.enbuild.2017.03.048.

Santangelo, A., & Tondelli, S. (2017). Occupant behaviour and building renovation of the social housing stock: current and future challenges. *Energy and Buildings, 145*, 276. https://doi.org/10.1016/j.enbuild.2017.04.019.

Schultz, P. W., Estrada, M., Schmitt, J., Sokoloski, R., & Silva-Send, N. (2015). Using in-home displays to provide smart meter feedback about household electricity consumption: a randomized control trial comparing kilowatts, cost, and social norms. *Energy, 90*, 351. https://doi.org/10.1016/j.energy.2015.06.130.

Shi, W., & Elsharkawy, H. *An innovative energy efficiency application development: through the evaluation of occupants' behavioural issues and its impact on domestic energy consumption in the UK. Proceedings of the International Conference for Sustainable Design of the Built Environment, SDBE 2017,* https://eprints.kingston.ac.uk/id/eprint/51170/1/ElSharkawy-H-EPrintID-VoR.pdf

Steemers, K., & Yun, G. Y. (2009). Household energy consumption: a study of the role of occupants. *Building Research & Information, 37*(5–6), 625. https://doi.org/10.1080/09613210903186661.

Steg, L. (2008). Promoting household energy conservation. *Energy Policy, 36*(12), 4449. https://doi.org/10.1016/j.enpol.2008.09.027.

Sun, K., & Hong, T. (2017). A framework for quantifying the impact of occupant behavior on energy savings of energy conservation measures. *Energy and Buildings, 146*, 383. https://doi.org/10.1016/j.enbuild.2017.04.065.

Sunikka-Blank, M., & Galvin, R. (2012). Introducing the prebound effect: the gap between performance and actual energy consumption. *Building Research & Information, 40*(3), 260. https://doi.org/10.1080/09613218.2012.690952.

Ürge-Vorsatz, D., Danny Harvey, L. D., Mirasgedis, S., & Levine, M. D. (2007). Mitigating CO_2 emissions from energy use in the world's buildings. *Building Research & Information, 35*(4), 379. https://doi.org/10.1080/09613210701325883.

Verplanken, B., Aarts, H., & Van Knippenberg, A. (1997). Habit, information acquisition, and the process of making travel mode choices. *European Journal of Social Psychology, 27*(5), 539–560. https://doi.org/10.1002/(SICI)1099-0992(199709/10)27:5<539::AID-EJSP831>3.0.CO;2-A.

Walker, S. L., Lowery, D., & Theobald, K. (2014). Low-carbon retrofits in social housing: interaction with occupant behaviour. *Energy Research & Social Science, 2*, 102. https://doi.org/10.1016/j.erss.2014.04.004.

Watts, C., Jentsch, M. F., & James, P. A. (2011). Evaluation of domestic energy performance certificates in use. *Building Services Engineering Research and Technology, 32*(4), 361. https://doi.org/10.1177/0143624411404486.

Wei, S., Hassan, T. M., Firth, S. K., & Fouchal, F. (2016). Impact of occupant behaviour on the energy-saving potential of retrofit measures for a public building in the UK. *Intelligent Buildings International, 9*(2), 97. https://doi.org/10.1080/17508975.2016.1139538.

Xu, X., Taylor, J. E., Pisello, A. L., & Culligan, P. J. (2012). The impact of place-based affiliation networks on energy conservation: a holistic model that integrates the influence of buildings, residents and the neighborhood context. *Energy and Buildings, 55*, 637. https://doi.org/10.1016/j.enbuild.2012.09.013.

Zhang, X., Shen, J., Yang, T., Tang, L., Wang, L., Liu, Y., & Xu, P. (2016). Smart meter and in-home display for energy savings in residential buildings: a pilot investigation in Shanghai, China. *Intelligent Buildings International, 11*, 4–26. https://doi.org/10.1080/17508975.2016.1213694.

Zhao, D., Mccoy, A. P., Du, J., Agee, P., & Lu, Y. (2016). Interaction effects of building technology and resident behavior on energy consumption in residential buildings. *Energy and Buildings, 134*, 223. https://doi.org/10.1016/j.enbuild.2016.10.049.

4
HOUSING TYPOLOGY IN THE UK

Introduction

The development of UK housing history began shortly before the Conquest, during the Norman occupation. First dwellings were showed Norman influence (Blackwell & Kohl, 2018). This continued with Roman and Saxon times. Roman occupation of Britain had little permanent effect on native architecture. Today, in the UK, most of the existing housing stock background goes back to the early 1850s (Newman & Newman, 2013). Many domestic buildings were designed during the remainder of the 19th century. The early 18th century was the time of rapid social change, and the British population was growing at a furious pace through the 19th century (Spyropoulos et al., 2022). UK housing can be categorized under four main house types, which are listed below:

1 Purpose-Built Flats
    - Purpose-Built High-Rise Flat
    - Purpose-Built Low-Rise Flat
    - Converted Flat
2 Bungalows
3 Detached Houses
    - Detached House
    - Semi-Detached House
4 Terraced Houses
    - Large Terrace House
    - Medium Terrace House
    - Small Terrace House

Purpose-Built Flat

A flat in a built block includes cases where there is only one flat with independent access in a building which is also used for non-domestic purposes (Satsangi & Dunmore, 2002). A built flat, which

DOI: 10.1201/9781003490364-6

Figure 4.1 High-rise flats; low-rise flats; converted flats.

is in a block at least six stories high is a high-rise flat and in a block less than six stories high is a low-rise flat. A flat resulting from the conversion of a house (generally other house types) or former non-residential building, which includes buildings converted into a flat as well as commercial premises (typically corner shops) is a converted flat, as shown in Figure 4.1.

Bungalow

It is a house with all of the habitable accommodation on one floor. Excludes chalet bungalows and bungalows with habitable loft conversions, which are treated as houses. The name stems from the Indian word "bangla" or "of Bengal" and it was a relatively common house type in India. Bungalows were particularly popular in Scotland (Flynn & Montalbano, 2023). Partly this may have been smaller and Scottish house sizes have traditionally been smaller than those in England, Wales and Northern Ireland (Griffith et al., 2022). Basically, it can be seen as a flat with a garden, as shown in Figure 4.2.

Detached House

It is a house where none of the habitable structures is joined to another building other than garages, etc. (Davey, 2019). There is a type of detached house, a semi-detached house, which is attached to one other house, as shown in Figure 4.3.

Terraced House

It is a house forming part of a block where at least one house is attached to two or more other houses (Ozarisoy & Elsharkawy, 2019). In the UK, especially at the beginning of the 19th century, terrace houses were the most common middle-class urban house type. There are small, terraced houses,

Figure 4.2 Bungalows.

Figure 4.3 Detached houses; semi-detached houses.

Figure 4.4 Small terraced houses; medium- and large-sized terraced houses.

which are relatively small in size compared to the medium- and large-sized terraced houses, which are fairly large compared to the small ones, as shown in Figure 4.4.

Today, in the UK as well as in the world, the development of a more sustainable, ecologically built environment is one of the major challenges for architects. Sustainable buildings are increasing and becoming more important every day (Sousa et al., 2017). Therefore, energy efficient housing should also be included under the main housing typology in view of the fact that it has become a significant part of future housing schemes in the UK (Mavrogianni et al., 2015).

Energy Efficient House Design

It is the recent house design scheme that has environmental concerns added to the design and also has a major impact on the world's approach to energy efficiency (Altan et al., 2015). An energy efficient house has many benefits for its habitants:

- The house is warmer in winter in general.
- The house has affordable fuel bills even for families on low incomes.
- The house minimises condensation possibility due to reduced heat loss via high insulation and better ventilation.
- The house is healthy and designed to have a long life and lower maintenance costs.
- The house reduces environmental pollution (e.g., carbon dioxide emissions) and this way has less impact on the environment (Taylor et al., 2015).

Existing Housing and the Profile of the Stock

According to the National Statistics and Regional Trends carried out in the years 1991 and 1994, there were 24,248,000 total existing dwellings in the UK (Lloyd & Gleeson, 2022). The number of dwellings with reference to each individually country – in England: 20,219,000, in Wales: 1,219,000, in Northern Ireland: 600,000, and in Scotland: 2,210,000 (Lloyd & Gleeson, 2022).

Furthermore, in England, there were some 20.4 million dwellings in 1996. The stock has increased by 649,000 dwellings since 1991, some 696,000 dwellings have been added and some 130,000 have been lost. Of those dwellings added, 19,000 were through conversion and/or changes of use from non-residential to residential accommodation (Loga et al., 2016). The remaining dwellings were newly built and almost all of the newly built ones have been for owner occupation or for registered social landlords. The housing typology profile varies within the UK. Figure 4.5 illustrates the composition of the housing stock within the UK by type and age.

As it can be seen above, there are significant differences in the profile of houses by type in the UK. Both in England and Wales, semi-detached and terraced houses are the most common types of dwelling, as shown in Figure 4.6. In England, semi-detached and terraced houses are each accounting for 30% of the stock. In Wales, terraced houses account for 34% and semi-detached houses account for 32% of the stock (Özer & Jacoby, 2023). In Scotland, the stock profile is substantially different compared to England and Wales. Only 60% of dwellings are houses. Semi-detached

Table 4.1 Total housing stock in the UK, 1981–1994.

Country	*1981*	*1986*	*1991*	*1992*	*1993*	*1994*	*Increase 1981–1994 (%)*	*Rates Per 1,000 Population 1994*
England	18,025	18,883	19,788	19,927	20,070	20,219	12.2	425
Wales	1,089	1,128	1,191	1,201	1,210	1,219	11.9	418
Northern Ireland	502	540	573	580	590	600	19.5	366
Scotland	1,970	2,050	2,160	2,175	2,193	2,210	12.2	431

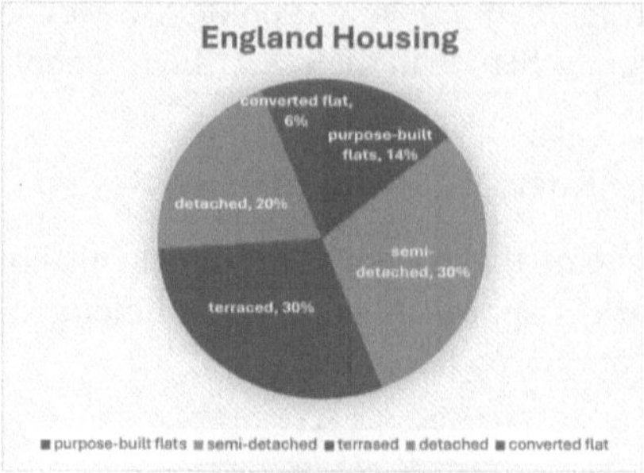

UK Housing Stock by Type, 1991

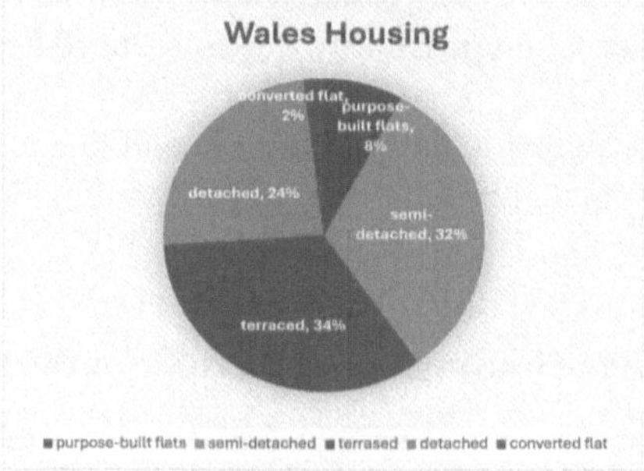

UK Housing Stock by Type, 1993

Figure 4.5 UK housing stock by type, 1991/1993.

houses are the most common type and account for 23% of all dwellings (Morrison, 2016). In Northern Ireland, dwelling stock is more similar to that in England and Wales, with relatively few flats (8%, of which the majority are purpose-built). Terraced dwellings are the most common (37%), followed by detached houses (31%) and semi-detached houses (24%) (Morgan & Cruickshank, 2014)

Then again, if we look at the profile of houses by age, we can easily see that there are considerable differences in dwelling stock. Both in England and Wales, almost half of the housing stock was built before 1944 with 26% and 33% respectively built before 1919. 34% of the dwellings stock in England was built after 1965 and, in addition, in Wales, recently built houses account for 33% of all the dwellings (Monahan & Powell, 2010). Scotland's housing stock is slightly newer on average, however in Northern Ireland the stock is much more recent, with only 32% built before 1944 (Aleksić et al., 2016). The recent statistics of the housing stock in England by dwelling type are also given in Table 4.2.

Conclusions

When focused on the English housing stock, the highest profiles of houses by type are terraced and semi-detached houses accounting for 30% of all the dwellings (Ji et al., 2014). In addition, 45% of the housing stock was built after 1944, which can also be seen as almost half of all the dwellings are relatively recent (Liu et al., 2024). Table 4.3 shows the relationship between dwelling age, tenure, type and energy rating (SAP).

As shown in Table 4.3, there is a variation in the overall energy efficiency (SAP ratings) between terraced and semi-detached houses in 1991 (Aragon et al., 2017). In general, the age

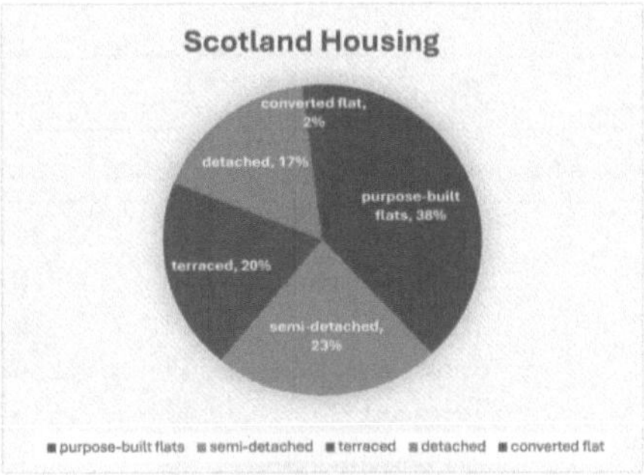

UK Housing Stock by Age, 1991

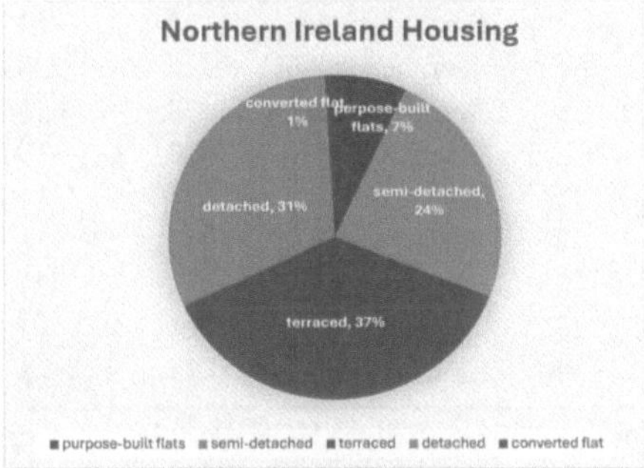

UK Housing Stock by Age, 1993

Figure 4.6 UK housing stock by age, 1991/1993 (DOE, 1996).

Table 4.2 UK housing stock – dwelling type by region, 1996

Government Office Region	*Terraced*	*Semi-detached*	*Detached*	*Purpose-built flat*	*Converted flat*
Northeast	36	38	12	13	2
Yorkshire and the Humber	30	40	15	11	3
Northwest	37	33	17	11	2
East Midlands	25	35	31	8	2
West Midlands	29	34	21	13	3
Southwest	28	27	29	11	4
East of England	28	29	27	12	3
Southeast	28	26	28	11	6
London	32	16	5	37	11
England (dwellings %)	30.5	29.8	20.5	14.9	4.3

of these dwelling types is the dominant influence together with tenure playing an important but secondary role. Moreover, the table not only shows the mean SAP rating for terraced and semi-detached houses but also shows the distribution from the worst to the best by age and tenure (Taylor, 2017). The final column gives the percentage of dwellings in each group, which have SAP ratings less than ten and thereby fall within the worst of the existing stock. Therefore, in order to understand the frequency distribution of SAP ratings by dwelling age and tenure, as shown in Table 4.4.

Table 4.3 England dwelling types, 1996

Dwelling age	*Tenure*			*Type*	*Energy rating (SAP)*			
	OO	*LA/HA*	*PR*		*Mean*	*Min*	*Max*	*<10 (%)*
Post-1980	OO			Terraced	56	21	72	0
Post-1980	OO			Semi-detached	46	18	63	2
Post-1980		LA/HA		Terraced	47	16	88	0
Post-1980		LA/HA		Semi-detached	47	16	88	0
1965–1980	OO			Terraced	47	−13	70	2
1965–1980	OO			Semi-detached	46	−20	68	1
1965–1980		LA/HA		Terraced	43	−3	69	1
1965–1980		LA/HA		Semi-detached	32	−14	62	7
Pre-1965	OO			Terraced	36	−23	71	5
Pre-1965		LA/HA		Semi-detached	28	−36	60	11
Pre-1965		LA/HA		Terraced	33	−19	65	8
Pre-1965			PR	Terraced	24	−19	56	18
Pre-1965			PR	Semi-detached	13	−38	47	40
1850–1964	OO			Semi-detached	33	−27	69	5
Post-1964			PR	Terraced	31	3	54	12
Post-1964			PR	Semi-detached	31	3	54	12

Note: The minimum and maximum values in the EHCS sample represent statically the range covered, on average, by over 99% of dwellings in the national stock of each type.

Categories listed and shaded are below-average SAP rating (35).

Owner occupied: OO, local authority/housing association: LA/HA, private rented: PR

To conclude, it could be said that for all tenures on both housing types, the dwelling age and form are the important factors, as shown in Figure 4.7. These factors also determine the heat loss through the building envelope. As the English House Condition Survey (EHCS) explains, in England, many older buildings have been upgraded by the addition of roof and wall insulation. Besides, since 1964, building regulations also require progressively higher insulation standards for all new housing (Cuéllar-Franca & Azapagic, 2012).

Overall, the age of the dwelling is the dominant influence together with tenure playing an important but secondary role. Both the table and figure show how the dwelling groups fall neatly into three blocks corresponding to the age bands 'post-1980', '1965–1980' and 'pre-1965' (Baiocchi et al., 2015). There are only four exceptions to this rule, which can be explained by owner occupied properties rising into a higher SAP rating block or dwellings from the rented sectors slipping into a lower SAP rating block. Also, within each block, there is a fairly clear pattern of owner-occupied properties being slightly more efficient than those owned by local authorities or housing associations, with private rented dwellings coming in a poor third. Moreover, SAP rating frequency distributed in Figure 4.7 shows that terraced houses are being superior to semi-detached houses.

Furthermore, to summarise, there are 21.1 million dwellings in England with an increase of 800,000 since 1996 and most additions were provided by new-build (averaging around 150,000 each year) although the level of new construction continued a longer-term trend of decline and was less than half the post-1945 high of the mid- to late-1960s. Consequently, there has been little change in the composition of the stock, and it remains relatively old with 8.1 million dwellings

Table 4.4 Comparison between terraced and semi-detached housing in England, 1996

Terraced housing			*Semi-detached housing*		
Dwelling age	Tenure	Mean SAP rating	Dwelling age	Tenure	Mean SAP rating
Post-1980	OO	56	Post 1980	OO	46
Post-1980	LA/HA	47	Post 1980	LA/HA	47
1965–1980	OO	47	1965–1980	OO	46
1965–1980	LA/HA	43	1965–1980	LA/HA	32
Pre-1965	OO	36	1850–1964	OO	33
Pre-1965	LA/HA	33	Pre-1965	LA/HA	28
Pre-1965	PR	24	Pre-1965	PR	13
Post-1964	PR	31	Post-1964	PR	31

Note: Categories listed and shaded are below average SAP rating (35).
Owner occupied: OO, local authority/housing association: LA/HA, private rented: PR.

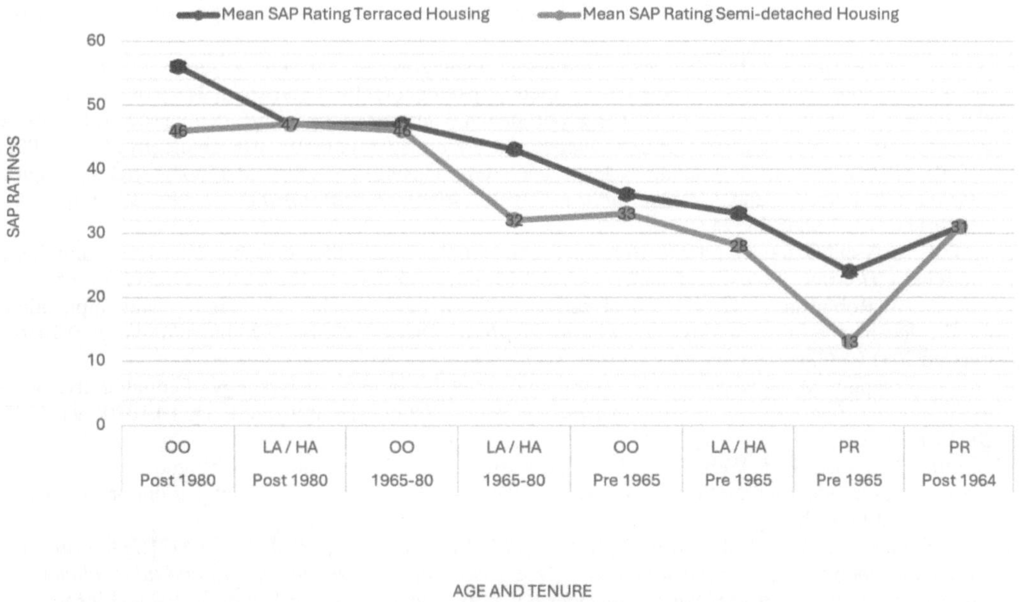

Figure 4.7 Frequency distribution of SAP ratings by age and tenure of two main house types in England, 1996.

(39%) built before 1945, including 4.4 million dwellings (21% of all) built before 1919 (Gallent & Robinson, 2011).

This again is the recognition that the existing housing profile is in poor condition and has a large impact on the environment given that it occupies the majority of the building stock in the country. It is therefore a serious concern that more and more housing is causing the increase in carbon dioxide emission and falling behind building required standards, which should make the actual point.

This chapter has reviewed the housing typology in the UK and looked at the conditions of the existing housing stock to determine the influences which will affect building designers when tackling issues concerning energy efficiency improvements and carbon dioxide emissions in these dwellings. In addition, undertaken improvements within the last decade, and how poor the conditions were across different types of dwellings, households and areas have also been summarised. There is more and more revised data being published every five years by the Office of the Deputy Prime Minister on housing conditions and undertaken improvements.

References

Aleksić, J., Kosanović, S., Tomanović, D., Grbić, M., & Murgul, V. (2016). Housing and climate change-related disasters: a study on architectural typology and practice. *Procedia Engineering, 165*, 869. https://doi.org/10.1016/j.proeng.2016.11.786.

Altan, H., Gasperini, N., Moshaver, S., & Frattari, A. (2015). Redesigning terraced social housing in the UK for flexibility using building energy simulation with consideration of passive design. *Sustainability, 7*(5), 5488. https://doi.org/10.3390/su7055488.

Aragon, V., Gauthier, S., Warren, P., James, P. A. B., & Anderson, B. (2017). Developing English domestic occupancy profiles. *Building Research & Information, 47*(4), 375. https://doi.org/10.1080/09613218.2017.1399719.

Baiocchi, G., Creutzig, F., Minx, J., & Pichler, P. (2015). A spatial typology of human settlements and their CO_2 emissions in England. *Global Environmental Change, 34*, 13. https://doi.org/10.1016/j.gloenvcha.2015.06.001.

Blackwell, T., & Kohl, S. (2018). *Historicizing housing typologies: beyond welfare state regimes and varieties of residential capitalism*. Informa UK Limited. https://doi.org/10.1080/02673037.2018.1487037.

Cuéllar-Franca, R. M., & Azapagic, A. (2012). Environmental impacts of the UK residential sector: Life cycle assessment of houses. *Building and Environment, 54*, 86. https://doi.org/10.1016/j.buildenv.2012.02.005.

Davey, R. (2019). Snakes and ladders: legal coercion, housing precarity, and home-making aspirations in southern England. *Journal of the Royal Anthropological Institute, 26*(1), 12. https://doi.org/10.1111/1467-9655.13175.

Flynn, L. B., & Montalbano, G. (2023). Housing regime typologies and their discontents: a systematic literature review. *Housing, Theory and Society, 41*(2), 127. https://doi.org/10.1080/14036096.2023.2265370.

Gallent, N., & Robinson, S. (2011). Local perspectives on rural housing affordability and implications for the localism agenda in England. *Journal of Rural Studies, 27*(3), 297. https://doi.org/10.1016/j.jrurstud.2011.05.004.

Griffith, E. J., Jepma, M., & Savini, F. (2022). Beyond collective property: a typology of collaborative housing in Europe. *International Journal of Housing Policy, 24*(1), 121. https://doi.org/10.1080/19491247.2022.2123272.

Ji, Y., Fitton, R., Swan, W., & Webster, P. (2014). Assessing overheating of the UK existing dwellings – a case study of replica Victorian end terrace house. *Building and Environment, 77*, 1. https://doi.org/10.1016/j.buildenv.2014.03.012.

Liu, Y., Grimmond, S., Luo, Z., Hertwig, D., Mcgrory, M., Lo Piano, S., & Smith, S. T. (2024). *Dwelling conversion and energy retrofit modify building anthropogenic heat emission under past and future climates: a case study of London terraced houses*. Elsevier BV. https://doi.org/10.1016/j.enbuild.2024.114668.

Lloyd, C. D., & Gleeson, J. (2022). Housing space in England at the neighbourhood-level, 1971 to 2017. *Applied Spatial Analysis and Policy, 15*(4), 1433. https://doi.org/10.1007/s12061-022-09458-y.

Loga, T., Stein, B., & Diefenbach, N. (2016). TABULA building typologies in 20 European countries—making energy-related features of residential building stocks comparable. *Energy and Buildings, 132*, 4. https://doi.org/10.1016/j.enbuild.2016.06.094.

Mavrogianni, A., Taylor, J., Davies, M., Thoua, C., & Kolm-Murray, J. (2015). *Urban social housing resilience to excess summer heat*. Informa UK Limited. https://doi.org/10.1080/09613218.2015.991515.

Monahan, J., & Powell, J. C. (2010). A comparison of the energy and carbon implications of new systems of energy provision in new build housing in the UK. *Energy Policy, 39*(1), 290. https://doi.org/10.1016/j.enpol.2010.09.041.

Morgan, M., & Cruickshank, H. (2014). Quantifying the extent of space shortages: English dwellings. *Building Research & Information, 42*(6), 710. https://doi.org/10.1080/09613218.2014.922271.

Morrison, N. (2016). Institutional logics and organisational hybridity: English housing associations' diversification into the private rented sector. *Housing Studies, 31*(8), 897. https://doi.org/10.1080/02673037.2016.1150428.

Newman, C., & Newman, R. (2013). Housing the workforce in 19th-century east Lancashire: past processes, enduring perceptions and contemporary meanings. *Post-Medieval Archaeology, 42*(1), 181. https://doi.org/10.1179/174581308x354047.

Ozarisoy, B., & Elsharkawy, H. (2019). Assessing overheating risk and thermal comfort in state-of-the-art prototype houses that combat exacerbated climate change in UK. *Energy and Buildings, 187*, 201. https://doi.org/10.1016/j.enbuild.2019.01.030.

Özer, S., & Jacoby, S. (2023). Space standards in affordable housing in England. *Building Research & Information, 52*(6), 611. https://doi.org/10.1080/09613218.2023.2253337.

Satsangi, M., & Dunmore, K. (2002). The planning system and the provision of affordable housing in rural Britain: a comparison of the Scottish and English experience. *Housing Studies, 18*(2), 201. https://doi.org/10.1080/0267303032000087720.

Sousa, G., Jones, B. M., Mirzaei, P. A., & Robinson, D. (2017). A review and critique of UK housing stock energy models, modelling approaches and data sources. *Energy and Buildings, 151*, 66. https://doi.org/10.1016/j.enbuild.2017.06.043.

Spyropoulos, K., Gidlow, C. J., & Ellis, N. J. (2022). Proposing a typology to examine the health impact of Housing First: a systematic review and meta-analysis. *Housing Studies, 39*(3), 766. https://doi.org/10.1080/02673037.2022.2084047.

Taylor, J., Wilkinson, P., Davies, M., Armstrong, B., Chalabi, Z., Mavrogianni, A., Symonds, P., Oikonomou, E., & Bohnenstengel, S. I. (2015). *Mapping the effects of urban heat island, housing, and age on excess heat-related mortality in London.* Elsevier BV. https://doi.org/10.1016/j.uclim.2015.08.001.

Taylor, S. P. (2017). What is innovation? A study of the definitions, academic models and applicability of innovation to an example of social housing in England. *Open Journal of Social Sciences, 05*(11), 128. https://doi.org/10.4236/jss.2017.511010.

5

ENERGY CONSUMPTION IN DOMESTIC BUILDINGS

Introduction

All societies, including both developed and less developed countries, are dependent on energy for their existence. Today, the consumption of energy is costly and degrades the environment, and therefore, its use should be reduced (Dong et al., 2020). In the UK, the domestic sector is one of the main consumers of energy and accounts for 29% of the total primary energy consumption in the country, as shown in Figure 5.1 (Ekins & Lees, 2008).

In view of the fact that buildings (both domestic and commercial buildings) account for the largest proportion of the UK's energy consumption, they should be considered one of the highest priorities when it comes to reducing the overall energy consumption in the country and meeting required targets (Kaveh et al., 2018). According to the statistics, the UK domestic sector energy consumption has increased by 32% since 1970 and by 19% since 1990. In addition, since 1990 factors affecting domestic energy consumption, such as the number of households, have increased by 10%, population has increased by 4% and household disposable income has increased by 30% (Johnson et al., 2023).

Environmental damage caused by carbon dioxide emitted during energy use, combined with the more basic issue of cost has made energy conservation the central theme of sustainable development (Penasco & Anadon, 2022). In this chapter, energy consumption in domestic buildings and strategies for energy efficiency design have been reviewed for the purposes of identifying the domestic sector energy use and the drivers of energy consumption in this sector, which can eventually help bring down the energy being consumed.

Domestic Sector Energy Use

Today in the UK, almost 30% of the carbon dioxide produced is caused directly or indirectly by the use of energy in dwellings and as a result, it is clear that the way we use energy is extremely inefficient. If all energy efficiency measures outlined in this research were to be carried out the amount of CO_2 produced could have been reduced by 60–80% (Rosenow, 2012; Ozarisoy, 2022). Moreover, if we were also to change our lifestyle and use renewable energy sources we could reduce the CO_2 emissions even further.

DOI: 10.1201/9781003490364-7

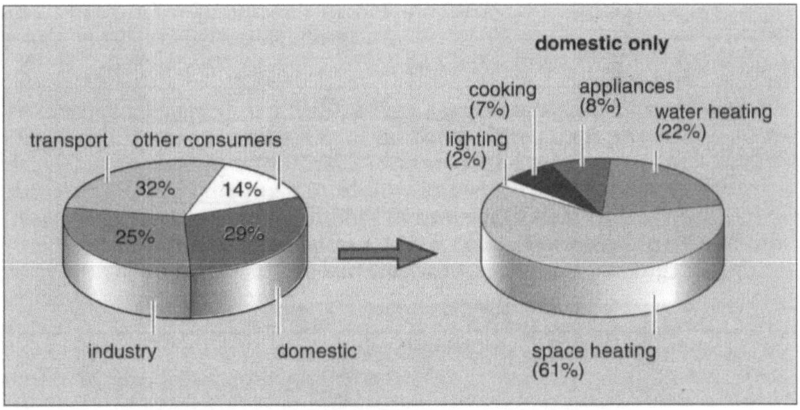

Figure 5.1 Breakdown of UK energy consumption and domestic only energy use.

Services and systems installed within dwellings use energy to create a comfortable, safe, healthy and productive environment (Ben & Steemers, 2014). These services and systems are as follows:

- Space Heating
- Water Heating
- Lighting
- Cooking
- Other Appliances

Each dwelling should be considered as a total system and by doing so, the way energy is being used can be studied and therefore the ways we can reduce CO_2 emissions can also be worked out, as shown in Table 5.1. The following steps can help save energy and therefore carbon dioxide emissions can be reduced in each dwelling:

- Choosing the sources of energy emitting the least amount of carbon dioxide.
- Finding the most efficient ways of using energy in heating systems, lighting and other appliances.
- Conserving the heat in dwellings through draught proofing, ventilation control and insulation (Wang et al., 2021).

The main issue is not improving the overall energy efficiency of a dwelling but the resources that can be directed to the task and how far to go with each measure to achieve the efficiency levels (Menconi et al. 2023).

The majority of energy consumed in the domestic sector is for space heating, which accounted for 58% of all delivered energy consumed in 2000, as shown in Figure 5.2 (Moran et al., 2012). Space heating in any year is largely dependent on outside temperatures, which explains the year-to-year fluctuations, although increases in internal temperatures, the growth in central heating and the increased number of households have all contributed to the increase over the period 1970–2000, despite the increased presence of insulation (Nieto et al., 2021). The other major areas of energy consumption in the domestic sector are heating water, lighting and appliances and cooking. Between 1970 and 2000, energy consumption in lighting and appliances increased by 157%, while energy use in cooking has fallen by 16% (Marshall et al., 2015).

Table 5.1 Dwelling as a total system

Input (sources of energy)		*Use (and misuse)*		*Output (loss)*
Fossil fuel	Gas	Space heating		Draught, ventilation
	Oil	Water heating		Hot flue gases
	Coal	Lighting		Heat loss through:
Electricity		Electricity		Walls and windows
Biomass:	Wood	Appliances:	Cooking	Roof
Solar energy:	Passive	Solar energy:		Floor
	Active	Human metabolism		Waste warm water

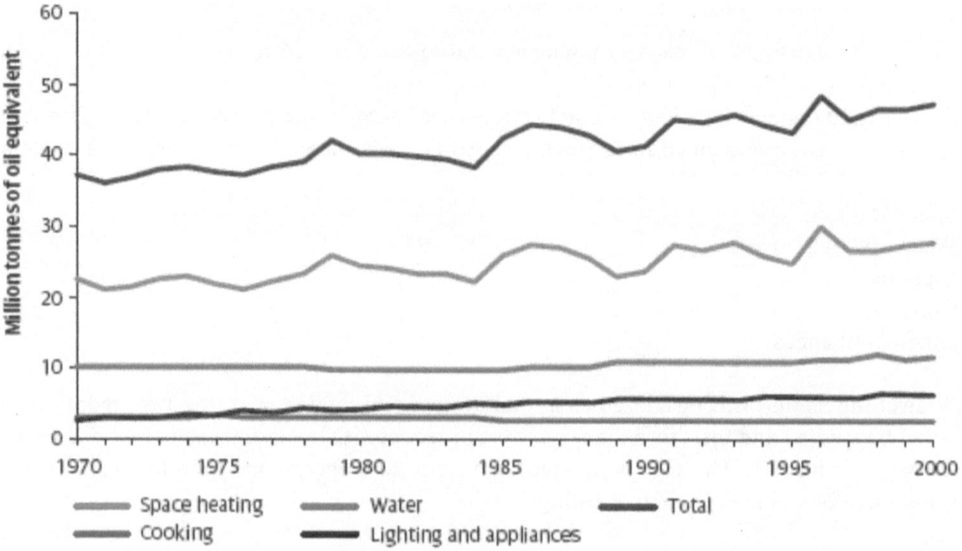

Figure 5.2 Domestic final energy consumption by end use, 1970–2000.

The fall in energy consumption in cooking is partly explained by a change in lifestyle (Keay, 2016). More convenience foods are now consumed, and people eat out more frequently than in the past. The average person in 2000 spent £7.36 per person per week on food and drink consumed out of the house. In addition, excluding alcoholic drink, expenditure on eating out has increased by 13% in real terms since 1994, whilst household food and non-alcoholic drink purchases fell by 2% (Kesicki, 2012).

Space and water heating require heat, which is produced by converting the chemical energy in gas, oil or coal to thermal energy by combustion in a heating appliance (Ozarisoy & Elsharkawy, 2019). The remaining services require electricity, which can also be used to provide space and water heating; however, this is somewhat costly (Zhang et al., 2012). In the UK, the majority of electricity is produced in power stations from steam or gas turbines driven by the combustion of fossil fuels (Warren, 2013). Smaller quantities are generated from nuclear power and renewable energy sources.

Embodied Energy and Materials

When considering energy consumption in buildings, it is important that not only the energy consumed during the life of a building has to be considered but also the extent of energy involved in the extraction, manufacture and transportation of building materials should be considered (Eyre & Baruah, 2015). This is known as the 'embodied energy' and directly relates to the gross carbon intensity of a material. The overall environmental credentials of a building are affected by a number of factors, which are as follows:

- Energy used over its estimated lifetime.
- Energy used in the construction process.
- The extent to which recycled materials have been used.
- The presence of pollutants in a material such as volatile organic compounds (VOCs).
- Toxic substance used in the production process.
- Energy used in demolition.
- Level of recyclable materials at demolition.
- Materials used in refurbishment.

Currently, it is known that a building consumes much more energy during its lifetime than is involved in extraction, manufacture and transportation. However, it is usually the case that the embodied energy will be a significant fraction of the total, as buildings become more energy efficient. As seen earlier in this chapter, domestic buildings account for almost 30% of the total primary energy consumption in the UK and therefore the embodied energy of a dwelling has to be considered in achieving the best results in overall energy efficiency (Foulds & Powell, 2014).

Energy Efficiency Basics

In housing, energy efficiency means using less energy for heating, cooling and lighting (Booth & Choudhary, 2013). It also means buying energy-saving appliances and equipment for use in a house. An important concept for energy efficiency in housing is the building envelope, which is everything that separates the interior of the dwelling from the outdoor environment: doors, windows, walls, foundation, roof and insulation. All the components of the building envelope need to work together to keep a dwelling warm in the winter and cool in the summer (Mcandrew et al., 2021). Insulation, for instance, will be less effective if the roof, walls and ceiling allow air to leak in or allow moisture to collect in the insulation.

Various approaches can help improve the building envelope, such as wall insulation can reduce heat loss through walls by up to 80% and draught stripping badly fitted windows and doors can also reduce heat loss. In addition, double-glazing increases comfort and reduces condensation (Ozarisoy & Altan, 2023). Even low-emissivity glass further enhances these benefits and can significantly improve a dwelling's energy efficiency.

Heating and cooling systems typically use the most energy in the domestic sector (Hamilton et al., 2013). In housing, the addition of efficient controls, like a programmable thermostat, can significantly reduce the energy use of these systems. Some dwellings can also use zone heating and cooling systems, which reduce heating and cooling in the unused areas of a house.

Both heating water more efficiently and reducing hot water use can reduce the energy consumed to heat water. A wide variety of fixtures, such as low-flow showerheads and faucet aerators, can also reduce hot water use. In a house, an older water heater can be replaced with a newer, more energy-efficient one, and the water heater and hot water pipes can be insulated to minimise heat loss (Tunstall, 2022).

Today, most common appliances and electronic devices are available in energy-efficient models, from washing machines and refrigerators to copiers and computers. Several energy-efficient lighting options, such as compact fluorescent light bulbs, are also available. There are ways to determine how energy efficient a dwelling really is and, if needed, what improvements can be made. Homeowners can conduct simple energy audits on their homes or have professional audits done (Sousa et al., 2017).

Energy efficiency standards in a house could be more desirable to the occupants than an equivalent higher energy-consuming dwellings. In addition, the maintenance costs are reduced due to less energy-consuming devices installed and the effect is seen as an immediate reduction in gas and electricity bills. However, the reductions in bills are often not as large as predicted. This is because the dwelling occupants use the opportunity to lower fuel bills to obtain an improvement in thermal comfort conditions (O'keeffe et al., 2016). It is well-known fact that homes can be better heated yet still have lower fuel bills. Improvements in temperatures also have an effect on health.

There is a well-known approach, which is to first reduce the demand for energy then make use of natural energy flows and finally supply energy that is needed in an efficient and well-controlled manner, as shown in Figure 5.3. The order to energy efficiency is therefore:

1 Site and orientate the house in a microclimate by choosing an appropriate site or using landscaping techniques to modify the climate.
2 Minimise fabric and ventilation heat loss rates.
3 Optimise any free sources of energy such as solar gains, daylight or natural ventilation.
4 Select the energy-consuming plant to be as energy efficient as possible.
5 Control the energy-consuming plant so that it operates only when required and to the level required.
6 Recover waste energy where viable.

Reducing Energy Consumption

While there are a number of important dimensions to the concept of sustainability, it is largely used in reference to sustain the balance and health of the environment; the greatest threat to our environment today is climate change associated with emissions of greenhouse gases. Under the Kyoto Protocol, the European Community agreed to an 8% cut in greenhouse gas emissions and under a burden-sharing agreement amongst its member states, the UK agreed a cut of 12.5% (Parkes & Spataru, 2017). Both the UK government and the devolved administrations also have a domestic goal to cut the emissions of carbon dioxide, the main greenhouse gas, to 20% below 1990 levels by 2010. It is accordingly believed that there will be clear benefits for the UK by moving beyond its Kyoto target towards its domestic goal. Taking early action and starting the transition to a low-carbon economy will therefore bring financial and economic gains. The UK's Climate Change Programme includes a flexible, cost-effective set of policies and measures that will help to achieve these objectives that will safeguard the UK's competitiveness and that will also deliver wider benefits for the environment and society (Xie et al., 2017).

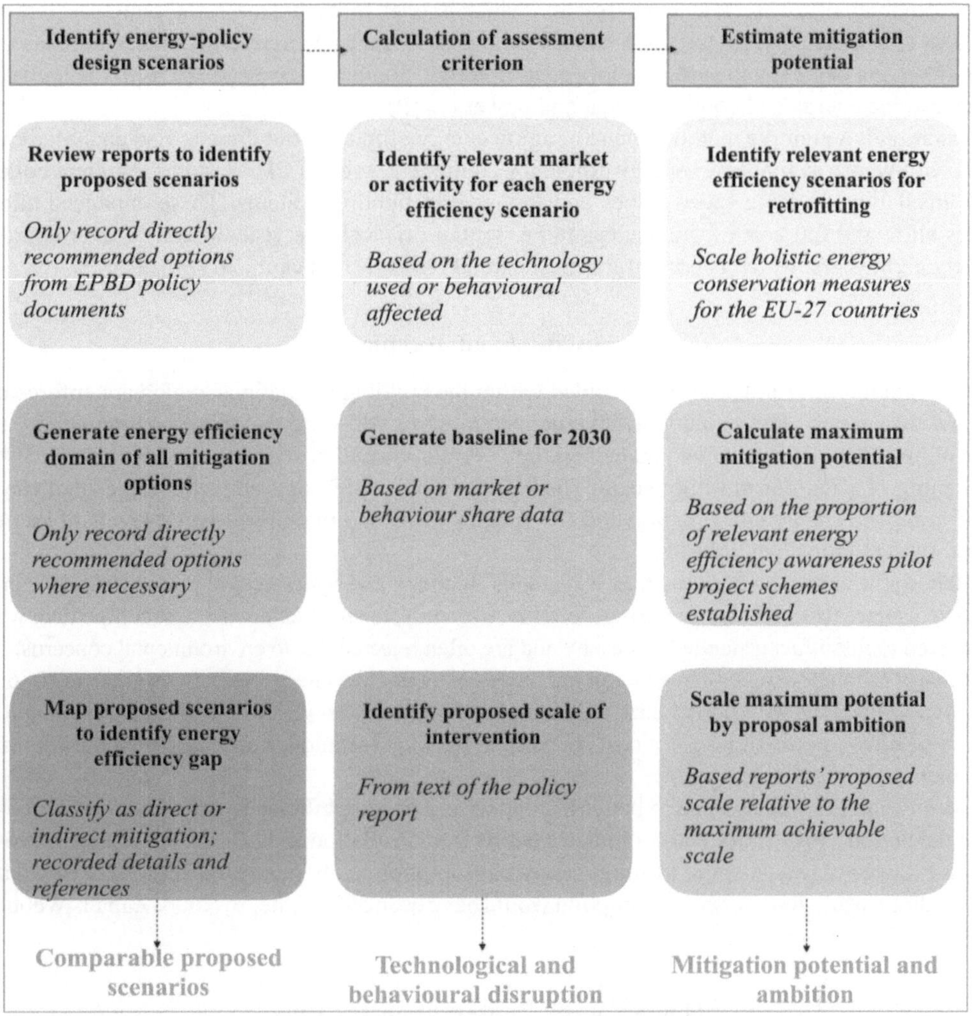

Figure 5.3 Energy efficiency basics.

Achieving emissions reductions associated with energy use in the domestic sector will inevitably mean reducing energy consumption. This itself must be considered as a significant challenge given that the domestic energy sector is a growing market (both gas and electricity consumption in the domestic sector has grown by 8–10% over the last five years and perhaps most importantly, a competitive supply market where unit prices have been steadily falling; reducing the incentive and economic benefits of energy efficiency (Johnson et al., 2023). The methods used by the UK government can be broadly categorised as fiscal, legislative and technical, and briefly explained below:

Fiscal

Taxation can be used to penalise the use of energy or reward the purchase of energy saving devices. The 'Climate Change Levy' (CCL) is a major new energy-related tax that came into operation in

2001 (Nieto et al., 2021). It is an energy tax, which applies to the business and public sectors and aims to discourage energy use by increasing the cost of fuels. Moreover, it makes purchases of energy-saving devices and systems more cost-effective. Some fuels are exempt from the levy such as renewable energy and combined heat and power (CHP) output.

Taxation is a punitive method of encouraging energy efficiency but there is also enchased capital tax allowance available for energy efficiency equipment, such as CHP systems, efficient boilers, pipe insulation, variable speed drives and low energy lighting systems. These enhanced allowances allow the full cost of the devices to be written off over one year instead of over approximately eight years, which is normal for most capital items (Sousa et al., 2017).

Legislation and Technical

Conservation of fuel and power is included within the building regulations, which are enforceable by law. In the past, the regulations have concentrated on reducing the space heating energy use of buildings, however, it now includes lighting, ventilation and air-conditioning with an excepted tightening of standards in future years. There are a number of improved technologies that can be used to reduce energy consumption and CO_2 output whilst maintaining existing levels of thermal comfort, ventilation and lighting.

Renewable energy sources, such as wind, solar, biomass and hydroelectric power stations, can be used to displace fossil fuel and nuclear generation. Currently, renewables are not cost competitive when compared to fossil fuel generated electricity and are often rejected due to environmental concerns.

Energy efficiency basics can generally achieve cost-effective energy savings as great as 40–60% in existing housing with poor fabric and services (Vagtholm et al., 2023). These savings can be achieved often at little or no extra cost if energy efficiency techniques are integrated during refurbishment or at the design stage.

As an example, savings due to better insulation and heating efficiency are shown in Figure 4.4 over the period 1970–2000. The combined savings from insulation and heating efficiency improvements reduced domestic space heating energy consumption to about 48% in 2000. It is calculated that without insulation energy consumption would have been 59% higher by 2000 (Zapata-Webborn et al., 2020).

Home Energy Conservation Act 1995

The Home Energy Conservation Act (HECA) 1995 set a target of 30% improvement in domestic energy efficiency to be achieved by 2010 (Wang et al., 2021). It came into effect in April 1996 and required every UK local authority with housing responsibilities – 'Energy Conservation Authorities' – to prepare, publish and submit to the Secretary of State an energy conservation report identifying practicable and cost-effective measures to significantly improve the energy efficiency of all residential dwellings in their area; and to report on progress made in implementing the measures, as shown in Figure 5.5.

Significant improvements are therefore possible in both public and private sectors, providing appropriate resources are made available to local authorities and registered social landlords/landladies. To a large extent, the financial resources required to meet this target need to come from private homeowners, since 70% of housing belongs to the private sector (Keay, 2016). There is local authority investment within the private sector, which will be targeted primarily at low income and other households at risk from fuel poverty and undertaken as part of the local authority's private sector renewal activity. There are also sources of external grants available, which could be actively

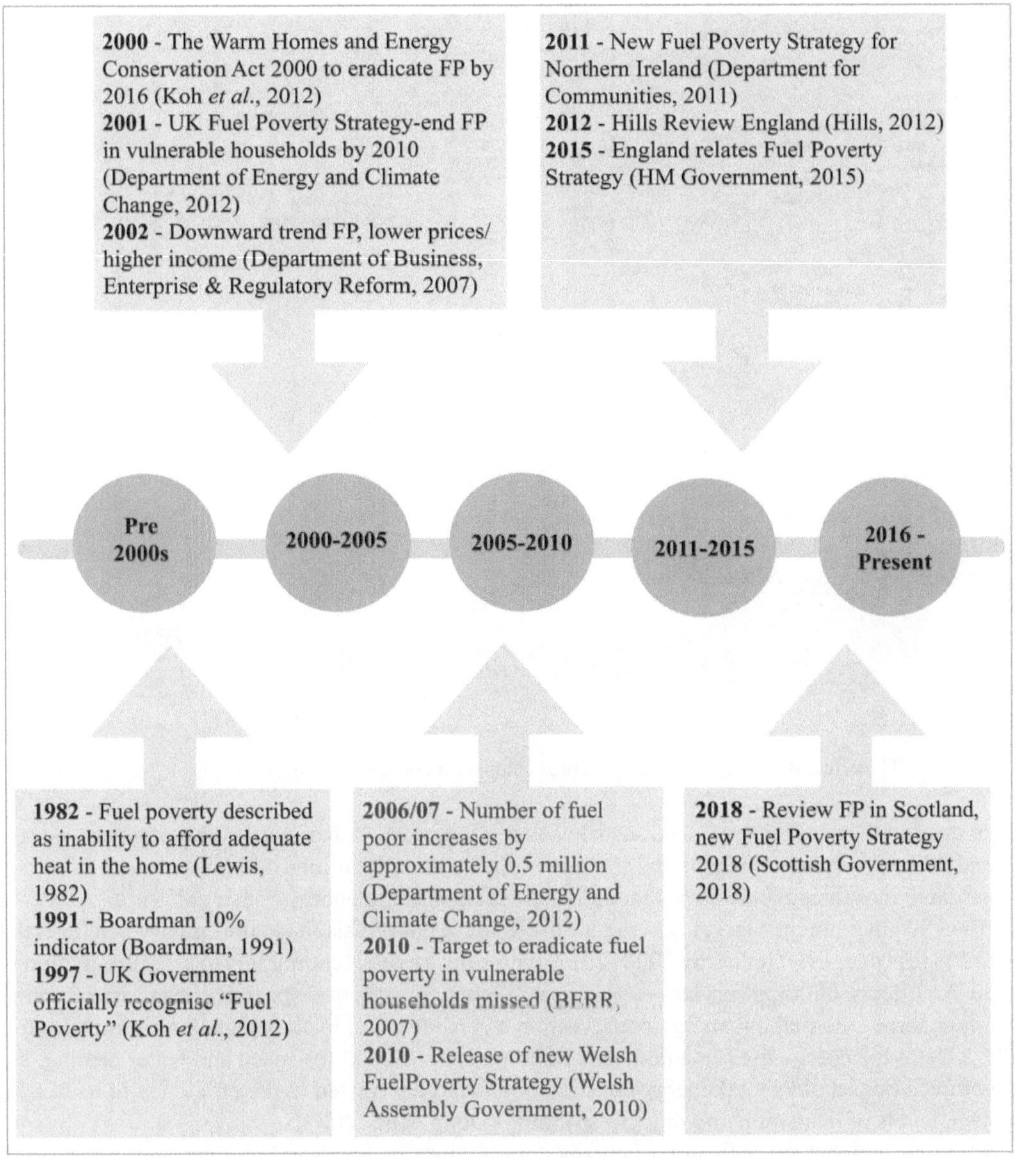

Figure 5.4 Savings due to better insulation and heating efficiency, 1970–2000.

promoted by the local authority. Furthermore, energy efficiency is an important priority for the local authorities in their investment and maintenance strategies for the public sector, and therefore, they will continue to support measures that help the provision of affordable heating to all households.

Conclusions

Given the value of energy efficiency basics and techniques for producing better buildings and reducing environmental impact, it is the duty of professionals in the construction industry to

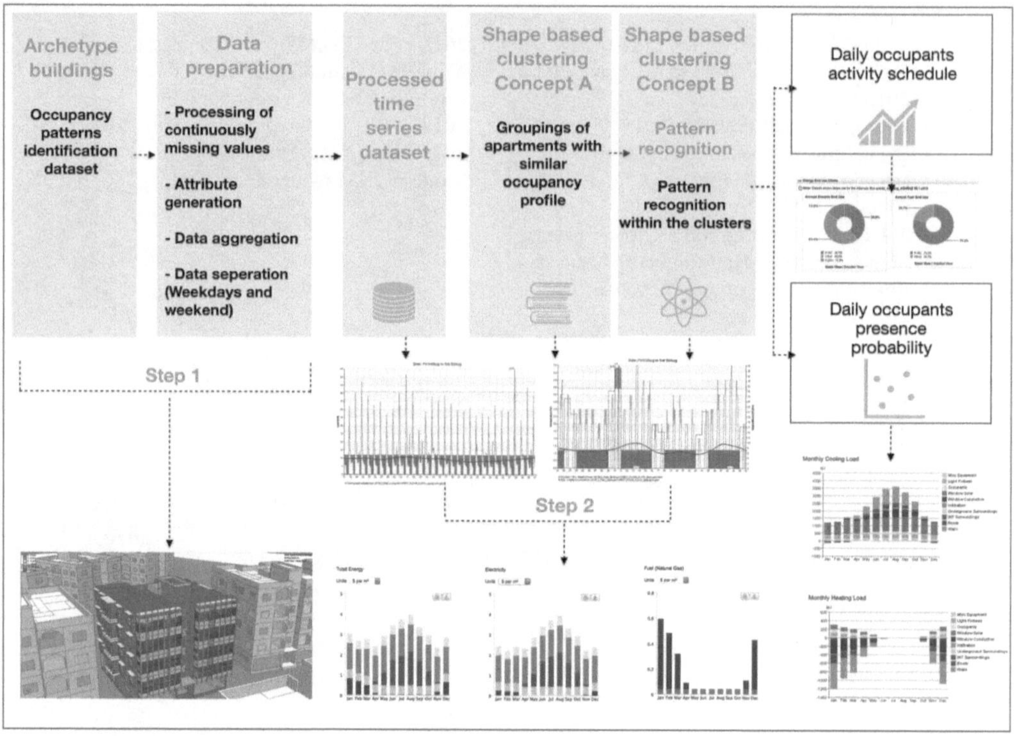

Figure 5.5 The effect of energy efficiency improvements on energy consumption.

recommend energy efficiency solutions whenever alternative choices occur. This must also be supported with increasing legislation by the government; in recognition of the importance of energy issues have introduced legislation that aims to reduce building energy consumption.

The UK government's target is to reduce greenhouse gas emissions in the country to a level that is 12.5% less than 1990 levels by 2008–2012, with a desire of extending this to a 60% reduction by 2050. Architects, building services engineers and other construction specialists can assist in meeting these targets by making energy conservation a consideration in all design decisions. Additionally, 83% of the energy used in a domestic building is consumed for space and water heating, and therefore, a household's total energy consumption is largely related to the efficiency of its heating system, levels of insulation and draught stripping (Dong et al., 2021).

It can be said that the reduction of energy consumption in domestic buildings and consequently the effective refurbishment of dwellings for energy efficiency relies on improvements in these key areas. Energy efficiency improvements in existing housing stock have made an impact on energy consumption since 1970, and it has been calculated that without these improvements, energy consumption would have been 46% higher by 2001 (Ozarisoy, 2022). However, opportunities to make much greater improvements to energy efficiency and therefore to reduce energy consumption remain.

This chapter outlines energy consumption in domestic buildings and determines potential improvements in this sector to help the UK meet its international commitment. If energy efficiency improvements in the domestic sector were to be accelerated both by professionals and the government, overall carbon dioxide emissions could be radically reduced. Thus, improving energy efficiency and reducing CO_2 emissions in this sector is not only going to help the country achieve its target but also tackle vital issues, such as fuel poverty, non-decent homes, excess winter deaths, etc. This will also raise awareness of sustainable and energy efficient building design in the UK.

Furthermore, sustainable design has become an important issue and to achieve this there must be a significant reduction in CO_2 emissions and the amount of energy consumed in buildings. The following chapter focuses on case studies carried out within the private housing sector that examine energy-related problems and determine potential improvements for energy efficiency in the domestic sector.

References

Ben, H., & Steemers, K. (2014). Energy retrofit and occupant behaviour in protected housing: a case study of the Brunswick Centre in London. *Energy and Buildings, 80*, 120. https://doi.org/10.1016/j.enbuild.2014.05.019.

Booth, A. T., & Choudhary, R. (2013). Decision making under uncertainty in the retrofit analysis of the UK housing stock: implications for the Green Deal. *Energy and Buildings, 64*, 292. https://doi.org/10.1016/j.enbuild.2013.05.014.

Dong, S., Kremers, E., Brucoli, M., Rothman, R., & Brown, S. (2020). Improving the feasibility of household and community energy storage: a techno-enviro-economic study for the UK. *Renewable and Sustainable Energy Reviews, 131*, 23–24. https://doi.org/10.1016/j.rser.2020.110009.

Ekins, P., & Lees, E. (2008). The impact of EU policies on energy use in and the evolution of the UK built environment. *Energy Policy, 36*(12), 4580. https://doi.org/10.1016/j.enpol.2008.09.006.

Eyre, N., & Baruah, P. (2015). Uncertainties in future energy demand in UK residential heating. *Energy Policy, 87*, 641. https://doi.org/10.1016/j.enpol.2014.12.030.

Foulds, C., & Powell, J. (2014). Using the Homes Energy Efficiency Database as a research resource for residential insulation improvements. *Energy Policy, 69*, 57. https://doi.org/10.1016/j.enpol.2014.01.015.

Hamilton, I. G., Steadman, P. J., Bruhns, H., Summerfield, A. J., & Lowe, R. (2013). Energy efficiency in the British housing stock: energy demand and the Homes Energy Efficiency Database. *Energy Policy, 60*, 462. https://doi.org/10.1016/j.enpol.2013.04.004.

Johnson, E., Betts-Davies, S., & Barrett, J. (2023). Comparative analysis of UK net-zero scenarios: the role of energy demand reduction. *Energy Policy, 179*, 25–27. https://doi.org/10.1016/j.enpol.2023.113620.

Kaveh, B., Mazhar, M. U., Simmonite, B., Sarshar, M., & Sertyesilisik, B. (2018). An investigation into retrofitting the pre-1919 owner-occupied UK housing stock to reduce carbon emissions. *Energy and Buildings, 176*, 33. https://doi.org/10.1016/j.enbuild.2018.06.038.

Keay, M. (2016). UK energy policy – Stuck in ideological limbo? *Energy Policy, 94*, 247. https://doi.org/10.1016/j.enpol.2016.04.022.

Kesicki, F. (2012). Costs and potentials of reducing CO_2 emissions in the UK domestic stock from a systems perspective. *Energy and Buildings, 51*, 203. https://doi.org/10.1016/j.enbuild.2012.05.013.

Marshall, E., Steinberger, J. K., Dupont, V., & Foxon, T. J. (2015). Combining energy efficiency measure approaches and occupancy patterns in building modelling in the UK residential context. *Energy and Buildings, 111*, 98. https://doi.org/10.1016/j.enbuild.2015.11.039.

Mcandrew, R., Mulcahy, R., Gordon, R., & Russell-Bennett, R. (2021). Household energy efficiency interventions: A systematic literature review. *Energy Policy, 150*, 13–15. https://doi.org/10.1016/j.enpol.2021.112136.

Menconi, M., Painting, N., & Piroozfar, P. (2023). An energy performance baseline scenario for 19thC listed dwellings in the UK. *International Journal of Building Pathology and Adaptation, 42*(1), 133. https://doi.org/10.1108/ijbpa-08-2022-0127.

Moran, F., Natarajan, S., & Nikolopoulou, M. (2012). Developing a database of energy use for historic dwellings in Bath, UK. *Energy and Buildings, 55*, 218. https://doi.org/10.1016/j.enbuild.2012.09.016.

Nieto, J., Pollitt, H., Brockway, P. E., Clements, L., Sakai, M., & Barrett, J. (2021). Socio-macroeconomic impacts of implementing different post-Brexit UK energy reduction targets to 2030. *Energy Policy, 158*, 32–33. https://doi.org/10.1016/j.enpol.2021.112556.

O'keeffe, J. M., Gilmour, D., & Simpson, E. (2016). A network approach to overcoming barriers to market engagement for SMEs in energy efficiency initiatives such as the Green Deal. *Energy Policy, 97*, 582. https://doi.org/10.1016/j.enpol.2016.08.006.

Ozarisoy, B. (2022). Energy effectiveness of passive cooling design strategies to reduce the impact of long-term heatwaves on occupants' thermal comfort in Europe: Climate change and mitigation. *Journal of Cleaner Production, 330*, Article 129675. https://doi.org/10.1016/j.jclepro.2021.129675.

Ozarisoy, B., & Altan, H. (2023). State-of-The-Art I: energy efficiency directives and policy aspirations in retrofit interventions. In: *Handbook of retrofitting high density residential buildings*. Springer (457–459). https://doi.org/10.1007/978-3-031-11854-8_2.

Ozarisoy, B., & Elsharkawy, H. (2019). Assessing overheating risk and thermal comfort in state-of-the-art prototype houses that combat exacerbated climate change in UK. *Energy and Buildings, 187*, 201. https://doi.org/10.1016/j.enbuild.2019.01.030.

Parkes, G., & Spataru, C. (2017). Integrating the views and perceptions of UK energy professionals in future energy scenarios to inform policymakers. *Energy Policy, 104*, 155. https://doi.org/10.1016/j.enpol.2016.11.019.

Peñasco, C., & Anadón, L. D. (2022). Assessing the effectiveness of energy efficiency measures in the residential sector gas consumption through dynamic treatment effects: Evidence from England and Wales. *Energy Economics, 117*, 47–49. https://doi.org/10.1016/j.eneco.2022.106435.

Rosenow, J. (2012). Energy savings obligations in the UK—a history of change. *Energy Policy, 49*, 373. https://doi.org/10.1016/j.enpol.2012.06.052.

Sousa, G., Jones, B. M., Mirzaei, P. A., & Robinson, D. (2017). A review and critique of UK housing stock energy models, modelling approaches and data sources. *Energy and Buildings, 151*, 66. https://doi.org/10.1016/j.enbuild.2017.06.043.

Tunstall, R. (2022). An empirical test of measures of housing degrowth: learning from the limited experience of England and Wales, 1981–2011. *Urban Studies, 60*(7), 1285. https://doi.org/10.1177/00420980221121198.

Vagtholm, R., Matteo, A., Vand, B., & Tupenaite, L. (2023). Evolution and current state of building materials, construction methods, and building regulations in the U.K.: implications for sustainable building practices. *Buildings, 13*(6). https://doi.org/10.3390/buildings13061480.

Wang, Y., Qu, K., Chen, X., Zhang, X., & Riffat, S. (2021). Holistic electrification vs deep energy retrofits for optimal decarbonisation pathways of UK dwellings: a case study of the 1940s' British post-war masonry house. *Energy, 241*, 33–35. https://doi.org/10.1016/j.energy.2021.122935.

Warren, P. (2013). A review of demand-side management policy in the UK. *Renewable and Sustainable Energy Reviews, 29*, 941. https://doi.org/10.1016/j.rser.2013.09.009.

Xie, Y., Gilmour, M. S., Yuan, Y., Jin, H., & Wu, H. (2017). A review on house design with energy saving system in the UK. *Renewable and Sustainable Energy Reviews, 71*, 29. https://doi.org/10.1016/j.rser.2017.01.004.

Zapata-Webborn, E., Mckenna, E., Pullinger, M., Cheshire, C., Masters, H., Whittaker, A., Few, J., Elam, S., & Oreszczyn, T. (2020). The impact of COVID-19 on household energy consumption in England and Wales from April 2020 to March 2022. *Energy and Buildings, 297*, 14–17. https://doi.org/10.1016/j.enbuild.2023.113428

Zhang, T., Siebers, P., & Aickelin, U. (2012). A three-dimensional model of residential energy consumer archetypes for local energy policy design in the UK. *Energy Policy, 47*, 102. https://doi.org/10.1016/j.enpol.2012.04.027.

6
UK HOUSING AND POLICIES

Introduction

In most countries in the world, buildings are responsible for at least 40% of energy use (Dolge & Blumberga, 2021). This figure is rising fast as construction booms in countries, such as China and India. In the UK, over 40% of energy consumption is building-related. The energy used in homes alone is responsible for more than a quarter of CO_2 emissions. CO_2 is the main greenhouse gas and the most significant cause of climate change (Dowson et al., 2012). Most of the energy used in homes is produced using processes that release CO_2 into the atmosphere. Challenging targets for emissions reductions have been set by the UK government. The UK government has recently committed to a minimum of 80% reductions in UK GHGs emissions by 2050 from 1990 level, as shown in Figure 6.1 (Day et al., 2009).

Energy is used in homes to provide useful services for householders, such as heat, hot water and power for appliances. As a result of the use of fossil fuels to provide this energy either directly, such as gas for central heating, or indirectly, such as coal for power stations, to generate electricity, this energy use results in carbon dioxide emissions (Li & Guo, 2021). The need to reduce these carbon dioxide emissions that result from energy use in housing is a key part of the UK's efforts to combat climate change (Coley & Kershaw, 2010). Energy use in homes is strongly affected by the population, the number of households, the type of homes, the level of insulation with heating systems, the age of a dwelling, householder income and the temperature difference between inside and outside. In terms of heating energy consumption, it usually correlates more strongly to the size of dwellings and type of dwellings, heating energy is related to external wall area and window area, and household size makes little difference to heating (Collins et al., 2010). The two most important determinants of heating energy use are insulation and the efficiency of heating systems. The age of a dwelling also affects its energy efficiency, and older homes typically have poorer insulation than modern homes (Gupta & Chandiwala, 2009). The difference in temperature between outside and inside homes in winter is the single most important factor shaping heating energy use at home as well. Hot water use and use of some appliances increase in proportion to household size.

This chapter shows the state of the UK housing sector such as improving the energy efficiency of the UK housing sector over the period of time, and how the UK government has introduced

DOI: 10.1201/9781003490364-8

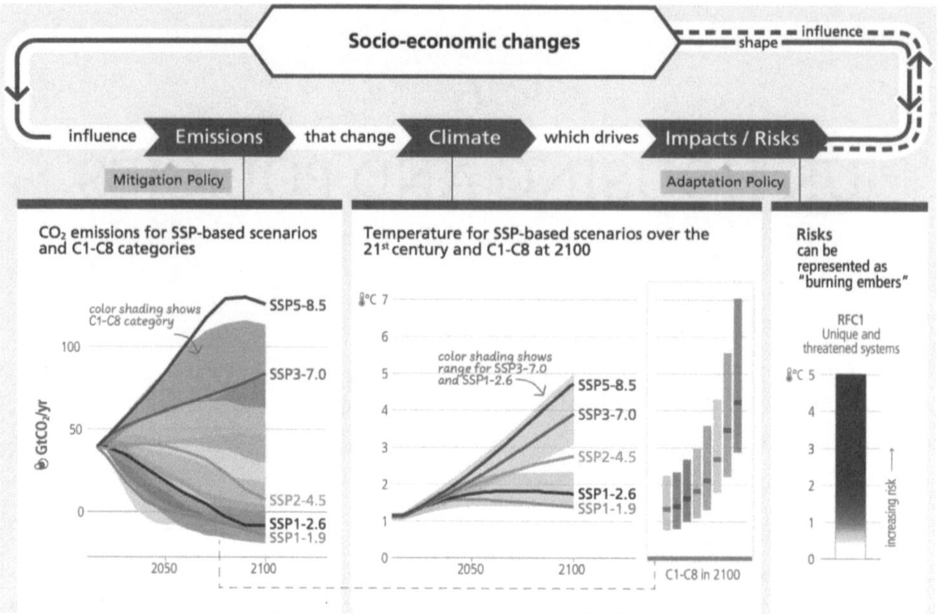

Figure 6.1 Carbon dioxide emissions by end user.

policies in order to improve the energy efficiency with indoor environments, especially thermal comfort. This chapter studied existing policies for 'New homes, existing homes and fuel poverty', in order to meet the UK government's ambitious target. For the existing home's study, user behaviour was also considered to have an effect on energy consumption in our homes. Also studied were UK's building standards, such as Building Regulation Part L, and UK's home rating systems, such as the Code for Sustainable Homes against Building Regulation, and Energy Performance Certificates (EPCs) with Standard Assessment Procedure (SAP). The direction of UK building standards for the future has been studied, in order to meet the government's challengeable CO_2 reduction targets. This chapter also examines the issue of fuel poverty in more detail by looking at the inter-relationship between social and environmental issues in housing. It describes the main fuel poverty programmes, especially those relating to energy efficiency, considers the relationship between fuel poverty and climate change policies, and critically reviews fuel poverty policies and the impact of and progress towards the government's statutory fuel poverty target. The historic development of UK homes is summarized in this short overview, based on several sources, and mainly considered from 'Energy Efficiency in OLD HOUSES'. (Holmans et al., 2007).

Most of the Roman villas were built around the countryside and it had underfloor heating and/ or hypocausts. Medieval timber-framed buildings were reasonably comfortable, and with enough straw, the thatched roof might even match the thermal properties of a modern roof. Even the infill of the timber framing, made of wattle, mud, straw and animal hair, might be built up enough to provide modest levels of thermal insulation.

The development of the English house is indebted to the Normans and the stone buildings they created, initially in the form of castles for defence. Massive stone constructions such as cathedrals and castles are pleasantly cool and comfortable in summer. The thick stone walls are able to absorb heat and store it to give it out again when it is needed such as during the cooler night. This thermal mass is less useful in winter, however, as it would absorb heat – stored in the stone walls, rather than heat the space and its occupants. From the Norman Conquest, the principle of internal environmental design changed. The major concern for the building was the heating in the UK climate. Daylight and ventilation could be provided by narrow-plan buildings.

In the 16th centuries, the chimney was invented but in the earlier part of the century, it would be used for showing off the owner's ability to others. But after time it was used for providing better air quality and heating. After the Great Fire in London, in 1667, a City of London Building Act was passed to increase fire resistance and party-wall construction between dwellings. Many problems of dampness in old houses were due to parapet walls. The rise of suburbs, the most ubiquitous form of British dwelling, with their semi-detached houses, began to be built in response to an increase in urban populations in the 19th century. After World War II, as increased demand for mass social and private housing supply, the traditional method of housing construction could not be maintained due to a lack of skill and skilled labourers. About a third of Britain's dwellings were built between 1800 and 1911, and a third of them between 1870 and 1911. Traditional domestic construction techniques did not change significantly over the period 1840–1919. Generally, houses built with solid walls need to 'breathe', as they will absorb moisture from rainfall but will also release it through evaporation if the originally designed ventilation paths and construction techniques are retained. Most modern houses operate on quite different principles, as the introduction of the cavity wall in about 1920 started the change in the construction technology to try to totally exclude moisture from dwellings and other building types. Measures to improve the energy efficiency of new homes were first introduced into Building Regulations in 1965, with the introduction of a requirement for loft insulation and this was followed by changes in 1971 to limit condensation. In 1981, Building Regulations sought to further reduce heat loss through the fabric although, at the time, the aim was to cut energy costs. Since then, they have been subject to regular review and update, the last change taking effect in April 2006.

Existing UK Homes Profile

Energy Consumption and CO_2 Emissions

Energy use in homes is just under a third of total energy use in Britain, up from a quarter in 1970 (Broer & Titheridge, 2010). The total energy consumption shows all energy used: transport, industry, public sector and housing. Total energy use in Britain rose and fell during the 40 years. However, it finished the period a little below the level of use in 1970: 1,600 TWh. Energy use in housing rose by 1.6% from 1970 to 2009 and an average increase of 0.4% per year (Brown et al., 2022). However, the number of homes also increased by two-fifths, and average household size has fallen. This means that average energy use per home has fallen from 22,235 to 18,639 kWh. The Department of Energy and Climate Change also provides a breakdown of domestic energy consumption, which has risen from 36,884,000 tonnes of oil equivalent in 1970 to 43,590,000 tonnes of oil equivalent in 2009 (Cerin et al., 2014). Breaking this down into types of fuel, between 1970 and 2009, total domestic electricity consumption increased by 59%, the use of solid fuels decreased by 96% and natural gas consumption increased by more than 300%.

Space heating is responsible for the majority of the domestic energy consumption, followed by lighting and appliances, and water heating. Out of a total of over 43 million tonnes of oil equivalent in 2009, space heating was responsible for 61%, lighting and appliances for 18% and water heating for 18% (Lee at al., 2018). The study shows the energy consumption by end-use in UK homes between 1970 and 2008. The energy consumed for space heating, lighting and appliances has increased while for cooking and water heating has decreased. Reducing the energy used for

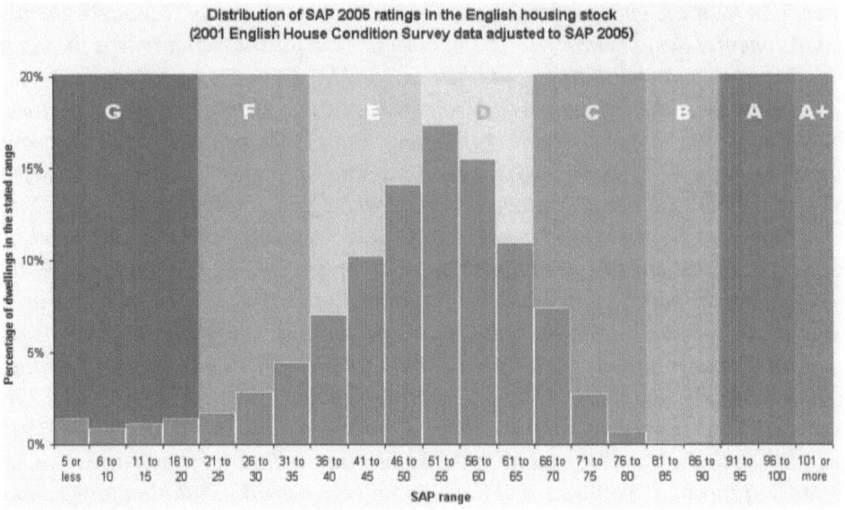

Figure 6.2 Distribution of the housing stock and EPC bands, UK.

space heating, lighting and appliances is important in order to achieve the government's carbon reduction target which is an 80% reduction by 2050 from the 1990 level (Sousa et al., 2017).

Energy Efficiency of Housing Stock

There were around 22.2 million dwellings in 2008 (Rosenow & Eyre, 2016). England has a relatively old housing stock with some 8.4 million homes built before 1945, of which 4.8 million were built before 1919 (Roberts, 2008). A fifth of homes have been built since 1980. About 60% of homes in the UK were built before introducing the modern Building Regulation, which was introduced in 1965 with U-value limits for walls and roofs. Around 19% of the dwelling stock was flats; most of these were purpose-built low-rise flats (O'malley et al., 2014). The most common types of dwellings were terraced houses, which accounted for 29% of the stock, followed by semi-detached houses at 26%, as shown in Figure 6.2.

Based on the SAP methodology's standard assumptions regarding occupancy and heating of the stock, mean carbon dioxide emissions per dwelling were 6 tonnes/yr. Some 15% of all dwellings were in the lowest EPCs Energy Efficiency Rating Bands F and G (SAP less than 39) (Miu et al., 2018). Over 40% of properties built before 1919 have an SAP rating of less than 41. Two-thirds of all properties have an SAP rating of 41–70, irrespective of age, whereas 60% of properties built since 1990 have an SAP rating greater than 70. Hard-to-treat homes generally feature characteristics such as solid walls, no gas supply, no loft space or being part of a high-rise block (Mcgilligan et al., 2008).

Space Heating

The pilot study shows that even with the improvements in energy efficiency of UK dwellings, heating's share of total energy use in homes has grown from 58% in 1970 to 66% in 2008 (Mempou et al., 2010). In the UK, the main types of insulation are loft insulation, cavity wall insulation, double glazing and hot water tank insulation. Improving building fabric efficiency by government's policies, almost a third of the UK's dwelling heat loss from building fabric has been reduced (Ozarisoy, 2022). This study also shows how improved insulation for building fabrics and improved heating systems' efficiency have reduced energy consumption for space heating in the UK housing sector. The combined savings from insulation and heating efficiency improvements reduced domestic space heating by an estimated 41.2 million tonnes of oil equivalent. Without these improvements, it is estimated that energy consumption would have been doubled (Coyne & Denny, 2021).

Heating and hot water account for around 84% of energy use in a typical home in the UK (Booth & Choudhary, 2013). Improving the energy efficiency of heating systems, or reducing demand, by even a small amount, can have a far bigger effect on energy use than making changes to, for example, lighting. This study shows that turning a room's thermostat down by just 1°C can reduce heating bills by 10% (Gupta & Gregg, 2012). People increasingly want to be comfortable in buildings. The average temperatures inside domestic dwellings are estimated by Building Research Establishment (BRE) to have increased from 12°C in 1970 to 17.3°C in 2008. The rise in temperatures has been helped by the increased energy efficiency of home such as use of central heating systems and improved fabrics' heat loss. The central heating ownership increased from 79% of households in 1990 to 96% in 2008 (Henderson et al., 2015). Temperature in a domestic dwelling is an important factor in energy consumption. According to Shorrock and Utley (2003), a comfortable living room temperature during occupancy hours is 21°C, in other parts of the house it is 2°C lower. The thermostat setting and length of time heating is on influence a dwelling's space heating energy use (Shipworth et al., 2010). A 1% increase in heating temperature is estimated to cause a 1.55% increase in CO_2 emissions. The same percentage increase in heating duration is likely to result in a 0.62% increase in CO_2 emissions. Therefore, increasing the temperature in a home results in more CO_2 emissions than having the heating on for a longer time period (Gaterell & Mcevoy, 2005).

The temperature that occupants set their thermostat to is affected by many different factors, such as social grade, type of dwelling, tenure and number of persons in the household (Marchand et al., 2015). From the Department for Environment, Food and Rural Affairs shows higher social grades reported setting their thermostat to around 21°C. People in lower social grades reported setting their thermostat at least 3°C higher. DEFRA found the two important factors which are as follow: *People living in flats set their thermostat to as high as 28°C, 8°C higher than people living in detached houses and, Private renters set their thermostat to 27°C on average, which is higher than other tenure groups (Jenkins et al., 2021).*

Further findings on internal temperature suggest that people living on their own tend to live in colder homes (Yohanis & Mondol, 2009). Meier & Rehdanz (2009) investigated that residential space heating behaviour of British households and accordingly, heating energy demand increases with households' size, average household age and number of children. Owners have higher consumption than renters (Hall & Purchase, 2006). The reasons for this may include housing type, with renters mainly living in flats and the majority of owners living in detached or semi- detached houses. In general flats are more energy efficient than houses.

Appliances and Lighting

The growth in appliance energy consumption had been sharply increased and it used less than 5% of total energy in 1970, now it is about 12% (Ozarisoy, 2022). The average annual growth in appliances energy consumption is nearly 3% a year. Since the 1990s people have been buying more energy efficient appliances but, at the same time, there has been an overall increase in the purchase of appliance, for example extra televisions and other consumer electronics (Poel et al., 2007). Between 1972 and 2002, electricity used in the household sector doubled, and further rose another 12% by 2010. The increase in the purchase of appliances has been influenced by cheaper prices and increased personal demands. In addition, increasing individualised lifestyles can lead to households having multiple of the same products. For example, in 2003 over 60% of households owned two or more television. Other reasons for increased energy usage include use of appliance as 'standby' modes and increased use of rechargeable portable devices such as laptop, tablet pc and mobile phones. The use of standby on some electrical goods is responsible for around 1% of total domestic energy consumption and 6% of domestic electricity consumption (Ozarisoy & Altan, 2023).

The presence of children in a household is also likely to affect the number of appliances per home and the resulting energy use. An example of use of energy by children is that 58% of children age between 12 and 14 have electric appliances, such as televisions and game consoles, in their bedroom and 23% of young people reported being likely to fall asleep with appliances left on. Lighting energy consumption is a small portion of total energy consumption in dwelling. The study shows that energy use for lighting increased every year until 2002 then it is stabilised about 3%. Overall, it increased by nearly 60% from 2.5% in 1970 to 3.3% in 2008 (Semple & Jenkins, 2009).

Since 1990, energy consumption for lighting has increased by around 1% per year, mainly due to the shift away from rooms lit by single ceiling bulbs to multi-source lighting. So, although lighting is now more energy efficient, more lighting is being used. Older people appear to use less lighting than younger people: household aged 20–64 years used 61% more energy for lighting than households aged 65 and over (Shrubsole et al., 2014). This may be due to older people are more likely considering the cost and bills than other age groups and switch off lights that are no use.

Time Spent at Home and Lifestyle Changing

The average person in the UK uses 150 litres of water every day and over 30% of this for hot water which use energy to generate heat and producing carbon dioxide from it (Gillich et al., 2019). A shower can use as little as one-third of the volume of a typical bath in the UK (70 litres). But power showers with pumps can use significantly more water than baths. The Office of National Statistics said that people in the UK spend 70% of their time at home and 27% of time away from home on primary activities. Lifestyle factors that may increase the length of time spent at home such as increase in working at home, unemployment and retirement, and therefore increased the likelihood that heating duration and energy consumption will be higher (Harrison & Reeve, 2002). During the winter months in particular, space heating may increase if occupants are at home for longer periods during the day. Ruiz and Walling (2005) found that the increase in the availability of technology in the home has contributed to an increase in the number of home workers, which has risen from 2.3 million in 1997 to 3.1 million in 2005. If this growth continues there could be around 30% of the UK workforce home working for at least by 2014 (Cairns et al., 2004). This would lead to increase in energy use in domestic dwellings but needs to be offset against energy savings at the place of employment and reduction in travelling.

Background on Policies

In its Energy White Paper of 2003, the government announced a long-term target of reducing the UK's emissions of greenhouse gases by 2050 by 60% compared with the level in 1997 (after that it was increased to 80% by 2050). Reducing energy use in housing is critical to the achievement of this target (Campbell, 2007). As mentioned above, 27% of the UK emissions of CO_2 are attributable to energy use in the domestic sector and the White Paper emphasised the need for action in both new and existing housing, including the introducing of Low-Zero carbon technologies and greater use of renewable energy sources (Alwan et al., 2016).

In December 2006, the Department for Communities and Local Government (DCLG) published its strategy for moving toward zero carbon development including progressive improvements in the energy standards required by building regulations in order to achieve zero carbon housing by 2016. Through changes to the building regulations, standards of thermal efficiency in new housing have been radically improved over the past three decades. Even so, the average dwelling constructed to the Part L 2006 Building Regulations standards would consume around 3,600 kWh each year. Clearly, significantly higher levels of thermal performance are required. It also published a Code for Sustainable Homes (Baba et al., 2012).

The measures required to achieve a 60~80% reduction in emissions from the housing sector were assessed in detail in a Tyndall Centre report published in 2005. This concluded that there would need to be significant investment to improve energy performance in the existing housing stock (two thirds of which would still be in existence in 2050) together with a higher rate demolition and new-build, and much greater use of low-carbon and renewable energy technologies on the supply side. Specifically, the report stated that new dwellings constructed after 2020 should consume on average no more than 2,000 kWh a year for space heating, in contrast to the 6,800 kWh that existing housing, on average, would consume even after the necessary improvements (Georgiadou et al., 2013).

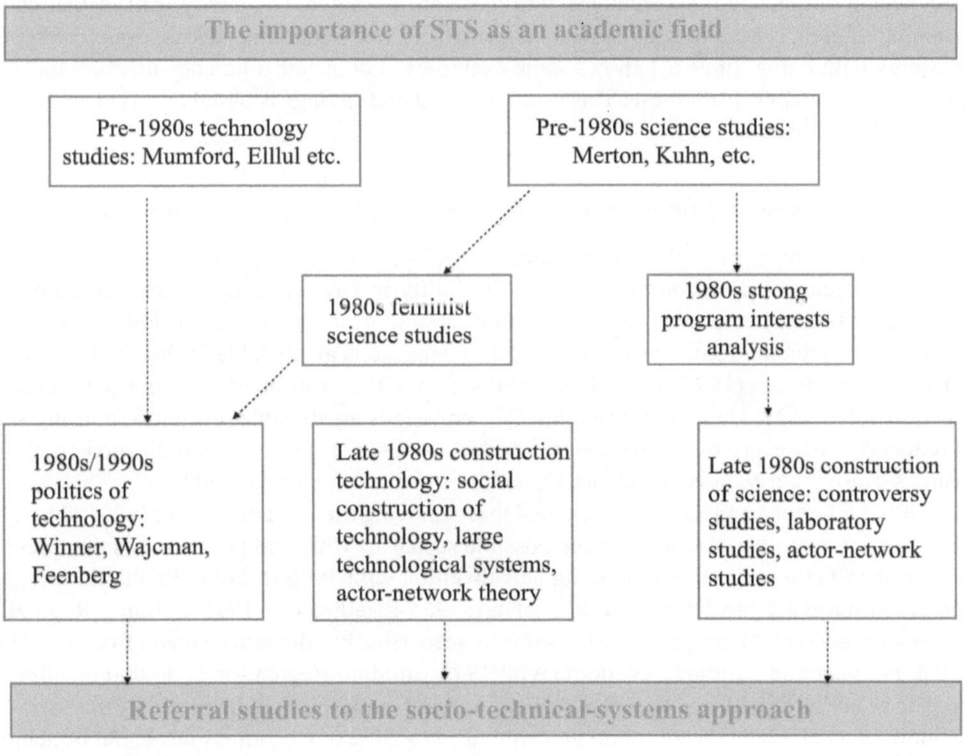

Figure 6.3 Major policies.

Table 6.1 Domestic efficiency measures: costs and savings

Measures	*Costs (£)*	*Bill saving (£/yr)*	*Carbon saving (kgC/year)*	*Pay-back (years)*
Cavity wall insulation	342	133	242	2.6
Solid wall insulation	3150	380	694	7.5
Loft insulation	284	104	190	2.7
A-rated boiler	1500	168	177	8.9
Hot water cylinder insulation	14	29	53	0.5
Improved heating controls	147	43	77	3.4
Draught proofing	100	23	43	4.3
Double glazing	4,000	41	26	97.6

In addition, there are large challenges in bringing the existing stock up to a suitable level of performance without excessive cost, reduction in internal space or unacceptable changes to external appearance. New materials and insulation technologies may offer improved performance and the incorporation of renewable sources of energy in refurbishment projects will be an essential component of an overall strategy for reducing emissions from existing housing (Jacoby et al., 2022). As shown in Figure 6.3, the UK government has announced many different policies to improve the UK homes' energy efficiency especially for fuel poverty. The measures promoted through policies and supported by grants with a simple payback of less than seven years, which is determined 'energy efficiency measurements' and these are considered as cavity wall insulation, loft insulation, efficient central heating systems and controls, hot water tank jackets, draught-proofing, low-energy light bulbs. Table 6.1 shows some examples of domestic efficiency measure for installation costs and savings from these. This estimated cost and savings is based on a typical three bed semi-detached property.

Research on Emissions Cuts from UK Housing Scenarios

Recent years, a number of studies have examined methods of reducing the carbon emissions from UK housing sector as based on the technical feasibility in UK dwelling in order to achieve UK government's target. Letherman and Samo (2000) and Shorrock et al. (2001). These studies have attempted to predict the future energy use and CO_2 emissions attributable to the UK housing sector. The study by Evans (1997) studied the stabilisation of the atmospheric CO_2 concentration over the period 1991–2021. He assumes that the CO_2 emissions attributable to the UK housing sector will reduce by 10% every five years over the period 1991–2021. From this simple method, the UK housing sector would achieve the 60% CO_2 emission reduction target by 2021.

In 2000, Letherman and Samo investigated that achieving the UK government's 20% CO_2 emissions reduction target by examining three possible scenarios which are; (1) demolishing pre-1965 dwelling at 180,000 per year and building new dwelling with 1996 Building Regulations requirements, (2) improving the fabric U-values of pre-1965 dwellings to 1996 Building Regulations requirements at 400,000 per year, and (3) same as scenario 2 but the rate changed from 400,000 to 200,000 per year and building new home with 1995 Building Regulation at 50,000 dwelling per year. The result indicated that the 20% reduction target could be only achieved by scenario two. The Shorrock et al. (2001) work found that around 17%–31% of CO_2 emissions could be achieved by introducing a range of energy efficient measures into the UK housing stock.

Table 6.2 Research on emissions cuts from UK housing scenarios

Research	*CO_2 reduction targets*	*Considering cooling energy demand*
Evans (1997)	60% by 2021	No
Letherman and Samo (2000)	20% by 2020	
Shorrock et al. (2001)	17%–31% by 2020	
BRE (2005)	60% by 2050	
Johnston et al. (2005)	60% by 2050	
Boardman et al. (2005)	60% by 2050	
Boardman et al. (2007)	80% by 2050	
Natarahan and Levermore (2007)	50% by 2030	
	60% by 2050	
	70% by 2030	

In 2005, the three studies were carried out and each study confirmed that 60% target could be achieved. From the BRE research, they suggested that to achieve emissions reduction over 60% is by 50% solar thermal, solar PV and heat pumps, and 25% of centrally heated homes with biomass boiler installation. Boardman et al. (2005) explained the use of an average of two 'low or zero carbon technologies' per home that could make homes to be achieved 60% emissions reductions, and the technologies are 60% of dwellings installed solar thermal, almost 40% heat by micro-CHP, 30% had solar PV and 20% of heat from community heating. In 2007, have reviewed the three studies and they found Johnston et al. were not able to achieve 60% emissions reductions. The most efficient suggestions from Natarajan and Levemore (Boardman et al., 2005) is that solar PV, solar thermal or micro-CHP each would need to be installed in around 45% of all UK homes. The WWF research found that 80% reductions could be achieved through the most extensive refurbishment with reduced energy demand and gird decarbonization. The Energy Saving Trust study found that 80% of reduction could be achieved by reducing energy demand, grid decarbonization, maximal use of technical interventions and further reductions in appliance usage, as shown in Table 6.2.

There is general agreement within the various studies to achieve emissions reductions over 60% in the UK housing sector (Fuerst & Mcallister, 2011). The essential premise to achieve these targets is an extensive government programmes for housing stock improvements then most extensive technologies considered. None of above studies, however, has indicated about energy consumption and CO_2 emissions increased from cooling energy demand due to climate change and global warming. Of course, they mentioned about improving energy efficiency of lightings and appliances but without considering of cooling energy demand. If adding cooling energy demand with these studies, the government target should be changed or be carefully re-planned to avoid dangerous level of climate change due to greenhouse gases concentration into atmosphere by human activities (Kelly et al., 2012).

New Homes

Little more than a few years ago, few of us had considered the possibility of the zero carbon homes. The concept of modern homes was self-sufficient in terms of their energy use had been limited to a handful of 'autonomous houses', not connected to mains services, designed, built and occupied by enthusiasts. That all changed in December 2006, with the launch of a government consultation which proposed that from 2016 all new homes should be built to a zero-carbon standard. The policy statement which followed in July 2007, *Building a greener future: policy statement* confirmed the policy and its timetable. The extent of change required to deliver zero carbon homes cannot be overstated (Mckee et al., 2016).

Critical to achieving zero carbon homes is having a definition that is realistic, workable and has sufficient flexibility to enable compliance on all developments and for all kinds of homes. Clearly, the challenge of developing a high-density urban infill site, where there may be less opportunity to exploit renewable energy resources is quite different from that of building a single detached home on a large south facing plot where there is scope for using a variety of renewable technologies (Mcmanus et al., 2010). The consequence of an inflexible definition of zero carbon, which did not recognize these limitations and allowed no scope for the use of renewable energy produced offsite, would be to prevent the development of a significant proportion of site. Clearly that would run counter to government policy of improving the supply of new homes in the UK (Ozarisoy & Elsharkawy, 2019).

The commitment to build zero carbon homes is one of several government strategies and initiatives which aim to reduce the causes, and adapt to the effects, of climate change. On 3 October 2008 the government announced the creation of a new Department of Energy and Climate Change as part of its commitment to a wide range of strategies to reduce energy consumption, conserve water, reduce waste, mange flood risk, provide renewable energy sources, reduce risk of overheating, reduce fuel poverty and so on (Sweetnam et al., 2013). On 17 December 2008, following on from work done by the UK Green Building Council (UKGBC) earlier in the year the government launched a second consultation which deals specifically with the definition of zero carbon and how it is assessed.

Zero Carbon New Homes by 2016

In order to reduce UK domestic carbon emissions, the government has set an ambitious target for new homes to be 'zero-carbon' by 2016. This means that new homes will need to reduce the energy that is used for activities such as space heating, water heating, lighting, cooking and running electrical appliances. Once that energy reduced to a minimum, homes will need to supply the remaining energy required from a renewable or zero-carbon energy sources, so that over a year there are no net carbon dioxide emissions as a result of the activities within the home. Government has also introduced stamp duty land tax exemption of new zero-carbon homes costing less than £500,000. From October 2007, in order to qualify, a zero-carbon home must be certified and calculated in accordance with the approved method, set out the Stamp Duty Land Tax (Zero Carbon Home Relief) Regulations 2007, as mentioned in Table 6.3 (Watts et al., 2011).

However, carbon emissions associated with production and transport of building materials are not considered in this definition, nor are the energy used during the construction process. With the construction of each new home thought to be responsible for 20–40 tonnes of carbon dioxide (Swan et al., 2017). There is growing concern that the present definition of a zero-carbon home does not reflect the 'true' carbon footprint. It may be argued that the definition of zero-carbon home should also address carbon emissions associated with the lifestyle of the occupant (Wolf et al., 2021). The location of the home will greatly influence the mode of transport, how and where food is bought from, travel to work and school, and interaction with their local community. All of these activities have associated carbon emissions. The three mechanisms currently in place for requiring or encouraging zero carbon standards for new homes are the Building Regulation, the Code for Sustainable Homes and Stamp Duty Land Tax Relief (Carter & Fortune, 2007).

Table 6.3 What does zero carbon home mean?

Building A Greener Future sets out that all new homes are to be built from 2016 in such a way that, after taking account of: *emissions from space heating, ventilation, hot water and fixed lighting expected energy use from appliances exports and imports of energy from the development (and directly connected energy installations) to and from centralised energy networks,* the building will have net zero carbon emissions over the course of a year.
The present consultation retains the approach of looking at net emissions (including from appliances) over the course of a year. It proposes that, to meet the zero carbon homes standard, homes should: *be built with high levels of energy efficiency and achieve at least a minimum level of carbon reductions through a combination of energy efficiency, onsite energy supply and/or (where relevant) directly connected low carbon or renewable heat; and choose from a range of (mainly offsite) solutions for tackling the remaining emissions.*

Building Regulation

The concept for zero carbon was first introduced in the *Building a green future: towards zero carbon development* consultation document, published by Communities and Local Government (CLG) in December 2006. It laid out a proposed timetable for changes to Building Regulation for all new homes to achieve zero carbon by 2016, and defined zero carbon as follows:

> For a new home and be genuinely zero carbon it will need to deliver zero carbon (net over the year) for all energy use in the home – cooling, washing and electronic entertainment appliances as well as space heating, cooling, ventilation, lighting and hot water.
>
> (Fuerst et al., 2015)

This consultation marked a turning point because it was the first time that it was suggested that the CO_2 emissions from appliances and cooking, and cooling within the home, might be brought under control using the Building Regulations. A policy statement was issued by CLG in July 2007, which confirmed the government's intention to press ahead with the policy to the following timetable in England and Wales. It proposed that by 2010, new housing achieves a 25% carbon improvement relative to 2006 Building Regulations and then this would be increased to a 44% improvement in 2013 and leading to zero carbon housing by 2016, as shown in Figure 6.4. Note that zero carbon represents an improvement in the order of 150% over AD L1A (Mcgilligan et al., 2009).

Table 6.4 shows a summary of the range of UK minimum building fabric standards need to the respective certification. From this, we could image how the UK home's fabric efficiency would be improved to achieve the Zero carbon home by 2016. It also makes it clear that the building fabrics thermal property will be more and more increased, which means less heat losses through the building fabric and the buildings will be more airtight.

Code for Sustainable Homes

Launched in December 2006, the Code for Sustainable Homes is an environmental assessment method for rating and certifying the performance of new homes. The Code has nine categories of sustainable design and credits are earned under these categories if specified performance targets are reached. Based on the number of credits achieved an overall rating is awarded, ranging from Level one (1 star) to Level six (6-star). Amongst the nine categories of sustainable design

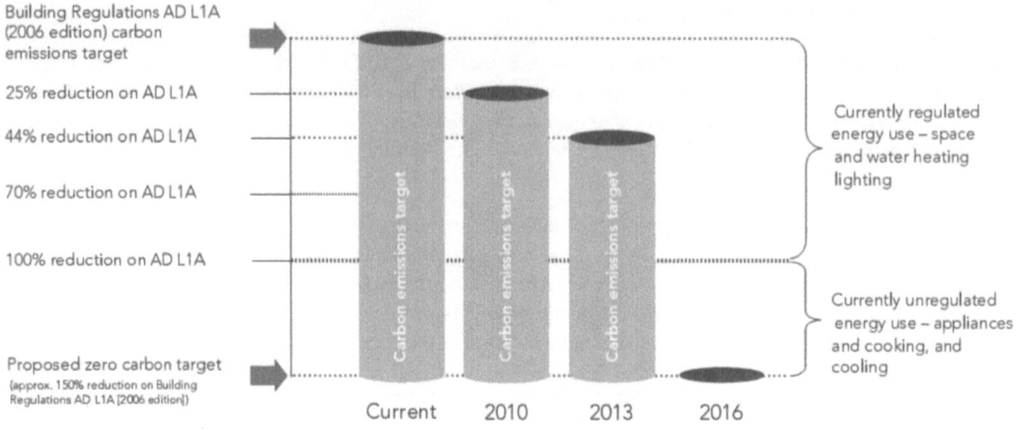

Figure 6.4 Reduction of CO_2 emissions from new housing.

Table 6.4 Building fabrics standards

U-value and airtightness	*Building standards*						
	Building Regulation 2006	*EST Good Practice*	*EST Best Practice*	*EST Advanced Practice*	*AECB Silver*	*AECB Gold*	*PassivHouse*
Roof	0.25	0.16	0.13	0.15	0.15	0.15	0.1
Walls	0.35	0.3	0.25	0.15	0.25	0.15	0.1
Floors	0.25	0.22	0.2	0.15	0.2	0.15	0.1
Windows	2.2	2.2	1.8	0.8	1.5	0.8	0.8
Airtightness (m^3/ m^2/hr@50)	10	10	3	0.75	3	0.75	0.6 (AC/hr)

is 'energy efficiency measures' and emissions (Pickvance, 2009). Due to the importance of this category, mandatory levels of performance for energy and emissions are prescribed for each level of the Code. Again, these are expressed as a percentage improvement in CO2 emissions over the current level of performance required by AD L1A, as indicated in Table 6.5.

The percentage improvement required over the target emissions rate (TER) required by Building Regulation Part L is shown in Table 6.5. Code level 3 needs a 25% TER improvement (CO2 emissions set at least 25% lower than basic Building Regulation 2006 Part L compliance). Similarly, the Code level 4 needs a 44% improvement on the Part L and requires 10.4 points. The design measures need to achieve the Code level 5 and 6 with regard to energy are essentially the same for all types of construction, requiring significant use of renewable energy conjunction with a very high standard of fabric performance. The definition of zero carbon that the Code uses is essentially the same as that used in the 2006 CLG consultation document (Pickvance, 2009), but two additional criteria are set as: *To ensure homes are built to excellent standards of fabric energy efficiency – this prevents homes heated using fuels with low or zero CO_2 emissions from being built with poor fabric performance, To allow the use of renewable energy generated offsite to be counted, only where it is directly connected by means of a 'private wire' – this prevents the*

Table 6.5 Anticipated mandatory improvement in CO_2 emissions

Code level	*Percentage improvement over TER required by Part L1A (2006)*	*Points required*	*Government target for regulatory introduction*
1	10% (EST 'Good Practice' level)	1.25	–
2	18%	3.77	–
3	25% (EST 'Best Practice' level)	6.28	2010
4	44% (similar to 'PassivHaus' standard)	10.04	2013
5	100% (zero emissions: heating, hot water, ventilation and lighting)	17.57	–
6	≈150% (zero net emissions from all energy use in the home)	18.83	2016

developer from simply paying towards a wind turbine away from the site that would have been constricted in any case. (Sodagar & Starkey, 2015).

From 1 May 2008, it has been mandatory for a Code rating certificate to be provided for all new homes built in England. For publicly funded new homes and those built on public land, achievement of Code level 3 is the minimum requirement. The study shows predicted carbon emissions from homes through various building standards and rating levels. The existing housing stock is responsible for an average of around 72 CO_2 kg/m²/yr. After Code level 5, carbon emissions from heating and hot water would be zero due to using renewable source to provide the heating demand by solar thermal, district heating, biomass and CHP, and only remaining would be from cooking and appliances. These would be also removed by energy provided from renewable energy sources which are mostly coming from PVs and wind turbines (Monahan & Powell, 2010).

Stamp Duty Land Tax Relief

A relief from stamp duty land tax (SDLT) form most new zero carbon homes built in the UK was announced in the 2006 pre-Budget report. To establish eligibility for this relief HM Revenue and Customs (HMRC) needed to adopt a definition of zero carbon. Like the Code, the HMRC definition follows that in the 2006 CLG consultation document but, again requires in addition as (HMRC): *Excellent standards of fabric energy efficiency, Stipulation that a 'private wire' shall be used to connect offsite renewable energy sources* (Mccord et al., 2020). It is understood that, to date, only a very small number of homes have qualified for relief from SDLT.

Why Aim for Zero Carbon?

It is estimated that the new homes built between now and 2050 could account for up to one third of the total stock of housing in 2050. Logic dictates, therefore, that if those new homes are built to high standards of energy efficiency, the amount of CO_2 they will emit will add less to total housing emission than if they are built to lower standards (O'keeffe et al., 2016). And so, it follows that, if homes can be built to zero carbon design standards, which they will not in theory add to overall emissions at all. Importantly, the zero-carbon target provides certainty to the industry on what is needed and when, allowing time for progressive up-skilling and development of the new approaches required. Homes built to zero carbon standards should be relatively future proof: They will not contribute towards climate change during their operation; Their occupants will be unlikely to suffer fuel poverty; Their occupants will be well cushioned from the effects of international fuel markets (Muir, 2013).

Construction of Zero Carbon Houses

A wide range of new technologies are appearing in an effort to meet low and zero-carbon standards, not just for creating renewable energy, but also in terms of constructing new houses. Many of these techniques are hybrids of tried and tested solutions but many are highly innovative. In the main, the new forms of construction are more expensive than traditional building, but the incorporation of much higher specification windows and doors, ventilation systems and substantially more insulation will inevitably increase costs. In an effort to increase efficiency and keep costs within limits there has been encouragement for more prefabrication, off-site construction and other new techniques. These range from 'high-tech' materials using steel and synthetic components, to timber frame and the use of low impact renewable crop-based materials (Williams & Dair, 2006). Many of these innovations have been encouraged by the government, and some are showcased at the BRE Innovation Park. These showcase buildings have highlighted major challenges facing housebuilders aiming to achieve zero-carbon performance. Only, 6% of homeowners surveyed by the National House Building Council Foundation (NHBC) felt that the £35,000 extra cost to achieve zero-carbon homes is reasonable. For example, the Lighthouse, in BRE Innovation Park, built by Kingspan costs an estimated £180,000, compared to the average £100,000 for a building regulation-compliant home. A slightly lower cost of £150,000 has been estimated for Bill Dunster's Rural Zed house to achieve Code level 6. However, this is a major investment that many homeowners are unwilling to make, even with possible savings in running costs of £400 per year. This is a serious concern for house builders surveyed by NHBC, who believe that Code level 6 homes cannot be built profitable by 2016 (Scanlon & Elsinga, 2013).

Fuel Poverty

The UK is one of the world's developed countries, yet a large proportion of its population suffers from fuel poverty. Despite numerous initiatives, the issues remain prevalent, enduring and serious. Today, one in five UK households (21%) are fuel poor, but around the constituent nations the picture varies: the figure is as high as 44% in Northern Ireland, one in three homes (33%) in Scotland, a quarter of households in Wales (26%) and 18% in England. Fuel poverty is the term applied to household that spend more than 10% of their income on energy bills (Smol, 2022). As such, the proportion of people in fuel poverty varies as incomes rise or fall, as energy prices fluctuate and as the amount of energy consumed changes. During the early 2000s, when competitiveness in the liberalised energy market led to falling prices, combined with a period of rising benefits for low-income households, the number of people in fuel poverty fell. From 2004 onwards, though, the trend has been upwards. In the year 2000, the government made a commitment to the low-income households of the UK. By 2016, it said, fuel poverty in the UK would be history. At that time, 2016 seemed a long way away; the target appeared reasonable and achievable. However, up-to-date figure shows that the government's target to eliminate fuel poverty by 2016 would not be met within their timeline (Gooding & Gul, 2016).

A New Definition of Fuel Poverty

The *Hills Fuel Poverty Review*, commissioned by the Department for Energy and Climate Change (DECC) published its interim report in October 2011. The review proposes a new definition of fuel poverty. The '10% of incomes spent on energy' criteria says that Hills should be replaced. Instead, households will be classified as fuel poor based on a new calculation that takes into account high bills and low incomes. A household will be in fuel poverty under the Hills definition, when *To keep their home reasonably warm, they need to spend more on fuel than the average household, and That required high spending on fuel pushes their household income*[10] *below the official poverty line.* The Eastern Standard Time (EST) carries out regular research with a recent telephone survey and found out the following: *28% find it difficult to heat their home, 62% worry about bills: of these 32% feel it is a real personal stress to them, 40% find their home too cold, 30% find their house too draughty, 35% find that the way their home is built and maintained has an effect on their health, 21% report spending more than 10% of their household incomes on fuel bills, and 11% spend more than 20% of their household income on fuel bills. (Fuerst et al., 2016)*

21% is slightly higher than the 18% of households the government identified in 2009 as officially in fuel poverty. However, 11% is significantly higher than 3% of households the DECC identified as needing to spend over a fifth of income on fuel in 2009. Reporting spending over 10% of household income on bills is not the same as being in fuel poverty (defined as needing to spend 10% on energy bills in order to adequately heat the home) (Colburn, 2019).

Fuel Poverty in the UK

A household's income, the price it pays for its energy and the quality of its housing, heating and other equipment will determine whether the household is in fuel poverty. The quality of housing and equipment will have a major impact on the energy bill of low-income households generally. Fuel poverty is largely a problem of the existing housing stock and is not an issue in newer housing because of the comparatively higher standards of energy efficiency. The largest concentration of fuel-poor households is to be found in pre-1975 dwellings.

At the beginning of the century, the previous government published the *Warm Homes and Energy Conservation Act 2000*, which promised to eliminate fuel poverty by 2016. The following year, *the Fuel Poverty Strategy* committed the government to removing vulnerable households from fuel poverty by 2010 and all households by 2016, as shown in Table 6.6. With 4 million households in the UK trapped in fuel poverty in 2009, that target had certainly been missed. Indeed, recent estimates show that fuel poverty today is now more than twice as prevalent as it was in 2009 when targets for 2010 and 2016 were set (Martiskainen & Kivimaa, 2019).

The study shows that the reduction in fuel poverty between 1996 and 2003 was very substantial. Since 2003, however, there have been marked increases in the numbers of fuel poverty, so that in 2006 there were about 2.4 million households in England. Reductions in the number of fuel-poor households over the period 1996–2005 were driven by energy price decreases, as the combined product of world trends and the introduction of competition in the UK energy market. Household incomes also increased significantly during this period, especially with the introduction of Pension Credit and Tax Credits. Furthermore, the programmes for improving the energy efficiency of low-income households also made an important contribution, such as the Warm front (Gallent et al., 2020).

Table 6.6 Statutory obligation to eradicate fuel poverty in the UK

	Vulnerable households	*All households*
England	2010	2016
Wales	2010	2018
Scotland	N/A	2016
Northern Ireland	2010	2016

The government, in its 2009 annual report on fuel poverty, estimated that – of the reduction in overall fuel poverty between 1996 and 2005 – three-quarters was due to income improvements and around a fifth to energy efficiency improvements, with the rest resulting from energy price reduction. It is generally recognised that the best way to protect people against fuel poverty is with energy efficiency improvement. But, in recent years, all the progress made in tackling fuel poverty through energy efficiency has been undone through rising fuel prices. Electricity bills rose by 13% in the five years to 2011, and gas bills rose by 25% over the same period (Kearns, 2022). The result is that levels of fuel poverty now are estimated to be about the same as they were when the government committed to abolish the problem. The need to tackle cold, unhealthy homes does not go away. The government's own Hills Review recently estimated that more people die because of cold homes than die on the nation's roads.

Impact of Reducing Fuel Poverty

A reduction in and elimination of fuel poverty, especially through improved energy efficiency, brings a range of benefits and helps with a series of government objectives. There will obviously be considerable similarities between the impacts of measures to reduce fuel poverty and others to improve housing more generally. Reducing fuel poverty will help to:

> Enable people to keep warm, increase an individual's disposable incomes, Reduce carbon dioxide emissions, Reduce excess winter mortality – in part related to cold homes, Reduce cold-related illnesses, Enable more elderly people to live independently, Make it possible to release people from hospital to their homes more quickly, Reduce costs to the National Health Service, and Recue health inequalities.
>
> (Rivas et al., 2021)

Dwelling Characteristics and Efficiency

The study shows the fuel poverty rate in households with varying numbers of bedrooms. In households with one, two or three bedrooms the fuel poverty rate is broadly similar at around 19–21%, whereas in households with four or more bedrooms the fuel poverty rate is lower at around 12%. This is due to the fact that households living in properties with four or more bedrooms tend to have a greater household income than those households living in smaller properties, as shown in Table 6.7.

The energy efficiency level of the dwelling is another key driver in the propensity to fuel poverty. The study shows the fuel poverty rate by different SAP ratings, illustrating again that the least energy efficiency households are the ones most likely to be fuel poor. Most of the fuel poverty dwellings rate in EPCs as F and G, and these are mostly built before introducing modern

Table 6.7 Average annual income by number of bedrooms in 2009

Number of bedrooms	*Average annual income (£)*
1	17,000
2	21,900
3	26,700
4 or more	44,200
All households	27,900

Building Regulations (Pryce & Sprigings, 2009). These dwellings are also commonly called 'hard to treat homes' and these homes commonly have solid walls, electric heating systems, old boilers, minimum/or no loft insulation, no floor insulation, no completed double glazing and even off from the grid systems. This study shows the distribution of SAP ratings for both the fuel-poor households and the non-fuel-poor households. The average SAP rating amongst fuel-poor households in 2009 was 43.9, lower than the average SAP rating of 55.3 amongst non-fuel-poor households. However, the distribution of SAP rating in the fuel-poor group is more spread than that of the non-fuel-poor group. The 5th and 95th percentile of the fuel-poor group is 8.7 and 67.8, respectively, whereas the 5th and 95th percentile of the non-fuel-poor group is 32.3 and 74.8, respectively (Pryce & Sprigings, 2009).

Existing Programmes and Policies

Our homes need to be heated to a comfortable temperature, but some people cannot afford to heat them adequately to achieve a comfortable temperature. To tackle fuel poverty, it is necessary to address its three causes as shown above: high fuel prices, low incomes and poor energy efficiency. Of these three, poor energy efficiency is the most cost-effective to address. Improving home energy efficiency has other benefits too: it makes homes comfortable, and it helps to tackle climate change due to less heating energy demanded and less burning fossil fuel.

Government Funded Programmes

There are very sizeable schemes that provide energy efficiency measures free of charge for low-income households and these measures can significantly reduce fuel bills. This programme provides free insulation and heating system upgrades to people in England on means-tested benefits who own their own homes or who rent from private landlords. It also checks that vulnerable households don't miss out on their full benefits entitlements. In 2006/2007, the Warm Front programme secured an average improvement in SAP ratings per household from 40 to 56, a fall in CO2 emissions from 6.97 MtC per annum per household to 6.16 MtC and the potential to save an average of £190 per annum for each household in energy running costs. However, having helped more than 2.3 million vulnerable households since its inception in 2000, Warm Front is ending in 2013 (Pretlove & Kade, 2015). This will mean that, for the first time in 35 years, no publicly funded programme to improve the energy efficiency of vulnerable households will exist in England. Grant schemes in Scotland (the Energy Assistance Package or EAP) and Nest in Wales will continue. This provides between £150 and £300 paid to approximately 12 million UK households that have residents over the age of 60. Costing £2.7 billion a year, as a universal benefit, it does not target the fuel poor. Additional Cold Weather Payments are made to households on specific means-tested benefits during periods of very cold weather lasting more than a week.

Programmes to Improve Rented Homes

Decent Homes in England, the Scottish Housing Quality Standard and the Welsh Housing Quality Standard have set minimum levels of thermal comfort to tackle social housing in poor conditions. Many homes in the private rented sector are highly energy inefficient. The Landlords Energy Saving Allowance (www.direct.gov.uk) gives landlords a £1,500 tax allowance each year for energy saving improvements. The Housing Health and Safety Rating System is a risk management system used by local authority environmental health officers in England that aims to minimise hazards in homes. It is used in particular to regulate standards in the private rented sector. Cold homes are seen as a serious hazard, and although it is loosely defined, some official guidance – and many local authorities – has classified all homes in the F and G banding on an EPC as representing a serious cold hazard. Two-thirds of G-rated properties are occupied by fuel-poor households. Supporting lower tariffs for vulnerable households: Warm Home Discount Introduced in April 2011, the Warm Home Discount scheme, which is paid for by energy providers at a cost of £250 million p.a., subsidies electricity bills of selected vulnerable households – usually pensioners – by up to £140 a year. By the time the initiative is due to end in March 2015, it is expected to have helped more than 2 million households, although not all of them will be fuel poor (Li et al., 2019).

The Carbon Emissions Reduction Target (CERT)

This £1 billion per year scheme obliges energy suppliers to fund energy efficiency improvements in homes. It is the government's main policy for improving energy efficiency across the UK housing stock. Investment in the scheme is considerable and, whilst it generally only funds basic cost-effective measures, it has provided insulation in more than 1.8 million cavity walls, 2.4 million lofts and 42,000 solid walls. CERT funds measures in all homes, but 40% of the target must be achieved in the homes of a group of 'priority' customers assumed to be at risk of fuel poverty: those in receipt of certain income-related benefits and pensioners aged over 70. Within that 40%, since 2010 suppliers have been further obliged to target 15% (of the total target) at a 'super-priority' group on a more restricted set of benefits, and who are deemed to be at particular risk of fuel poverty. Some energy suppliers have struggled to identify customers in this super-priority group (Poon & Garratt, 2012).

The Community Energy Saving Programme (CESP)

CESP brings energy companies, local authorities and third-sector organisations together to introduce a 'whole house' approach to energy efficiency measures in low-income areas. Introduced in 2009, it obligates energy suppliers to install energy-saving measures in harder-to-treat, especially solid-wall, properties – the homes that they have not found it cost-effective to target under CERT. By the time the scheme ends in October 2012, it is expected to have delivered more than £350 million of energy efficiency measures to about 90,000 homes: equivalent to 1.6% of the total number of households in fuel poverty (Muir, 2013).

The Future: ECO and Green Deal

This year the Energy Company Obligation (ECO) will replace supplier obligations under CESP and CERT. Costing the suppliers around £1.3 billion per year, ECO has two elements: the affordable warmth obligation and the carbon-saving obligation. The first, affordable warmth, proposes to target the current CERT super-priority group but only in privately owned or privately rented accommodation. It will help low-income and vulnerable households to heat their homes adequately. The second element, carbon saving, will work with hard-to-treat homes where householders cannot fund energy efficiency improvements fully through the Green Deal. The government predicts that, by 2023, ECO will alleviate fuel poverty in between 350,000 and 550,000 households – around 10% of those currently affected – and, at the most optimistic level, 1.7 million cavity walls will be insulated: less than 30% of the potential (O'neill & Gibbs, 2019).

The Green Deal is the government's flagship energy efficiency policy. It will provide householders with access to finance for energy efficiency improvements. They will make repayments through the electricity meter. But there is a 'golden rule' for eligibility: annual repayments must be no more than the savings achievable by installing the improvements. The golden rule is based on the 'required' fuel use of the property, rather than the actual use. As many fuel-poor households use less than the required amount of fuel and are thus likely to achieve smaller savings, this poses a risk, because the golden rule may not apply to them.

Climate Change Effect on UK Homes

Previous Studies

Adaptation of the built environment to climate change is becoming increasingly as important as mitigation, as shown in Table 6.8. Adaptation, a responsive adjustment to reduce or eliminate possible risk, will be critical given the earth is committed to at least 1.8°C global average surface warming by the end of the century, even in the most optimistic projection presented by the Intergovernmental Panel on Climate Change. In the UK, it is very likely that the degree of change in mean summer temperature increase will be above 1.4°C for the southeast of England by the 2080s (Mitchell & Natarajan, 2020).

The housing sector, in particular, is not only contributing to a significant proportion (27%) of the UK's CO_2 emissions as a result of energy consumed for heating, lighting, cooking and use of electrical appliances, but it is also recognised to be inadequate in capacity to adapt to future climate change or even variation in the current climate (Ozarisoy & Elsharkawy, 2019). A warming climate is projected to change the comfort conditions and energy use patterns of existing UK homes to a considerable degree throughout this century. This is why emerging policy on climate change adaptation is overwhelmingly focused on tackling overheating in new and existing housing.

Climate change mitigation can be both synergistic and opposable to adaptation, so it is in our best interest to ensure that the retrofits do not lock-in overheating risk through measures that are not tested for future change in climate. It is also important that the potential to adapt in simple ways is not missed in the process of retrofitting the existing building stock. In order to most effectively reduce the risk to future generations, it is essential that neither mitigation nor adaptation have negative consequences nor become contradictory.

Climate change adaptation has a clear role to play in mitigation-oriented refurbishment approaches. One clear example is that a large majority of dwellings in the UK have no mechanical cooling. Modelling has shown that building performance in current 'heatwave condition' can function as a performance benchmark for a typical hot summer in the 2050s. Increases in summer temperatures like this have the potential to increase occupant vulnerability to overheating and the potential to increase energy use through the widespread responsive installation of air-conditioning for occupant comfort (Collins et al., 2010). Collins, Natarajan and Levermore modelled the impact of climate change on the overall UK housing stock assuming a 50% uptake of air-conditioning with no passive alterations and assuming heating methods would continue as they are currently and found that there would be a 10% decrease in CO2 emissions from the drop in heating demand. They, however, note later in

Table 6.8 Mitigation and adaptation

Mitigation: is to act to try to limit the amount of future change by reducing the emissions that are understood to be its root cause, notably from our use of fossil fuels.
Adaptation: is to recognise that change is inevitable and to design differently in response to that change.
The two are inextricably linked – the better we are at the former, the less the need for the latter.

this paper that another study suggests that one in four homes will be mechanically cooling by 2030 and that heat pumps (with cooling capacity) could become popular, placing pressure on the electricity supply. This study summarises the details and key findings of a list of previous studies which focus on or consider the necessity for adapting homes for climate change in the UK. In recent years, very remarkable projects, such as 'The COPSE' and 'SCORCHIO', were carried out. The COPSE (Coincident Probabilistic Climate change weather data for a Sustainable built Environment) project was funded by the Engineering and Physical Sciences Research Council between 2008 and 2011. The aim of this project was to develop new methodologies for producing weather data files to assess building designs in future climates, with particular focus on comfort and energy use. To generate future climate weather files, it adopted new methods that account for the Urban Heat Island (HUI), external temperature, wind data and relative humidity for investigating overheating/comfort studies and energy consumption (Mavrogianni et al., 2012).

The SCORCHIO (Sustainable Cities: Options for Responding to Climate Change Impacts and Outcomes) project was carried out between 2007 and 2010 and looked to develop methodologies and tools to help end-users' analysis adaptation options for urban areas with heat and comfort. From this project, a PC-based decision-making tool SCHEEME (Sustainable Cities: Heat, Energy and Emissions Evaluation) was developed to assess current and future heat scenarios and vulnerability at various scales ranging from city to building scale, to help support climate risk assessment activities. Table 6.9 delineates the studies to investigate the Sustainable Codes of Homes in the UK.

Other Considerations for the Housing Sector

CO_2 Coming from the Life Cycle of Home

The influence of successively more demanding Building Regulations means that new dwellings are now more energy efficient than ever before. Consumers living in new properties enjoy lower bills and better comfort levels as well as lower carbon dioxide emissions. The expected path of the Building Regulations means that these positive effects should continue. However, alongside home operational carbon dioxide emissions, the life cycle of a dwelling also includes a number of other stages in which carbon dioxide emissions are generated, including materials extraction, dwelling construction and ultimate demolition. These other stages are collectively known as embodied CO_2 emissions, and they are not currently accounted for within.

Building Regulations

This study shows the total CO_2 emissions arising from the life cycle of a dwelling, with operational carbon in purple and embodied carbon emissions in blue. Embodied CO_2 has historically been a niche area of investigation, due to both the difficulty in analysing it, and its perceived lack of importance when compared with operational CO_2. In general, the typical split between operational and embodied CO_2 in new build housing has been taken as 80% operational, 20% embodied (Porritt et al., 2012). However, within the context of future Building Regulations requirements – which are expected to be tightened to the point that new homes will be significantly lower in CO_2 from 2016 – operational carbon dioxide emissions are set to fall radically. This means that embodied CO_2 emissions will become increasingly significant in terms of the percentage they contribute to the overall CO_2 impact of newly built homes. In addition, typically the more energy efficient a given house type becomes, the greater the quantity of additional materials required to construct it such as additional insulation and more services. There is also potential that such additional materials like

Table 6.9 Review of literature studies: the impact of increased summer temperatures as a result of climate change on the UK housing sector

Study and methodology	*Location*	*House typologies*	*Weather data and climate change scenario*	*Key findings*
Computer modelling (Orme et al., 2003)	Not given	Detached house, semi-detached house, top floor flat and town house	Not given	– No combination of measures entirely eliminated overheating. – Night-time cooling with thermal mass is most effective in reducing overheating.
Ove Arup and Partners Ltd (2005), Computer modelling (Energy 2) (Arup Research and Bill Dunster Architects, 2005)	London, Manchester and Edinburgh	Four-bedroom detached house and two-bedroom semi-detached house	UKCIP 02 (DSY)	– Large areas of room exposed heavyweight thermal mass to absorb excess heat gains. – Generous volumes of cooler air ventilation particularly during nighttime with the increasing diurnal temperature swing. – Good opaque shading to block direct solar gain, ideally outside the occupied rooms. – Limiting ventilation during the hot period of the day. – Positioning bedrooms north facing away from direct solar gain. – Domestic appliances with reduced heat output, particularly on tand-by'.
Hacker, Belcher and Connell (2005), Computer modelling. (Hacker et al., 2005)	London, Manchester and Edinburgh	19th century semi-detached house and new build detached house	Not given	– Solar shading and ventilation are most effective. – New build, due to airtightness and insulation levels, shows better thermal and comfort performance than older home.
Gaterell and Mcevoy (2005), Computer modelling (TAS) (Gaterell & Mcevoy, 2005)	Southeast England	1968 Detached house	UKCIP 02	– Replacing single glazing to double glazing would be the best option to reduce heating energy demand and overheating as climate changes. Loft insulation would be the least option to saving energy and overheating terms.

(*Continued*)

Table 6.9 (Continued)

Study and methodology	*Location*	*House typologies*	*Weather data and climate change scenario*	*Key findings*
Simon Roberts (2008), literature study (Simon Roberts, 2008)	Not given	Not given	Not given	– Looked at the opportunities using passive and active energy efficiency measurements to reduce heating and cooling demand for UK buildings. – Overheating, rather than excessive cold, is set to become a growing problem for householders and employees in existing UK buildings.
Three Regions Climate Change Group (2008), Computer modelling (Three Regions Climate Change Group, 2008)	Southeast England, Ease England and London	1930s house, 1960s flat and an entire block of flats	Not given	– Increased façade reflectivity, solar control, ventilation, enhanced air movement and cooler floors would be effective in reducing overheating and the potential to eliminate the air conditioning demanding.
Day, Jones and Maidment (2009), Statistical modelling (Day et al., 2009)	London	Housing stock	Not given	– The growth in air conditioning systems in London could lead to a doubling of CO_2 emissions by 2030.
Day, Jones and Maidment (2009), Statistical modelling (Day et al., 2009)	London	Housing stock	Not given	– The growth in air conditioning systems in London could lead to a doubling of CO_2 emissions by 2030.
Cloey and Kershaw (2010) Computer modelling (IES) (Cloey & Kershaw, 2010)	London	Typical new built house and apartment	UKCIP 02	– The general form of the relationship between increases in external temperature due to climate change and increased in internal temperatures and it shows the relationship is linear.

Peacock, Jenkins and Kane (2010), Computer modelling (ESP-r) (Peacok et al., 2010)	Edinburgh and London	Detached houses (1988 timber frame construction, 2002 twin leaf insulated masonry construction and pre 1900 solid wall construction)	CIBSE TRY	– Warming climate and varying internal heat gains are estimated to examine weather UK domestic buildings are likely to be prone to overheating in the future, and therefore require air conditioning systems. – The ability, or inability, of the occupant to – adapt to bedroom temperature is paramount in the understanding of the conditions for overheating. – The effect of changing the building construction and geographical location can result in significantly different thermal conditions. – The problem appears most noticeable for buildings in the south of the UK and with lightweight constructions. – In the North of the UK for solid wall dwellings' cooling nights problem diminishes significantly.
Collins, Natarajan and Levemore (2010) Computer modelling (IES) (Collins et al., 2010)	Cardiff, Edinburgh, Manchester and London	UK housing stock (detached, semi-detached, end-terraced, mid-terraced, converted flat and purposed built flat)	UKCIP 02 (TRY)	– Heating demand will still be dominant over cooling demand in the UK dwellings by the 2080s.
Zero Carbon Hub (2010), Computer modelling (IES) and SAP (Zero Carbon Hub, 2010)	UK locations	Semi-detached house and top floor apartment	UKCIP 09	– Additional insulation coupled with reduced air leakage enhanced solar gain, decreasing space heating and decreasing summer comfort.
Porritt, Shao, Cropper and Goodier (2011), Computer modelling (IES) (Porritt et al., 2011)	London	Victorian terraced house	UKCIP 02 (TRY)	– The most effective interventions for reducing overheating were found to be wall insulation and measures to reduce solar heat gain. – A lighter coloured roof and increased levels of loft insulation would be less effective.

(Continued)

Table 6.9 (Continued)

Study and methodology	*Location*	*House typologies*	*Weather data and climate change scenario*	*Key findings*
Porritt, Cropper, Shao and Goodier (2012), Computer modelling (IES) (Porritt et al. 2012)	London	End and mid-terraced house	UKCIP 02 (TRY)	– External wall insulation and solar reflective coatings would be effective to reduce overheating. – Internal wall insulation would be less effective for reducing overheating.
Gupta and Gregg (2012), Computer modelling (IES) (Gupta & Gregg, 2012)	Oxford	Detached, semi-detached, mid-terraced and purpose built flat	UKCIP 09	– The most effective way to reduce overheating is insulation and controlled solar gain. – No measures were able to entirely eliminate the risk of overheating in the home.
Kendrick, Ogden, Wang and Baiche (2012), Computer modelling (TAS) (Kendrick et al., 2012)	London	Semi-detached house		– Heating demand is reduced by approximately 25% between 1990 and 2050 as a result of the milder conditions. – Overheating in the living area during occupied hours is reduced by small amount in the heavyweight case. – Thermal mass is effective in lopping peak temperatures during the daytime but can be detrimental to thermal comfort in the evenings and at night when lower ambient temperatures can cool a lightweight building more quickly. – Heavyweight buildings can store daytime heat gains and be slower to cool in the evening due to their increased thermal inertia.
Mavrogianni, Wilkinson, Davies, Biddulph and Oikonomou (2012) (Mavrogianni et al., 2012)	London	London Housing stock	UKCIP 09	– Under the current climate, the insulation levels had a considerable impact on indoor temperatures, with combined retrofitting of roof insulation and window upgrades reducing daytime living room temperatures during the warmest period. – Internally retrofitted walls and floors tended to increase daytime living room temperatures.

renewable generation installations may have particularly high embodied carbon dioxide levels. Both of these considerations suggest that as operation CO_2 emissions reduce, embodied CO_2 emissions will increase.

Considering User Behaviour

The rebound effect is the reduction of the potential energy saving caused by the user offsetting some of the savings through changed behaviour. It is what happens when more energy-efficient technologies lower the cost of using a particular product, allowing people to use it more extensively without increasing cost. Potential energy savings are reduced because the user changes their behaviour and so offsets some of the savings from improving energy efficient measurements. Energy saving could lead to additional activity through either an increased use of the same product or for another energy using action. For example, leaving lights on for longer because they are energy-saving bulbs or heating the house to a higher temperature after insulating the walls and loft. It is not only the energy use in the home that can be affected; savings can affect energy use in other ways. For example, people saving money on energy bills may decide to spend the savings on a holiday overseas by flight, buying more electrical appliances or replacing a TV with a bigger screen (Porritt et al., 2011).

Energy Consumption for Cooling

If climate change leads to a trend in higher temperatures in the UK, this could have an effect on thermal comfort, particularly in the summer. Cooling needs should be taken into account in design, construction and refurbishment to avoid high-energy demand for cooling in the future. Sales of air conditioners are increasing in the UK, especially during heat waves, and seem likely to increase even more if temperatures rise due to climate change. The recent extremely hot summers in the UK have seen a growing interest in mechanical cooling in the home and a rapidly expanding residential market in the UK. In one week during the hot summer of 2003, UK DIY hardware store, B&Q, reported air conditioner sales 176% higher than in the corresponding week of the previous year. Building Services Research and Information Association (BSRIA) UK reports suggest that over the 15 years from 2000 to 2015, sales of portable air conditioning units will increase by over 450%, while sales of split systems will increase by over 150% (Mitchell & Natarajan, 2020).

According to Boardman et al. (2005) in *The 40% House*, cooling is likely to become an increasingly important issue in the future. Consumer focus group research carried out in 2006 by BRE for the Department for Communities and Local Government, *the impact of societal change on the building regulations* (DCLG 2007), identified that just over 57% of householders would like to see air conditioning fitted as standard and even a necessity in the future, in spite of its potential environmental impact. Furthermore, once people have air conditioning available, they tend to use it more than is necessary to avoid dangerous overheating (Jones, 2007).

Observed Overheating in the UK Homes

Overheating was not a major concern in the heating dominated climate of the UK. There are very few published summer thermal monitoring studies in the UK. Six recent such studies are: the analysis of temperature measurements from five houses in London and four houses in Manchester during the 2003 heat wave, 15 low energy houses in Milton Keynes (Summerfield et al., 2007); four houses in Stamford Brook in 2006 (Wingfield et al., 2008); 62 houses in Leicester during the

2006 heat wave, 36 houses from a total of 120 dwellings in the overall study in London in 2009 and a large monitoring campaign involving 224 nationally representative houses across the UK during July and August 2007 (Peacock et al., 2010). From their studies, it is indicated that there could be a substantial problem of overheating in the existing stock. They also pointed out that increasing risks of summer overheating are likely in the future in new housing developments, especially in lightweight structures, as well as heavyweight structures that have been internally insulated. Dwellings that were recently built or refurbished to high-efficiency standards have the potential to face a significant risk of summer overheating.

Conclusions

This section shows the summary and findings from this chapter. UK homes are responsible for more than a quarter of CO2 emissions of total emissions from the UK. This figure has gradually increased, and this trend will continue in the future. Heating is responsible for the majority of domestic energy consumption and reducing heating in the home would have the most significant impact on the total domestic energy use. People in lower social grades report heating homes to a higher temperature than those in higher social grades. The average indoor temperatures of homes have increased due to the increased energy efficiency such as the use of central heating systems with high-efficiency boilers, improving the fabric's heat loss through added insulation for cavity walls, loft insulation, and changing expectations of occupants with regard to comfort temperature for their homes.

Since the 1990s, people have been buying more energy-efficient appliances and energy-saving light bulbs but, at the same time, there has been an overall increase in their usage. This leads to an increase in energy consumption from lighting and appliances. It is likely that trend will continue in the future. Time spent at home is an important issue as the more time people spend at home the more likely they are to have their heating on, and use lights and appliances. The government has announced the target of reducing the UK's emissions of GHGs by 80% by 2050 compared with the level in 1990s and line with this target, the government set the various target emissions through the building regulations and the Code for Sustainable Homes. According to these, after 2016 the UK homes will be zero carbon homes (Orme & Palmer, 2003). Improvements to the fabric of the building and the technologies installed should mean new homes use significantly less energy than existing homes. The UK government has addressed various policies to improve the energy efficiency of UK dwellings, and it seems to reduce energy demand and carbon emissions from our homes. These policies have helped various people, such as vulnerable households, elders, fuel poverty and even landlords to improve their rented homes' energy efficiency. Improved energy efficiency from new and existing homes, while eliminating fuel poverty, also seems to be the UK government's strategy to meet its own reduction target of CO_2 emissions. However, the subsequent sharp increase in the number of households in fuel poverty was primarily the result of energy price increases since 2004 (Mullins, 2017). Although the energy efficiency programmes have reduced the impact of rising prices, they have not been able to prevent its recent increasing trends.

Building standards in the UK are expected to be more airtight and energy efficient of building fabrics and it will help reduce the energy demand during the heating, but it could increase the chance of having an overheating home in summer then it will affect occupant health and cooling energy consumption. After improving the home's energy efficiency then it will require looking at the user behaviour which is sometimes a more important issue for effects on energy consumption than installing the energy efficiency measurements. There could be a substantial problem of overheating in the existing stock and houses that were recently built with high-efficiency standards lightweight construction or refurbished as heavyweight form that have been internally insulated have the potential to face a significant risk of summer overheating. It is likely that housing, both existing and new build, will need to contribute more to the overall strategy to reduce CO_2 emissions in the future. The UK government has stated that 80% of GHG emissions will be reduced by 2050 from 1990 level in the Climate Change Act 2008 (Slater, 2018). If CO_2 cannot be cut sufficiently from other sectors, such as shipping and aviation, then higher standards will be required elsewhere to make up the shortfall. In order to combat climate change, the energy efficiency of the housing stock will need to be increased, and carbon emissions reduced. The various national policies to improve the energy efficiency of our homes seem to be working for existing and new build homes, and it has helped to reduce heating demand from homes and gives better internal comfort for occupancy. However, the number of lighting and appliances has more than doubled in terms of electricity consumption by households which has led to an increase in CO_2 emissions. Re-educating people's behaviours to adapt their lifestyle provides greater consideration for the environment rather than for their convenience.

It has been confirmed that the UK government will increase their homes' energy efficiency through the Building Regulations and EPCs for existing homes and the Code level for new homes. It was also pointed out that the fabric standards for new builds and existing dwellings would be significantly increased so that they are more airtight. These strategies help to reduce heating energy consumption. However, in terms of climate change, the summer will be hotter and more frequent heat waves such as those of 2003 and 2006 will result. This means that increasing the energy efficiency of our homes through the building fabric efficiency measurements could be disastrous for future summer overheating in our homes. It will lead to installation the air conditioning systems in every home in the UK and contribute to increases in electricity consumption and CO_2 emissions. With the result that the government's CO_2 reduction target will fail. Therefore, the direction of improving energy efficiency for our existing and newly built homes is right, and we should do more for our energy-inefficient homes such as fuel poverty and vulnerable households. However, when we design our home, we should consider that our homes will be future proofed which means considering climate change and better performance of heating and cooling.

References

Alwan, Z., Jones, P., & Holgate, P. (2016). Strategic sustainable development in the UK construction industry, through the framework for strategic sustainable development, using Building Information Modelling. *Journal of Cleaner Production, 140*, 349. https://doi.org/10.1016/j.jclepro.2015.12.085.

Baba, A., Mahdjoubi, L., Olomolaiye, P., & Booth, C. (2012). Insights of architects' knowledge of the Code for Sustainable Homes (CSH) in relation to low carbon housing design and delivery in the UK. *Structural Survey, 30*(5), 443. https://doi.org/10.1108/02630801211288215.

Booth, A. T., & Choudhary, R. (2013). Decision making under uncertainty in the retrofit analysis of the UK housing stock: implications for the Green Deal. *Energy and Buildings, 64*, 292. https://doi.org/10.1016/j.enbuild.2013.05.014.

Broer, S., & Titheridge, H. (2010). Enabling low-carbon living in new UK housing developments. *Management of Environmental Quality: An International Journal, 21*(1), 90. https://doi.org/10.1108/14777831011010883.

Brown, D., Brisbois, M., Lacey-Barnacle, M., Foxon, T., Copeland, C., & Mininni, G. (2022). The Green New deal: historical insights and local prospects in the United Kingdom (UK). *Ecological Economics, 205*, 121–123. https://doi.org/10.1016/j.ecolecon.2022.107696.

Campbell, K. (2007). Energy performance and building regulations. *Journal of Building Appraisal, 3*(3), 231. https://doi.org/10.1057/palgrave.jba.2950077.

Carter, K., & Fortune, C. (2007). Sustainable development policy perceptions and practice in the UK social housing sector. *Construction Management and Economics, 25*(4), 399. https://doi.org/10.1080/01446190600922578.

Cerin, P., Hassel, L. G., & Semenova, N. (2014). Energy performance and housing prices. *Sustainable Development, 22*(6), 404. https://doi.org/10.1002/sd.1566.

Colburn, G. (2019). The use of markets in housing policy: a comparative analysis of housing subsidy programs. *Housing Studies, 36*(1), 46. https://doi.org/10.1080/02673037.2019.1686129.

Coley, D., & Kershaw, T. (2010). Changes in internal temperatures within the built environment as a response to a changing climate. *Building and Environment, 45*(1), 89. https://doi.org/10.1016/j.buildenv.2009.05.009.

Collins, L., Natarajan, S., & Levermore, G. (2010). Climate change and future energy consumption in UK housing stock. *Building Services Engineering Research and Technology, 31*(1), 75. https://doi.org/10.1177/0143624409354972.

Coyne, B., & Denny, E. (2021). Mind the energy performance gap: testing the accuracy of building energy performance certificates in Ireland. *Energy Efficiency, 14*(6), 23–25. https://doi.org/10.1007/s12053-021-09960-1.

Day, A. R., Jones, P. G., & Maidment, G. G. (2009). Forecasting future cooling demand in London. *Energy and Buildings, 41*(9), 942. https://doi.org/10.1016/j.enbuild.2009.04.001.

Dolge, K., & Blumberga, D. (2021). Economic growth in contrast to GHG emission reduction measures in Green Deal context. *Ecological Indicators, 130*. https://doi.org/10.1016/j.ecolind.2021.108153.

Dowson, M., Poole, A., Harrison, D., & Susman, G. (2012). Domestic UK retrofit challenge: Barriers, incentives and current performance leading into the Green Deal. *Energy Policy, 50*, 294 (45-47). https://doi.org/10.1016/j.enpol.2012.07.019.

Fuerst, F., & Mcallister, P. (2011). The impact of Energy Performance Certificates on the rental and capital values of commercial property assets. *Energy Policy, 39*(10), 6608. https://doi.org/10.1016/j.enpol.2011.08.005.

Fuerst, F., Mcallister, P., Nanda, A., & Wyatt, P. (2015). Does energy efficiency matter to homebuyers? An investigation of EPC ratings and transaction prices in England. *Energy Economics, 48*, 145. https://doi.org/10.1016/j.eneco.2014.12.012.

Fuerst, F., Mcallister, P., Nanda, A., & Wyatt, P. (2016). *Energy performance ratings and house prices in Wales: an empirical study*. Elsevier BV. https://doi.org/10.1016/j.enpol.2016.01.024.

Gallent, N., De Magalhaes, C., & Freire Trigo, S. (2020). Is zoning the solution to the UK housing crisis? *Planning Practice & Research, 36*(1), 1. https://doi.org/10.1080/02697459.2020.1829283.

Gaterell, M. R., & Mcevoy, M. E. (2005). The impact of climate change uncertainties on the performance of energy efficiency measures applied to dwellings. *Energy and Buildings, 37*(9), 982. https://doi.org/10.1016/j.enbuild.2004.12.015.

Georgiadou, M. C., Hacking, T., & Guthrie, P. (2013). *Future-proofed energy design for dwellings: case studies from England and application to the Code for Sustainable Homes*. Sage Publications. https://doi.org/10.1177/0143624412463016.

Gillich, A., Saber, E. M., & Mohareb, E. (2019). Limits and uncertainty for energy efficiency in the UK housing stock. *Energy Policy, 133*, 93–97. https://doi.org/10.1016/j.enpol.2019.110889.

Gooding, L., & Gul, M. S. (2016). Energy efficiency retrofitting services supply chains: a review of evolving demands from housing policy. *Energy Strategy Reviews, 11–12*, 29. https://doi.org/10.1016/j.esr.2016.06.003.

Gupta, R., & Chandiwala, S. (2009). Achieving low carbon buildings using Code for Sustainable Homes in the UK. *International Journal of Low-Carbon Technologies, 4*(3), 187. https://doi.org/10.1093/ijlct/ctp018.

Gupta, R., & Gregg, M. (2012). Using UK climate change projections to adapt existing English homes for a warming climate. *Building and Environment, 55*, 20. https://doi.org/10.1016/j.buildenv.2012.01.014.

Hall, M., & Purchase, D. (2006). Building or bodging? Attitudes to sustainability in UK public sector housing construction development. *Sustainable Development, 14*(3), 205. https://doi.org/10.1002/sd.265.

Harrison, M., & Reeve, K. (2002). *Social welfare movements and collective action: lessons from two UK housing cases*. Housing Studies, 17(5), 755–771. https://doi.org/10.1080/0267303022000009790.

Henderson, C., Ganah, A., & John, G. A. (2015). Achieving sustainable homes by 2016 in the UK: the current status. *Environment, Development and Sustainability, 18*(2), 547. https://doi.org/10.1007/s10668-015-9664-8.

Holmans, A., Stephens, M., & Fitzpatrick, S. (2007). Housing policy in England since 1975: an introduction to the special Issue. *Housing Studies, 22*(2), 147. https://doi.org/10.1080/02673030601132714.

Jacoby, S., Arancibia, A., & Alonso, L. (2022). Space standards and housing design: typological experimentation in England and Chile. *The Journal of Architecture, 27*(1), 94. https://doi.org/10.1080/13602365.2022.2045340.

Jenkins, D., Mccallum, P., Patidar, S., & Semple, S. (2021). *Alternative modelling approaches to energy performance certificates*. KU Leuven. https://doi.org/10.26868/25222708.2021.30173.

Johnston, D., Lowe, R., & Bell, M. (2005). An exploration of the technical feasibility of achieving CO_2 emission reductions in excess of 60% within the UK housing stock by the year 2050. *Energy Policy, 33*(13), 1643. https://doi.org/10.1016/j.enpol.2004.02.003.

Jones, G. A. (2007). Assessing the success of the sale of social housing in the UK. *Journal of Social Welfare and Family Law, 29*(2), 135. https://doi.org/10.1080/09649060701666606.

Kearns, A. (2022). Housing space and occupancy standards: developing evidence for policy from a health and wellbeing perspective in the UK context. *Building Research & Information, 50*(7), 722. https://doi.org/10.1080/09613218.2021.2024756.

Kelly, S., Crawford-Brown, D., & Pollitt, M. G. (2012). Building performance evaluation and certification in the UK: Is SAP fit for purpose? *Renewable and Sustainable Energy Reviews, 16*(9), 6861. https://doi.org/10.1016/j.rser.2012.07.018.

Kendrick, C., Ogden, R., Wang, X., & Baiche, B. (2012). Thermal mass in new build UK housing: a comparison of structural systems in a future weather scenario. *Energy and Buildings, 48*, 40. https://doi.org/10.1016/j.enbuild.2012.01.009.

Lee, H., Lee, M., & Lim, S. (2018). Do consumers care about the energy efficiency of buildings? Understanding residential choice based on energy performance certificates. *Sustainability, 10*(11), 47–49. https://doi.org/10.3390/su10114297.

Li, F., & Guo, Z. (2021). How does an expansion of mandatory inclusionary housing affect housing supply? *Journal of the American Planning Association, 88*(1), 83. https://doi.org/10.1080/01944363.2021.1928533.

Li, Y., Kubicki, S., Guerriero, A., & Rezgui, Y. (2019). Review of building energy performance certification schemes towards future improvement. *Renewable and Sustainable Energy Reviews, 113*, 7–9. https://doi.org/10.1016/j.rser.2019.109244.

Marchand, R. D., Koh, S. C. L., & Morris, J. C. (2015). Delivering energy efficiency and carbon reduction schemes in England: lessons from green deal pioneer places. *Energy Policy, 84*, 96. https://doi.org/10.1016/j.enpol.2015.04.035.

Martiskainen, M., & Kivimaa, P. (2019). Role of knowledge and policies as drivers for low-energy housing: case studies from the United Kingdom. *Journal of Cleaner Production, 215*, 1402. https://doi.org/10.1016/j.jclepro.2019.01.104.

Mavrogianni, A., Wilkinson, P., Davies, M., Biddulph, P., & Oikonomou, E. (2012). Building characteristics as determinants of propensity to high indoor summer temperatures in London dwellings. *Building and Environment, 55*, 117. https://doi.org/10.1016/j.buildenv.2011.12.003.

Mccord, M., Haran, M., Davis, P., & Mccord, J. (2020). Energy performance certificates and house prices: a quantile regression approach. *Journal of European Real Estate Research, 13*(3), 409. https://doi.org/10.1108/jerer-06-2020-0033.

Mcgilligan, C., De Wilde, P., & Goodhew, S. (2008). An assessment of the potential returns of energy certificates for the UK household sector. *Journal of Financial Management of Property and Construction, 13*(3), 187. https://doi.org/10.1108/13664380810913421.

Mcgilligan, C., Sunikka-Blank, M., & Natarajan, S. (2009). Subsidy as an agent to enhance the effectiveness of the energy performance certificate. *Energy Policy, 38*(3), 1272. https://doi.org/10.1016/j.enpol.2009.10.068.

Mckee, K., Muir, J., & Moore, T. (2016). Housing policy in the UK: the importance of spatial nuance. *Housing Studies, 32*(1), 60. https://doi.org/10.1080/02673037.2016.1181722.

Mcmanus, A., Gaterell, M. R., & Coates, L. E. (2010). The potential of the code for sustainable homes to deliver genuine 'sustainable energy' in the UK social housing sector. *Energy Policy, 38*(4), 2013. https://doi.org/10.1016/j.enpol.2009.12.002.

Meier, H., & Rehdanz, K. (2009). Determinants of residential space heating expenditures in Great Britain. *Energy Economics, 32*(5), 949. https://doi.org/10.1016/j.eneco.2009.11.008.

Mitchell, R., & Natarajan, S. (2020). UK Passivhaus and the energy performance gap. *Energy and Buildings, 224*. https://doi.org/10.1016/j.enbuild.2020.110240.

Miu, L. M., Wisniewska, N., Mazur, C., Hardy, J., & Hawkes, A. (2018). A simple assessment of housing retrofit policies for the UK: what should succeed the energy company obligation? *Energies, 11*(8) (5-6). https://doi.org/10.3390/en11082070.

Monahan, J., & Powell, J. C. (2010). A comparison of the energy and carbon implications of new systems of energy provision in new build housing in the UK. *Energy Policy, 39*(1), 290. https://doi.org/10.1016/j.enpol.2010.09.041.

Muir, J. (2013). The dynamics of policymaking under UK devolution: social housing in Northern Ireland. *Housing Studies, 28*(7), 1081. https://doi.org/10.1080/02673037.2013.803045.

Mullins, D. (2017). Achieving policy recognition for community-based housing solutions: the case of self-help housing in England. *International Journal of Housing Policy, 18*(1), 143. https://doi.org/10.1080/19491247.2017.1384692.

O'keeffe, J. M., Gilmour, D., & Simpson, E. (2016). A network approach to overcoming barriers to market engagement for SMEs in energy efficiency initiatives such as the Green Deal. *Energy Policy, 97*, 582. https://doi.org/10.1016/j.enpol.2016.08.006.

O'malley, C., Piroozfar, P. A. E., Farr, E. R. P., & Gates, J. (2014). Evaluating the efficacy of BREEAM code for sustainable homes (CSH): a cross-sectional study. *Energy Procedia, 62*, 210. https://doi.org/10.1016/j.egypro.2014.12.382.

O'neill, K., & Gibbs, D. (2019). Sustainability transitions and policy dismantling: zero carbon housing in the UK. *Geoforum, 108*, 119. https://doi.org/10.1016/j.geoforum.2019.11.011.

Ozarisoy, B. (2022). Energy effectiveness of passive cooling design strategies to reduce the impact of long-term heatwaves on occupants' thermal comfort in Europe: climate change and mitigation. https://doi.org/10.1016/j.jclepro.2021.129675.

Ozarisoy, B., & Altan, H. (2023). State-of-The-Art I: energy efficiency directives and policy aspirations in retrofit interventions. In: *Handbook of retrofitting high density residential buildings*. Springer. https://doi.org/10.1007/978-3-031-11854-8_2.

Ozarisoy, B., & Elsharkawy, H. (2019). Assessing overheating risk and thermal comfort in state-of-the-art prototype houses that combat exacerbated climate change in UK. *Energy and Buildings, 187*, 201. https://doi.org/10.1016/j.enbuild.2019.01.030.

Peacock, A. D., Jenkins, D. P., & Kane, D. (2010). Investigating the potential of overheating in UK dwellings as a consequence of extant climate change. *Energy Policy, 38*(7), 3277. https://doi.org/10.1016/j.enpol.2010.01.021.

Pickvance, C. (2009). The construction of UK sustainable housing policy and the role of pressure groups. *Local Environment, 14*(4), 329. https://doi.org/10.1080/13549830902764712.

Poel, B., Van Cruchten, G., & Balaras, C. A. (2007). Energy performance assessment of existing dwellings. *Energy and Buildings, 39*(4), 393. https://doi.org/10.1016/j.enbuild.2006.08.008.

Poon, J., & Garratt, D. (2012). Evaluating UK housing policies to tackle housing affordability. *International Journal of Housing Markets and Analysis, 5*(3), 253. https://doi.org/10.1108/17538271211243599.

Porritt, S., Shao, L., Cropper, P., & Goodier, C. (2011). Adapting dwellings for heat waves. *Sustainable Cities and Society, 1*(2), 81 (23-27). https://doi.org/10.1016/j.scs.2011.02.004.

Porritt, S. M., Cropper, P. C., Shao, L., & Goodier, C. I. (2012). *Ranking of interventions to reduce dwelling overheating during heat waves*. Paper presented at the, *55* 16. https://doi.org/10.1016/j.enbuild.2012.01.043.

Pretlove, S., & Kade, S. (2015). Post occupancy evaluation of social housing designed and built to code for sustainable homes levels 3, 4 and 5. *Energy and Buildings, 110*, 120. https://doi.org/10.1016/j.enbuild.2015.10.014.

Pryce, G., & Sprigings, N. (2009). Outlook for UK housing and the implications for policy. *International Journal of Housing Markets and Analysis, 2*(2), 145. https://doi.org/10.1108/17538270910963081.

Rivas, S., Urraca, R., Bertoldi, P., & Thiel, C. (2021). Towards the EU Green Deal: local key factors to achieve ambitious 2030 climate targets. *Journal of Cleaner Production, 320.* https://doi.org/10.1016/j.jclepro.2021.128878.

Roberts, S. (2008). Altering existing buildings in the UK. *Energy Policy, 36*(12), 4482. https://doi.org/10.1016/j.enpol.2008.09.023.

Rosenow, J., & Eyre, N. (2016). A postmortem of the Green Deal: austerity, energy efficiency, and failure in British energy policy. *Energy Research & Social Science, 21*, 141. https://doi.org/10.1016/j.erss.2016.07.005.

Scanlon, K., & Elsinga, M. (2013). Policy changes affecting housing and mortgage markets: how governments in the UK and the Netherlands responded to the GFC. *Journal of Housing and the Built Environment, 29*(2), 335. https://doi.org/10.1007/s10901-013-9390-1.

Semple, S., & Jenkins, D. (2019). Variation of energy performance certificate assessments in the European Union. *Energy Policy, 137.* https://doi.org/10.1016/j.enpol.2019.111127.

Shipworth, M., Firth, S. K., Gentry, M. I., Wright, A. J., Shipworth, D. T., & Lomas, K. J. (2010). Central heating thermostat settings and timing: building demographics. *Building Research & Information, 38*(1), 50. https://doi.org/10.1080/09613210903263007.

Shorrock, L. D., Henderson, J., Utley, J. L., & Walters, G. A. (2001). Carbon emission reductions from energy efficiency improvements to the UK housing stock, BRE Report 435. Garston, Watford, Building Research Establishment.

Shorrock, L. D., & Utley, J. I. (2003). Domestic Energy Fact File 2003. BR 457. Bracknell, IHS BRE Press. Available at: https://bregroup.com/store/bookshop/domestic-energy-fact-file-2003 [Accessed 30 June 2013].

Shrubsole, C., Macmillan, A., Davies, M., & May, N. (2014). *100 Unintended consequences of policies to improve the energy efficiency of the UK housing stock.* SAGE Publications. https://doi.org/10.1177/1420326x14524586.

Slater, T. (2018). The invention of the 'sink estate': consequential categorisation and the UK housing crisis. *The Sociological Review, 66*(4), 877. https://doi.org/10.1177/0038026118777451.

Smol, M. (2022). Is the green deal a global strategy? Revision of the green deal definitions, strategies and importance in post-COVID recovery plans in various regions of the world. *Energy Policy, 169.* https://doi.org/10.1016/j.enpol.2022.113152.

Sodagar, B., & Starkey, D. (2015). The monitored performance of four social houses certified to the code for sustainable homes level 5. *Energy and Buildings, 110*, 245. https://doi.org/10.1016/j.enbuild.2015.11.016.

Sousa, G., Jones, B. M., Mirzaei, P. A., & Robinson, D. (2017). A review and critique of UK housing stock energy models, modelling approaches and data sources. *Energy and Buildings, 151*, 66. https://doi.org/10.1016/j.enbuild.2017.06.043.

Swan, W., Fitton, R., Smith, L., Abbott, C., & Smith, L. (2017). Adoption of sustainable retrofit in UK social housing 2010-2015. *International Journal of Building Pathology and Adaptation, 35*(5), 456. https://doi.org/10.1108/ijbpa-04-2017-0019.

Sweetnam, T., Spataru, C., Cliffen, B., Zikos, S., & Barrett, M. (2013). PV system performance and the potential impact of the green deal policy on market growth in London, UK. *Energy Procedia, 42*, 347. https://doi.org/10.1016/j.egypro.2013.11.035.

Watts, C., Jentsch, M. F., & James, P. A. (2011). Evaluation of domestic energy performance certificates in use. *Building Services Engineering Research and Technology, 32*(4), 361. https://doi.org/10.1177/0143624411404486.

Williams, K., & Dair, C. (2006). What is stopping sustainable building in England? Barriers experienced by stakeholders in delivering sustainable developments. *Sustainable Development, 15*(3), 135. https://doi.org/10.1002/sd.308.

Wolf, S., Teitge, J., Mielke, J., Schütze, F., & Jaeger, C. (2021). The European green deal — more than climate neutrality. *Intereconomics, 56*(2), 99. https://doi.org/10.1007/s10272-021-0963-z.

Yohanis, Y. G., & Mondol, J. D. (2009). Annual variations of temperature in a sample of UK dwellings. *Applied Energy, 87*(2), 681. https://doi.org/10.1016/j.apenergy.2009.08.003.

7
CONCLUSION

Energy Efficiency Gap

The aim of this research has been to investigate one important aspect of the built environment, which is energy efficiency improvements and reduction of carbon dioxide emissions in the domestic sector. This research has reached the conclusion that there is great potential for improving energy efficiency and reducing carbon dioxide emissions in existing housing in the UK, particularly in the privately rented sector, such as the case studies carried out in Sheffield.

The results of this research have helped to develop a database of the 'local' current situation. By choosing university-controlled properties, this has assisted to utilise the university authority to act effectively and play a key role in guiding energy efficiency improvements in these properties. This will not only help to encourage energy efficiency standards throughout the private housing sector but also accelerate the necessity of improvements and consultancies in the domestic sector, particularly in the privately rented sector (Fylan et al., 2016). To summarize some of the important conclusions defined from this research in response to the original research aims and objectives: To provide a database of the current conditions and energy standards of privately rented properties.

As explained, feedback reports have been compiled for each property owner to show the current conditions and energy standards of their property and therefore suggest potential energy efficiency improvements when considering refurbishments. The case study analysis and results have helped to develop a database of the 'local' current situation. They provide a benchmark figure of possible improvements which if carried out could contribute significantly to the overall improvement programme within the privately rented sector in the UK (Banfill & Peacock, 2007). To investigate potentials for energy efficiency improvements and reduction of carbon dioxide emissions in the privately rented sector.

As one of the main aims, this research has investigated potentials for energy efficiency improvements and reduction of carbon dioxide emissions in the privately rented sector, and as a result, it has been concluded that there is great potential for improvements particularly in this sector, such as the case studies undertaken in Sheffield.

- To evaluate and review energy and environmental assessment methods, and their applications in terms of energy efficiency and carbon index (CI).

DOI: 10.1201/9781003490364-9

As explained, existing building energy and environmental assessment methods have been evaluated and reviewed. Furthermore, overall comparison was presented to be more familiar with these tools and techniques currently available and their applications concerning energy efficiency and carbon dioxide index (Kelly et al., 2012).

- To identify a way of cost-effective modifications to dwelling base designs which can improve energy efficiency within privately rented properties.

  As described in more detail, modifications were made to base designs to identify potential cost-effective improvements in these properties. These have been analysed to estimate the full savings on the increased energy standards of these dwellings (Rosenow, 2012; Reid et al., 2015; Salem et al., 2019). The modifications were identified as follows: fabric (improve wall and roof insulation); solar gain (PVC-U frames and Low e glazing), ventilation (additional mechanical vents), space heating and water heating (new high-efficiency boiler and radiators).

- To develop case study results that can be used by landlords/landladies to encourage energy conservation.

  This research is not only a reminder of energy conservation for those who were involved in the built environment, but also a database developed of the 'local' current situation providing a checklist for both university housing services and property owners to help increase energy standards of these properties examined (Mallaburn & Eyre, 2013).

- To promote general recommendations for immediate energy savings throughout privately rented properties.

  The research in a form of a comprehensive report will be submitted to the Sheffield City Housing Unit with an aim to help keeping up to date records of privately rented properties, i.e., conditions and energy standards in Sheffield. This information can also be useful for the Higher Education Colleges Association (HECA) team to help tackle energy-related problems in other existing dwellings in the city and meet its overall target.

As chapter stated, research results according to the research hypothesis indicates that privately rented properties in Sheffield have significant potentials for energy efficiency improvements. In the UK, more than 25 years of research efforts have produced broad understanding of the implications of the energy use and developed energy strategies and technologies with significant potential for energy savings (Ko & Fenner, 2007). These strategies and technologies have not been effectively transferred to the building design community and as a result, there has been a knowledge gap, which provided property developers designing without any energy-related considerations or environmental concerns (Norouzi et al., 2023). This again is one of the main reasons why the majority of domestic buildings in the UK are still designed with no further concerns about environmental issues or energy efficiency beyond those required by national building regulations (Shrubsole et al., 2014).

According to the recently published English House Condition Survey, there is considerable scope for improvements in energy efficiency and thermal performance of the housing stock (Alabid et al., 2022). The vast majority of the existing housing falls well below modern building regulations and standards with the average SAP rating being 51 (Dowson et al., 2012). The housing stock is inefficient due to designing without any energy-related considerations and other problems mentioned above. Many existing dwellings are not only physically capable of being made energy efficient but also not physically capable of being brought up to a high SAP rating with heating and insulation measures available. A large number of households are lacking affordable warmth, especially households in council and privately rented housing across the country (Gillich et al., 2019).

Energy Efficiency Measures

Energy efficiency programmes often form a core element of local strategies for sustainable development as they deliver social, economic and health benefits as well as reduce housing management costs and help the environment (Clarke et al., 2008). Looking at some of the energy efficiency design schemes carried out in the UK in the last ten years, it is anticipated that these low-energy houses will influence the design of housing in the near future (Ozarisoy, 2022). Domestic buildings will therefore have less impact on the environment, considering that sustainable green design issues are incorporated at the design stage. This should be also enforced by the building regulations and supported through existing energy and environmental assessment methods for buildings.

As sustainable design has become an important issue to achieve this there must be a significant reduction in carbon dioxide emissions and the amount of energy consumed in buildings (Sunikka-Blank, 2012). In this research, properties examined can be utilised to illustrate the potential for energy efficiency improvements and by some means encourage property owners to increase the energy standards of these dwellings.

An important incentive to use building energy and environmental assessment methods more effectively is the need to improve energy efficiency and reduce carbon dioxide emissions in the country (Mcevoy et al., 1999). In housing, this could contribute a significant proportion of the total energy use and therefore help tackle crucial issues, such as affordable warmth and fuel poverty in existing stock, as well as help the UK meet its international target (Boardman, 2004).

It is obvious from the case study results that the privately rented properties are below energy standards required by the building regulations and implementing energy efficiency improvements suggested within feedback reports compiled can therefore reduce the energy consumed in these dwellings. Overall, it should be emphasised that most of the energy efficiency improvements suggested are very cost effective and the payback time on these dwellings is significantly short. It is an important fact that if these improvements were to be implemented, both property occupiers and owners would benefit equally; especially when considering the forthcoming Government's Home Information (Sellers/Buyers) Pack and EU's Energy Performance of Buildings Directive.

On one hand, for the occupier, since the property would be consuming less energy there will be significant energy savings, which would be beneficial for longer tenancy considering affordable warmth and utilities. On the other hand, for owners, the property energy rating would be higher. The higher the energy rating, the better the energy performance of the property, which is again lower carbon dioxide emission and therefore less impact on the environment; it could well be used as a marketing strategy in the near future (Tovar, 2012).

Moreover, the government is determined to ensure that the thermal efficiency of housing (and also non-domestic building) is improved, and their intention is to introduce a Home Information (Sellers/Buyers) Pack in which the energy efficiency of the dwelling would have to be given (Mcgill et al., 2015). This can be therefore seen as a boost for improvements in the private sector and an argument for why potential improvements can be determined.

It is evident from the previous discussions in this thesis that further work is necessary, particularly in the private sector where there is no legislation to enforce retrospective energy standards of housing. This is an exceptional opportunity which would not only increase the overall energy standards of existing housing stock in the UK but also help achieve targets set by the Home Energy Conservation Act 1995 and enhance the overall rate for energy efficiency improvements in the country accordingly. This will consequently contribute to reducing carbon dioxide emissions consumed by domestic buildings (Mcmichael & Shipworth, 2012).

Work in this area may be expanded in a number of different ways. Some suggestions for further research emerge from this research: The focus of this study was on the privately rented housing sector. The extension of the study for improving energy standards and suggesting guidelines for other tenures of housing (owner-occupied, local authority/housing association and council housing) are also important. Parameters, such as housing conditions, both space and water heating and other issues, can also be looked at in other tenures. Energy efficiency measures could be particularly included within presently undertaken schemes, such as Responsible Landlord and Accreditation Schemes to accelerate the energy efficiency improvements in the private sector and therefore help achieve national targets more rapidly. Similar case studies can also be carried out such as undertaken in this research in Sheffield within properties registered with similar schemes in the country.

References

Alabid, J., Bennadji, A., & Seddiki, M. (2022). A review on the energy retrofit policies and improvements of the UK existing buildings, challenges and benefits. *Renewable and Sustainable Energy Reviews, 159*, 23–27. https://doi.org/10.1016/j.rser.2022.112161.

Banfill, P. F. G., & Peacock, A. D. (2007). Energy-efficient new housing – the UK reaches for sustainability. *Building Research & Information, 35*(4), 426. https://doi.org/10.1080/09613210701339454.

Boardman, B. (2004). New directions for household energy efficiency: evidence from the UK. *Energy Policy, 32*(17), 1921. https://doi.org/10.1016/j.enpol.2004.03.021.

Clarke, J. A., Johnstone, C. M., Kelly, N. J., Strachan, P. A., & Tuohy, P. (2008). The role of built environment energy efficiency in a sustainable UK energy economy. *Energy Policy, 36*(12), 4605. https://doi.org/10.1016/j.enpol.2008.09.004.

Dowson, M., Poole, A., Harrison, D., & Susman, G. (2012). Domestic UK retrofit challenge: barriers, incentives and current performance leading into the Green Deal. *Energy Policy, 50*, 294. https://doi.org/10.1016/j.enpol.2012.07.019.

Fylan, F., Glew, D., Smith, M., Johnston, D., Brooke-Peat, M., Miles-Shenton, D., Fletcher, M., Aloise-Young, P., & Gorse, C. (2016). Reflections on retrofits: overcoming barriers to energy efficiency among the fuel poor in the United Kingdom. *Energy Research & Social Science, 21*, 190. https://doi.org/10.1016/j.erss.2016.08.002.

Gillich, A., Saber, E. M., & Mohareb, E. (2019). Limits and uncertainty for energy efficiency in the UK housing stock. *Energy Policy, 133*. https://doi.org/10.1016/j.enpol.2019.110889.

Kelly, S., Crawford-Brown, D., & Pollitt, M. G. (2012). Building performance evaluation and certification in the UK: Is SAP fit for purpose? *Renewable and Sustainable Energy Reviews, 16*(9), 6861. https://doi.org/10.1016/j.rser.2012.07.018.

Ko, J., & Fenner, R. (2007). Adoption of energy efficiency innovations in new UK housing. *Proceedings of the Institution of Civil Engineers – Energy, 160*(4), 151. https://doi.org/10.1680/ener.2007.160.4.151.

Mallaburn, P. S., & Eyre, N. (2013). *Lessons from energy efficiency policy and programmes in the UK from 1973 to 2013*. Springer Science and Business Media LLC. https://doi.org/10.1007/s12053-013-9197-7.

Mcevoy, D., Gibbs, D. C., & Longhurst, J. W. S. (1999). The prospects for improved energy efficiency in the UK residential sector. *Journal of Environmental Planning and Management, 42*(3), 409. https://doi.org/10.1080/09640569911163.

Mcgill, G., Oyedele, L. O., Mcallister, K., & Qin, M. (2015). Effective indoor air quality for energy-efficient homes: a comparison of UK rating systems. *Architectural Science Review, 59*(2), 159. https://doi.org/10.1080/00038628.2015.1078222.

Mcmichael, M., & Shipworth, D. (2012). The value of social networks in the diffusion of energy-efficiency innovations in UK households. *Energy Policy, 53*, 159. https://doi.org/10.1016/j.enpol.2012.10.039.

Norouzi, M., Haddad, A. N., Jiménez, L., Hoseinzadeh, S., & Boer, D. (2023). Carbon footprint of low-energy buildings in the United Kingdom: effects of mitigating technological pathways and decarbonization strategies. *Science of the Total Environment, 882*. https://doi.org/10.1016/j.scitotenv.2023.163490.

Ozarisoy, B. (2022). Energy effectiveness of passive cooling design strategies to reduce the impact of long-term heatwaves on occupants' thermal comfort in Europe: climate change and mitigation. https://doi.org/10.1016/j.jclepro.2021.129675.

Reid, L., Mckee, K., & Crawford, J. (2015). Exploring the stigmatization of energy efficiency in the UK: an emerging research agenda. *Energy Research & Social Science, 10*, 141. https://doi.org/10.1016/j.erss.2015.07.010.

Rosenow, J. (2012). Energy savings obligations in the UK—a history of change. *Energy Policy, 49*, 373. https://doi.org/10.1016/j.enpol.2012.06.052.

Salem, R., Bahadori-Jahromi, A., Mylona, A., Godfrey, P., & Cook, D. (2019). Investigating the potential impact of energy-efficient measures for retrofitting existing UK hotels to reach the nearly zero energy building (nZEB) standard. *Energy Efficiency, 12*(6), 1577. https://doi.org/10.1007/s12053-019-09801-2.

Shrubsole, C., Macmillan, A., Davies, M., & May, N. (2014). *100 Unintended consequences of policies to improve the energy efficiency of the UK housing stock*. Sage Publications. https://doi.org/10.1177/1420326x14524586.

Sunikka-Blank, M., Chen, J., Britnell, J., & Dantsiou, D. (2012). Improving energy efficiency of social housing areas: a case study of a retrofit achieving an "A" energy performance rating in the UK. *European Planning Studies, 20*(1), 131. https://doi.org/10.1080/09654313.2011.638494.

Tovar, M. A. (2012). The structure of energy efficiency investment in the UK households and its average monetary and environmental savings. *Energy Policy, 50*, 723. https://doi.org/10.1016/j.enpol.2012.08.019.

PART 2

Methods and Tools

8
INTRODUCTION

Background

Climate change is the most important environmental problem facing the world. Climate change is caused by the emissions of greenhouse gases (GHGs) into the atmosphere. The single most important GHG is carbon dioxide, which is mostly emitted by burning fossil fuels. The carbon dioxide could stay in the atmosphere for up to 200 years, and there is no easy way to remove its concentration from the atmosphere. The most efficient way to reduce the concentration in the atmosphere is to start cutting emissions down as soon as possible and as much as we can (Averchenkova et al., 2021).

The evidence for climate change in the UK and global level is clear enough. Global surface temperatures have risen by 0.8°C, with those over central England rising by almost 1°C over the last century. In the summer of 2003, a new temperature record of 38.5°C was recorded in the south-western part of the UK. These heat waves were responsible for an additional 35,000 deaths in Europe in 2003 (Carter, 2008). The degree of climate change expected in future is highly dependent upon the level of GHG emissions in the atmosphere, with higher emissions leading to greater climate change. The UK could expect warmer and wetter winters along with warmer and drier summers, with the average annual temperature rising between 2°C and 3.5°C by 2080. Globally, temperatures could rise by up to 6°C by 2100 (Carter, 2008).

The UK contributes 1.5% of global GHG emissions and according to the Kyoto Protocol, the UK government has set the very ambitious target to reduce GHG emissions which is 80% of GHG emissions reduction by 2050 in 1990 level. The UK has the oldest housing sector across Europe, and it is responsible for almost 30% of carbon dioxide emissions in the UK. Two-thirds of these homes would still be standing in 2050 (Carter, 2014). Previous studies pointed out that the UK residential sector can deliver a 60–80% reduction in carbon dioxide emissions by 2050, in line with the UK Government's target of zero-carbon homes by 2016 (Carter, 2014). The focus on the role of households in securing emission reductions includes improving building fabric standards and replacing old boilers with high-performing boilers. Improved insulation standards for buildings are a significant benefit for the on-going UK government's aim of reducing carbon dioxide emissions due to increasing heating demand in buildings, especially within the housing sector (Ozarisoy, 2022). This increase in insulation has led to some concern that it may result in increased summer overheating and a consequent increase in the use of domestic air conditioning systems.

DOI: 10.1201/9781003490364-11

Consideration has also to be given to the likelihood of global warming increasing the summertime temperatures and the increasing consumer demand and acceptance of air-conditioning in homes (Morey et al., 2020). Both these factors may increase the installation of air-conditioning in houses if measures are not taken to reduce the level of overheating.

In the near future and in adaptation terms, it will become increasingly important to give occupants control over their indoor environments during periods of potential overheating, periods of technical changes in construction, climate change, and the 'urban heat island effect', which are also likely to increase summertime external temperatures (Ekins & Etheridge, 2005). The net effect will be to place increasing importance on limiting the overheating of housing. In the past, the focus has been on comfort during the heating season, but as occupants' expectations changed, it is probable that they will also seek solutions to improve comfort during warm periods (Ozarisoy & Altan, 2023a). Notwithstanding these pressures, it is important to recognise that the benefits of increased thermal insulation will provide greater periods of comfort for the occupants during the heating season and reduced energy use in heating. However, to address these concerns, this research has investigated the likely impact of high insulation standards, such as the best practice standard on overheating in houses with insulation to varying levels and this research has been studied to see how well-insulated houses could adapt to the future UK climate, such as 2020s (2010–2039), 2050s (2040–2069) and 2080s (2070–2099) (Lockwood, 2013). Research has also been carried out into homes that would be adapted to the air conditioning system under given UK climate change scenarios, with various construction methods, where the cooling energy demand was predicted. Furthermore, current overheating standards which are under Chartered Institution of Building Services Engineers (CIBSE) Guide A, and the newly suggested criteria, EN15251, have been studied to compare the difference between these standards and in terms of the number of overheating hours using two different overheating criteria under the UK future climate change scenarios with various construction methods.

Literature Review

As the climate warms due to anthropogenic climate change, higher than average daily and monthly temperatures are expected to occur in the UK (Hanlon et al., 2021). In this case, of particular concern is the likelihood of more frequent occurrences of sustained high temperatures (heat waves), which have been shown to significantly increase human morbidity and mortality rates. In addition to heat waves, it is expected that due to higher ultraviolet (UV) radiation and more frequent days of dangerous ozone levels, people will be encouraged to spend more time indoors.

In addition to global warming, the issues of increased urbanisation and an aging population contribute to the impact of possible future scenarios. In the past 50 years, there has been a 30% increase in the ratio of those living in urban areas to those in rural areas. Due to the urban heat island (UHI) effect and the unique microclimate generated by cities, this means that more people may be affected by adverse health outcomes from rising temperatures (Mourkos et al., 2020). Population demographics are also changing globally, and particularly within the UK. As the UK population aging this means that more of the population will be exposed to higher outdoor/indoor temperatures.

Northern European cultures are not familiar with very warm weather, not having measures in place, such as those in Southern Europe, developed over time, to deal with the heat like taking of afternoon siestas (Parker, 2020). Moreover, the difference in the physical make-up of the building design which has developed to deal with a cool Northern European climate designed to maximise daylighting and passive solar heating with a defensive envelope to retain heat during winter. On the other hand, cooling in summer, if needed, is provided by ventilation through openable windows, utilising the fact that the external air is generally lower than that required for comfort within the building. These two factors mean that the UK buildings are not necessarily well suited to the climatic conditions now being experienced in summer and projected for the future (Ozarisoy & Elsharkawy, 2019).

To reduce the heating energy consumptions from domestic sector, which is responsible for more than a quarter of carbon dioxide emissions of the UK, the government has improved building fabric standards, such as increased insulation levels, replacing windows with high-performance one, boilers with high energy efficiency ones, for new built homes and even existing homes as well (Ozarisoy, 2022).

Newer, more highly insulated houses have been shown to have a higher indoor temperature than older buildings due to high levels of insulation and airtightness. They are warmed significantly by solar energy and energy gains from indoor heat sources, where this heat may remain trapped, and the temperature is less likely to decrease during the night when compared with older dwellings As building regulations increasingly incorporate higher levels of airtightness and thermal insulation, the potential risk of increased overheating in housing is evident. For example, such types of houses may overheat because the excessive heat gains from solar radiation and internal gains are not removed or absorbed due to fewer heat losses from the building fabric and the lack of thermal mass. However, increasing the insulation of the fabric will increase the average indoor temperatures and improving the airtightness will reduce heat losses during winter and summer through the fabric and infiltration. During the winter this is a desirable effect, but during the summer this situation can lead to overheating. It will increase the temperature of the air in the space and fabric of the building unless they are exceeded by the losses of heat from the house. As building regulations in the UK are expected to be revised in 2013 and the development agenda is strongly emphasising more airtight and energy-efficient homes, it is important to look at the potential for hotter homes and their effect on occupants' health (Gupta & Gregg, 2020).

The literature review has been carried out to discuss the effect of improving standards and changing construction methods on indoor thermal comfort and overheating. Furthermore, very recent research projects which are importantly related, such as COPSE and SCORCHIO, were reviewed to improve the quality of this research. The following literature review has focused only on UK homes' energy consumption and a summer overheating study using UKCIP 02 and UKCP 09 future climatic data with more common building simulation tools, in this case, IES, Tas, etc.

Thermal Comfort Study

The thermal interaction between people and their surrounding environment is highly complex and has been researched with a great deal of study, involving not only a study of heat transfer between the human body and their environment but also a study of human body responses to the environment and consideration of the social factors, which can also determine the way human body reacts to the environment.

During the 1920s, some initial work on thermal indices was carried out, and some major studies were done by Bedford in 1936 and continued over the next 20 years (Parkinson et al., 2019). The main idea of the research on thermal comfort in buildings has taken one of two main approaches: Laboratory studies – the studies based on experiment work and carried out in a climate chamber or laboratory; Field studies – the studies based on surveys and people are asked about their feeling of comfort with comfort scales. These two research approaches (laboratory and field studies) have led to two different approaches to specifying comfort conditions: *Deterministic methods* – which relate given environmental conditions, such as temperature, humidity and air speed, and given clothing and activity levels, to the likely level of occupant comfort; *Adaptive methods* – which are based on the surveys and aim to define the variations in comfort expectations with different climates. (Ozarisoy & Altan, 2023b).

Adaptive Models

The adaptive methods are based on field studies and an important premise of the adaptive model is that the person is no longer a passive recipient of the given thermal environment. The aim of these studies is to provide guidance that is related to ordinary living conditions. The difference between the deterministic methods and the adaptive methods is the adaptive methods do not require knowledge of the clothing level and the metabolic rate of occupants in order to establish the temperature required for thermal comfort (Gauthier et al., 2020). The adaptive method is based on the observation people take various actions to adjust their environment in order to achieve thermal comfort. To adapt to their changed condition and to achieve thermal comfort, people take various actions, such as shivering, sweating, changing clothing level, activities and closing/opening windows even adjusting blinds. However, in practice, they could not always possible to achieve comfort due to the dress code or lack of control/operation such as non-openable windows then the mechanical/air-conditioning system will be introduced to improve comfort.

The concept of adaptability has only recently been included in comfort standards, such as ASHRAE 55 and CIBSE Guide A, due to increased interest in naturally ventilation buildings rather than closely controlled air-conditioned buildings in order to reduce carbon emissions (Ozarisoy & Altan, 2021). As people adapt to changing conditions, the indoor comfort temperature in free-running buildings (naturally ventilated buildings) will vary with the outdoor temperature, and from person to person. ASHRAE and CIBSE have been used for this to predict/determine thermal comfort temperature bands but slightly different approaches. ASHRAE: the relationship between indoor comfort and outdoor temperature has usually been expressed in terms of the average monthly outdoor temperature.

CIBSE: a running mean of outdoor temperatures is used as research shows that UK weather can give considerable variations of outdoor temperature at much shorter than monthly intervals. According to the adaptive approach, people's comfort response depends on their very recent thermal experience, i.e., they are more affected by temperatures from the past few days. Therefore, when considering comfort temperatures, the daily running mean outdoor temperature is more appropriate than the monthly mean outdoor temperature (Gauthier et al., 2020).

If a hot spell has been experienced recently, then it will have a greater effect on comfort temperatures for the next few days or next week than the cooler temperatures earlier in the month giving a higher value than a straight average would do. Looking at the deterministic and the adaptive approach, the ideal comfort temperature band can be given which is in the range of around 20°C to 28°C and could be acceptable to most people (Ozarisoy & Elsharkawy, 2019). However, the degree of comfort within the bands is easily affected by other environmental factors, such as solar radiation gains, surrounding buildings or surfaces, air velocity, air quality and humidity and the occupants' activities.

Thermal Comfort and Overheating

Heat Related Definition

The Heatwave Plan for England defines forecast regional external day and night temperature thresholds that trigger four escalating action levels. In connection with this, a 'Heat-Health Watch'

Table 8.1 Forecast regional external temperature thresholds

Region	*Day (°C)*	*Night (°C)*
London	32	18
Southeast	31	16
Southwest	30	15
Eastern	30	15
West Midlands	30	15
East Midlands	30	15
Northwest	30	15
Yorkshire and Humber	29	15
Northeast	28	15

system operates in England from 1 June to 15 September each year. The Plan gives the following explanation of the thresholds:

> Although excess seasonal deaths start to occur at approximately 25°C [outdoors], for practical reasons the health heatwave alert system is based upon temperature thresholds where the odds ratio is above 1.15–1.2 (a 15–20% increased risk). The different trigger temperatures are summarised (see Table 8.1), with regional variations due to the relative adaptation to heat. However, a significant proportion of excess summer deaths occur before the health heatwave alert is triggered, which emphasises the importance of long-term planning actions by local authorities and the health sector.
>
> (Gupta & Gregg, 2012)

NHS heatwave guidance for care home managers and staff advises:

> Create cool rooms or cool areas. High risk groups that are vulnerable to the effects of heat are physiologically unable to cool themselves efficiently once temperatures rise above 26°C. Therefore, every care, nursing and residential home should be able to provide a room or area that maintains a temperature at 26°C or below.
>
> (Gupta & Gregg, 2012)

This guidance forms part of the implementation of the National Heatwave Plan for England. Table 8.1 shows the regional external temperature thresholds for day and night. When the external temperature is over the suggested temperature in Table 8.1 in related regions then it could be defined as "Heatwave" and overheating.

Thermal Comfort-Related Definition

For dwellings, CIBSE Guide A sets the following criteria for overheating, calculated based on the CIBSE Design Summer Years: Living areas – 1% of annual occupied hours over an operative temperature of 28°C, and Bedrooms – 1% of annual occupied hours over an operative temperature of 26°C. In addition to the overheating criteria, CIBSE Guide A also provides general summer indoor comfort temperatures for non-air-conditioned dwellings (Ozarisoy & Elsharkawy, 2019). These are that living areas should be at an operative temperature of 25°C and bedrooms at an operative temperature 23°C, nothing that sleep may be impaired above an operative temperature of 24°C.

It is necessary to use dynamic thermal simulation software to assess a design against the CIBSE Guide A criteria. The CIBSE criteria are not mandatory for housing. They are typically only applied by developers when they are required for new housing by client contractual specifications.

Care Quality Commission guidance for service users mentions they should be able to control temperature but does not set limits. For the design of new healthcare buildings, Health Technical Memorandum 03-01 states: *Calculations and thermal modelling should be undertaken to ensure that, during the summertime, internal temperatures in patient areas do not exceed 28°C (dry bulb) for more than 50 hours per year.* Concerning overheating, the English Partnerships Quality Standards stated:

> In order to ensure homes shall not be susceptible to overheating in rising summer temperatures, English Partnerships adopts the CIBSE standard. CIBSE Guide A (2006) required that: For living areas, less than 1% of occupied hours are over an operative temperature of 28°C, for bedrooms, less than 1% of occupied hours are over 26°C. This must be proven using appropriate simulation software in the design process, and adequate measures must be introduced to ensure it is maintained within the completed dwelling.
>
> (Gupta & Gregg, 2020)

A certification criterion set by the International Passive House Association requires that the indoor temperature should exceed 25°C for no more than 10% of the hours each year. This should be demonstrated at the design stage using the Passive House Planning Package software. Department for Communities and Local Government (DCLG) have a published expectation for dwelling temperatures:

> It is recommended that indoor operative temperatures (taken as a mean of air and radiant temperatures) should not exceed 26°C when RH is 30% or 24°C at RH 60%.
>
> (Ozarisoy, 2022)

It is likely that the CIBSE Guide A definition is the most widely used in terms of peak temperatures and the avoidance of overheating. CIBSE also provides general summer indoor comfort temperatures for non-air-conditioned dwellings.

Findings from, and Gaps in the Literature Review

A considerable number of UK-focused thermal modelling studies have been published in recent years, which have investigated the impact of climate change on indoor overheating levels and energy demand. A common feature of these studies is the use of dynamic thermal simulation packages, which are able to explore the effects related to indoor overheating/temperatures at a various resolution, such as daily, hourly or even sub-hourly time steps. Whilst these studies usually examine a limited number of dwelling types, most authors have tried to use some of the typical dwelling types that are broadly representative of the UK housing stock or the parts of the stock in the region under examination. However, the lack of these types of input parameters across the various studies does not allow the direct comparison of their results and even most of them are not using monitoring data for validation of their simulation models.

The general finding from the literature review is: It is impossible to eliminate overheating without mechanical cooling systems but the most effective ways to reduce overheating hours without an active cooling system are control of the solar gain and thermal mass with night ventilation.

Light-weight construction forms of houses would experience more overheating than heavy-weight construction houses. However, during the night-time, heavy-weight construction houses would have some problems, such as slow cooling during the evening and during the mild summer, over cooling then would lead to turning on the heating system, especially in the northern part of the UK, heating demand would be reduced due to external temperature increased by climate change, new build, due to airtightness and insulation levels, shows better thermal and comfort performance than older home during the winter. However, it would not provide any benefits for comfort during the summer, external insulation shows better thermal comfort performance than internal insulation, most of the energy-efficient measurements would be improved thermal comfort and reduced energy demand during the heating season. However, these measurements have not always given benefits for thermal comfort and cooling energy consumption during the summer (Semenov, 2007). The present climate change scenario and weather file generator have not accounted for considerable weather events, such as heat wave, UHI effect and it also could not provide high confidence of wind data for natural ventilation simulations analysis and relative humidity (RHs) for comfort study (Jowkar et al., 2019).

From the literature review, there are some gaps in current studies that may need to be filled which are: None of the simulation model studies have mentioned calibrating the study with actual monitoring data (Nicol et al., 2020). But some of them used real homes' physical data to build the simulation models. None of the studies have done comparison studies for heating, cooling and overheating with UKCIP 02 and UKCP 09 future weather data (it may need to be mentioned here that as UKCP09 data was not available at the early simulation studies, UKCIP02 projections were used with an aim to substitute them in the future. Later 2011, after being available to use the UKCP09 data, some of the simulation studies were carried out in order to make some comparison studies (UKCIP02 and UKCP09). None of the studies have adapted different thermal comfort criteria, such as EN15251, which uses the running mean outdoor temperature rather than CIBSE Guide A, which uses a fixed upper limit temperature for bedroom and living room (Rupp et al., 2021).

Aims and Objectives

Faced with the problems that are likely to be posed by climate change and/or global warming, the overall aim of this thesis has been to study energy consumption and overheating of selected case study homes in the UK under a range of possible future climate change scenarios. This research would share findings with previous/recent works presented by focusing on typology, location and construction methods with various building standards and defined climate projections. This work, however, differs notably in that it presents the modelling of each individual home at low and high emissions scenarios with UKCIP 02 and high emissions scenario UKCIP 09 with 10% probabilities where change is very unlikely to be greater than 10% projections for up to 2,100 and two different methods to define the upper limit of overheating temperature under futuristic climate data.

The literature review shows that the majority of previous studies using UKCIP 02 future weather data where the probability was fixed at a central estimate (50%) rather than UKCP 09 fixed at 10% probability. This is not a limitation of the previous work but simply where the data had progressed at the time of research. None of the studies have compared with Post Occupancy Evaluation (POE) but, in this study, there has been a monitoring of indoor environments, such as temperatures and RH levels, and energy consumption for more than one year after occupying the case study homes. In addition, a thermal comfort study has been completed, looking especially at the overheating sensitivity under different thermal comfort criteria, such as CIBSE Guide A and

EN15251. Furthermore, this research has studied whether the reduction on heating demand by climate change can offset the increase in carbon emissions from cooling systems by global warming in case study homes. Set out below is the summary of the objectives for this study which are

- To show how to use the POE data in order to reduce uncertainty of the input data and to increase the accuracy of the simulation study, which is almost one of the first attempts using such methods in the simulation field.
- To study future climate change predictions, climate change and global warming with a focus on an analysis of the Intergovernmental Panel on Climate Change (IPCC), UKCIP 02 and UKCP 09 scenarios undertaken to determine through analysis and discussion of the results obtained from the simulations and monitoring data, how much of heating demand reduction and cooling energy consumption is increasing and overheating may occur by climate change, different building standards and construction methods
- To identify, based on previously completed UK housing studies and policies (see Chapter 3), the direction of the UK government policy and adapting the policy such as improved building fabrics' thermal properties and the airtightness level, in the case study homes under future climate scenarios then investigate the potential for demanding energy consumption and CO_2 emissions, and overheating from case homes with better building fabrics and various level of airtightness.
- To develop various scenarios, in order to determine the most energy-efficient method of construction, such as lightweight construction with a better insulation level or heavyweight construction with a low insulation level.
- To investigate location sensitivity studies, such as London, Manchester, Edinburgh and building types of sensitivity studies; here it would call house typology studies, such as end-terraced, mid-terraced, ground floor flat and top floor flat, which will be investigated to find out the energy consumption and thermal comfort under present climatic weather data and future climatic data.

Contribution to the Field

The work that has been undertaken within this thesis will make an important contribution to the research field in a number of different ways. These include: This work represents the detailed attempt that has been undertaken to project the energy consumption and overheating consideration in the UK housing up to 2080s. To do this, this work has brought two different UK future climatic data from UKCIP 02 and UKCP 09 and this could be the first study to make a direct comparison study for such two different weather data with low-energy homes in order to predict their energy consumption and thermal comfort.

This work could be one of the first modelling exercises to be simulated with post-occupancy study to collaborate with a computer simulation model rather than using typical UK housing model input data without observation. When using a computer simulation tool for this kind of study, the most important considering factor is the accuracy and validation of the simulation model. This study shows how to improve accuracy and to reduce uncertainty of input data for simulation models using post-occupancy data. The results will give insights into a better understanding of construction methods to reduce energy consumption and overheating in the UK housing stock. These insights will enable rational and considered responses to problems, such as overheating and energy consumption increased by installing air-conditioning systems, which are likely to be faced in the future due to climate change. The simulation results of this study should be valuable to referenced studies and to the direction of government policy or strategies for UK housing in their effort to reduce energy consumption and overheating under a range of future climate change scenarios. The results could also be used to explain the benefit of having a heavyweight structure rather than a lightweight structure. In terms of cooling and overheating, this research also provides an explanation for the disadvantageous effect of using modern insulation standards with improved airtightness in UK homes.

References

Averchenkova, A., Fankhauser, S., & Finnegan, J. J. (2021). *The influence of climate change advisory bodies on political debates: evidence from the UK Committee on Climate Change*. Informa UK Limited. https://doi.org/10.1080/14693062.2021.1878008.

Carter, N. (2008). Combating climate change in the UK: Challenges and obstacles1. *The Political Quarterly, 79*(2), 194. https://doi.org/10.1111/j.1467-923x.2008.00913.x.

Carter, N. (2014). The politics of climate change in the UK. *WIREs Climate Change, 5*(3), 423. https://doi.org/10.1002/wcc.274.

Ekins, P., & Etheridge, B. (2005). The environmental and economic impacts of the UK climate change agreements. *Energy Policy, 34*(15), 2071. https://doi.org/10.1016/j.enpol.2005.01.008.

Gauthier, S., Bourikas, L., Al-atrash, F., Bae, C., Chun, C., De Dear, R., Hellwig, R. T., Kim, J., Kwon, S., Mora, R., Pandya, H., Rawal, R., Tartarini, F., Upadhyay, R., & Wagner, A. (2020). The colours of comfort: from thermal sensation to person-centric thermal zones for adaptive building strategies. *Energy and Buildings, 216*, 1–3. https://doi.org/10.1016/j.enbuild.2020.109936.

Gupta, R., & Gregg, M. (2012). Using UK climate change projections to adapt existing English homes for a warming climate. *Building and Environment, 55*, 20. https://doi.org/10.1016/j.buildenv.2012.01.014.

Gupta, R., & Gregg, M. (2020). Assessing the magnitude and likely causes of summertime overheating in modern flats in UK. *Energies, 13*(19), 21–23. https://doi.org/10.3390/en13195202.

Hanlon, H. M., Bernie, D., Carigi, G., & Lowe, J. A. (2021). Future changes to high impact weather in the UK. *Climatic Change, 166*(3–4). https://doi.org/10.1007/s10584-021-03100-5.

Jowkar, M., De Dear, R., & Brusey, J. (2019). Influence of long-term thermal history on thermal comfort and preference. *Energy and Buildings, 210*, 13–15. https://doi.org/10.1016/j.enbuild.2019.109685.

Lockwood, M. (2013). The political sustainability of climate policy: the case of the UK climate change act. *Global Environmental Change, 23*(5), 1339. https://doi.org/10.1016/j.gloenvcha.2013.07.001.

Morey, J., Beizaee, A., & Wright, A. (2020). An investigation into overheating in social housing dwellings in central England. *Building and Environment, 176*, 5–7. https://doi.org/10.1016/j.buildenv.2020.106814.

Mourkos, K., Mcleod, R. S., Hopfe, C. J., Goodier, C., & Swainson, M. (2020). Assessing the application and limitations of a standardised overheating risk-assessment methodology in a real-world context. *Building and Environment*, 181, 51–53. https://doi.org/10.1016/j.buildenv.2020.107070.

Nicol, F., Bahadur Rijal, H., Imagawa, H., & Thapa, R. (2020). The range and shape of thermal comfort and resilience. *Energy and Buildings, 224*, 37–38. https://doi.org/10.1016/j.enbuild.2020.110277.

Ozarisoy, B. (2022). Energy effectiveness of passive cooling design strategies to reduce the impact of long-term heatwaves on occupants' thermal comfort in Europe: Climate change and mitigation. *Journal of Cleaner Production, 330*, Article 129675. https://doi.org/10.1016/j.jclepro.2021.129675.

Ozarisoy, B., & Altan, H. (2021). Regression forecasting of 'neutral' adaptive thermal comfort: a field study investigation in the south-eastern Mediterranean climate of Cyprus. *Building and Environment, 202*, 11–17. https://doi.org/10.1016/j.buildenv.2021.108013.

Ozarisoy, B., & Altan, H. (2023a). Thermal comfort survey I: a field study investigation to assess on households' thermal discomfort and overheating risk of European buildings. In *Handbook of retrofitting high density residential buildings*. Springer (378–379). https://doi.org/10.1007/978-3-031-11854-8_6.

Ozarisoy, B., & Altan, H. (2023b). Thermal comfort survey II: a field study investigation on the regression forecasting of neutral adaptive thermal comfort. In *Handbook of retrofitting high density residential buildings*. Springer (545–547). https://doi.org/10.1007/978-3-031-11854-8_7.

Ozarisoy, B., & Elsharkawy, H. (15 March 2019). Assessing overheating risk and thermal comfort in state-of-the-art prototype houses that combat exacerbated climate change in UK. *Energy and Buildings, 187*, 201–217. https://doi.org/10.1016/j.enbuild.2019.01.030.

Parker, J. (2020). The Leeds urban heat island and its implications for energy use and thermal comfort. *Energy and Buildings, 235*, 23–27. https://doi.org/10.1016/j.enbuild.2020.110636.
Parkinson, T., De Dear, R., & Brager, G. (2019). Nudging the adaptive thermal comfort model. *Energy and Buildings*, 206. https://doi.org/10.1016/j.enbuild.2019.109559.
Rupp, R. F., Parkinson, T., Kim, J., Toftum, J., & De Dear, R. (2021). The impact of occupant's thermal sensitivity on adaptive thermal comfort model. *Building and Environment, 207*, 45–47. https://doi.org/10.1016/j.buildenv.2021.108517.
Semenov, M. A. (2007). Development of high-resolution UKCIP02-based climate change scenarios in the UK. *Agricultural and Forest Meteorology, 144*(1–2), 127. https://doi.org/10.1016/j.agrformet.2007.02.003.

9

ENERGY AND ENVIRONMENTAL ASSESSMENT METHODS FOR BUILDINGS IN THE UK

Introduction

As a global matter, environmental issues and good building design have also been increasingly important in the UK for the last 20 years (Adedoyin & Zakari, 2020). Therefore, in order to accomplish good building design that could contribute considerably to reducing pollution and improving the built environment, building energy and environmental assessment methods have been developed. These assessment methods for buildings identify criteria for a range of issues also concerning the global, national and indoor environments. Furthermore, using these methods can help to achieve the following:

- To provide recognition for buildings, which are friendlier to the global environment than normal practice, and therefore help simulate a market for them.
- To improve the indoor environment quality and occupants' health.
- To raise awareness of the dominant role which buildings play in global warming through the greenhouse effect; in the production of acid rain, through the burning of fossil fuels for energy; and in the depletion of the ozone layer.
- To reduce today's long-term impact that buildings have on the global environment.
- To reduce energy consumption and, thus, CO_2 emissions caused by buildings, while encouraging savings of energy.
- To provide a common set of targets and standards and accordingly avoid the false claims of environmental friendliness.
- To encourage designers to achieve environmentally eco-sensitive and low-energy building designs (Lomas & Porritt, 2016).

Existing Energy and Environmental Assessment Methods

In the UK, there are nine main methods that have been developed since the early 1980s (Gupta & Gregg, 2020). Some of these methods are relatively in use today and have also been added to the building regulations to improve building energy usage and increase efficiency in the country. The

DOI: 10.1201/9781003490364-12

main building energy and environmental assessment methods including simulation tools, which have been developed since 1980s, are listed below:

1 BREDEM
2 SAP
3 Environmental Standard – BREEAM
4 NHER
5 STARPOINT
6 ENVEST
7 The Toolkit – The Office Toolkit
8 APACHE
9 TAS

BREDEM

The Building Research Establishment Domestic Energy Model (BREDEM) provides a framework for the calculation of energy use in buildings. It is an energy model developed in the 1980s by the Building Research Establishment (BRE) and adopted by the Energy Efficiency Office of the Department of Energy and the Department of the Environment to estimate energy costs and potential savings in housing (Mckenna et al., 2018). This model assesses the energy use in occupied dwellings and shows that the physical characteristics of a dwelling and the lifestyles of the occupants are about equally important in determining energy consumption (Natarajan et al., 2011).

BREDEM exists in a number of standardised versions, which differ in technical detail and in the precise requirements for data related to a particular application (Ozarisoy, 2022). Some versions are designed to be implemented as computer programs while others are defined in such a way that do not require the use of a computer. The latest version is a particular development which is suitable for calculating SAP ratings in both new and existing dwellings, which can also be separated as one of the main energy and environmental assessment methods (Ozarisoy & Elsharkawy, 2019). BREDEM is designed to be easy to use to provide reliable results and based on practical experience gained through measurements made in many occupied buildings, as shown in Figure 9.1. The main aims of this scheme are:

- To identify the various uses of energy in dwellings.
- To estimate the annual requirement for each use.

When space-heating requirements are calculated, all the physical details of the dwelling, the heating system, internal and external temperatures and the living pattern of its occupants are all considered. Other energy uses are estimated from average consumptions derived from surveys, which are also appropriate to the composition and activities of households (Sousa et al., 2017).

SAP

Standard Assessment Procedure (SAP) is the government's standard system for home energy rating, which was published in 1993 (Strachan et al., 2008). The latest edition is SAP 2001, published in December 2001 and operational in the UK. The SAP is used to calculate the rating based on the BREDEM, which provides a framework for the calculation of energy use in dwellings (Summerfield et al., 2010). The first edition was included as one means to satisfy the Building Regulations Part L

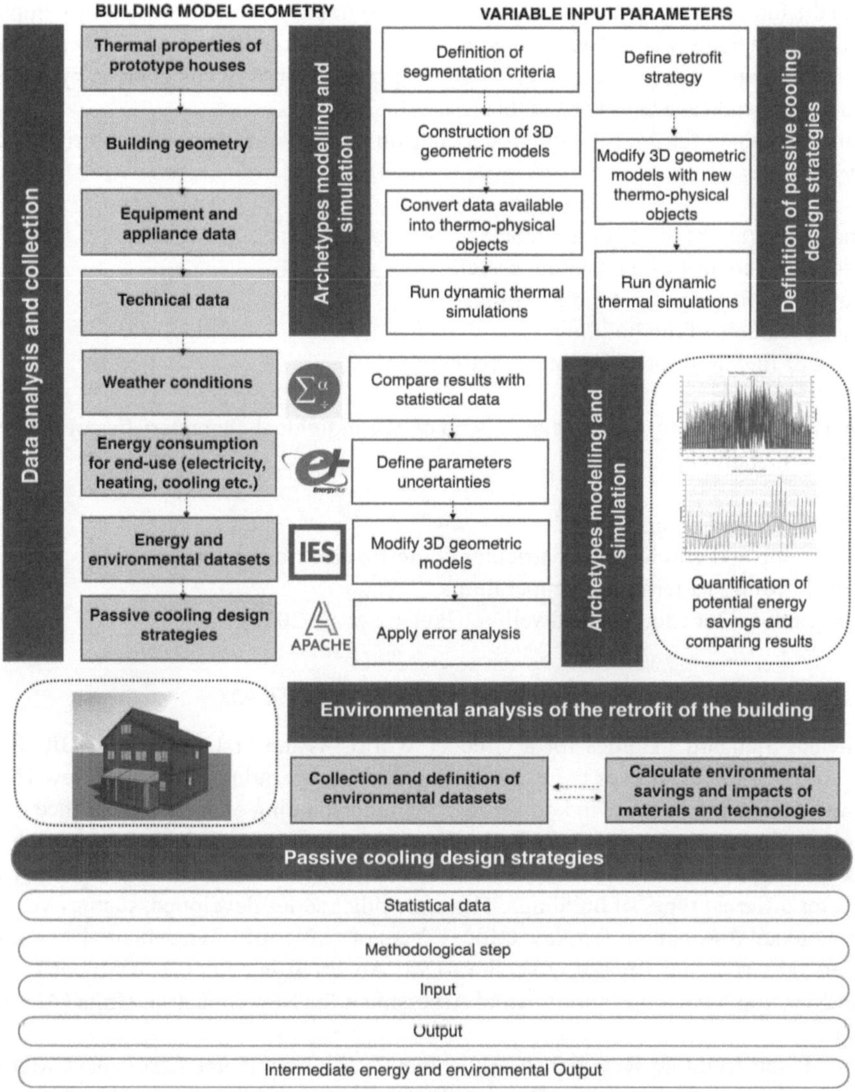

Figure 9.1 List of widely used environmental design software suites.

Domestic Buildings since July 1995. The latest edition, on the other hand, calculates the carbon index (CI), which can also demonstrate compliance with Approved Document L1 (England and Wales) and Technical Part J (Scotland). SAP provides a manual guide for energy rating of dwellings and energy efficiency in buildings (Rosenow et al., 2017). This manual guide describes the Government's SAP for producing an energy rating for a dwelling, which is based on calculated annual energy cost for space and water heating. The rating is normalised for floor area and therefore the size of the dwelling does not strongly affect the result, which was presented on a scale of 1–100: the higher the number, the better the standard. This rating has been changed in the latest edition (SAP 2001) to a scale from 1 to 120 with an additional scale of 0.0–10.0 for the CI, which is based on the annual CO_2 emissions associated with space and water heating (Simpson et al., 2021). The method of calculating the rating is set out in the form of a worksheet, accompanied by a series of tables. Therefore, by completing the numbered boxes in the worksheet and using the data in tables as indicated, the SAP calculations can be made. In addition, there is a computer program approved for SAP calculations by the BRE that could also be applicable, such as used in this research.

The rating obtained for the SAP rating depends on a range of factors that contribute to energy efficiency such as:

- Thermal insulation of the building fabric.
- Efficiency and control of the heating system.
- Ventilation characteristics of the dwelling.
- Solar gain features of the building.
- The price of fuel used for space and water heating.

The rating is not affected by factors that depend on the individual characteristics of the household inhabitants in the building such as:

- Household size and composition.
- The ownership and efficiency of particular domestic electric appliances.
- Individual heating patterns and temperatures.
- The geographical location of the dwelling (Preston et al., 2020).

Environmental Standard – BREEAM

Environmental Standard, 'Homes for a Greener World', is the first revision of BREEAM Version 3/91 for new homes. It uses the experience gained in operating BREEAM/New Homes and replaces and reflects the increase in knowledge and understanding of early issues since it launched in 1991 (Hall & Buckley, 2016). BREEAM/New Homes resulted from joint research between BRE and Municipal Mutual Insurance Ltd. It is one of the series of five assessment methods developed for different types of buildings. There are other series developed, such as Version 1/93: An Environmental Assessment for New Offices, Version 2/91: An Environmental Assessment for New Superstores and Supermarkets, Version 4/93: An Environmental Assessment for Existing Offices and Version 5/93: An Environmental Assessment for New Industrial Units (Ahern & Norton, 2020).

BREEAM, the Building Research Establishment's Environmental Assessment Method, was developed by Equivalent Circulating Density (ECD) Energy and Environment with the UK BRE and introduced in 1990 (Taylor et al., 2014). One of its most important aims is to provide guidance on ways of minimising the adverse effect of buildings on the global and local environments while promoting a healthy and comfortable indoor environment for inhabitants. This assessment method, which is carried out at the design stage, is based on readily available and generally accepted information (Kane et al., 2014). As it is stated in one of 'Building Research Establishment Reports', the method identifies and credits designs where specific targets are met. It is not expected that any single design will meet all of the target requirements. The main aims of this scheme are:

- To distinguish dwellings of reduced environmental impact in the marketplace,
- To encourage best environmental practice in building design, operation, management and maintenance,
- To establish criteria and standards going beyond those required by law and regulation,

- To increase awareness of owners, occupants, designers and operators of the benefit of environmentally sensitive buildings.

Since its launch in 1990, BREEAM has become the UK industry standard for specifying and benchmarking the environmental performance of buildings (Oliveira et al., 2016).

NHER

The National Home Energy Rating (NHER) is a policy tool for Authorities and Housing Associations, which calculates the energy performance of dwellings. It is the first scheme to receive full governmental approval for the delivery of SAP. This method was designed in 1990 by the BRE in the function of a comprehensive system for evaluating the energy efficiency of housing stock. In addition to SAP rating, it also delivers accurate running cost rating, CO_2 emission, building fabric rating and condensation analysis (Sandberg et al., 2016).

The NHER has three steps towards establishing and implementing an energy policy:

- Step 1: *A Stock Profile* – Step 1 sets the energy performance of the stock of 0 (poor) to 10 (excellent) and is based on the total annual running cost.
- Step 2: *Priorities and Targets* – Once an 'NHER Stock Profile' is obtained, the following priorities and targets could be used:
  - Identify priority estates or house types – those with the worst energy performance will often show the best return on investment.
  - Establish targets for refurbishment, e.g., an NHER of over 6.0.
  - Establish overall policy goals, e.g., all dwellings should have an NHER of 4.0 or over by the year 2000.
- Step 3: *Implementation* – Once the goals of the energy policy are established, the practicalities of implementing the energy policy need to be addressed. The main aims of this scheme are:
  - To aid energy efficient design of new buildings.
  - To check affordable warmth targets can be met.
  - To calculate accurate running costs based on actual living patterns if known.
  - To detail costs and savings of refurbishment packages.

The software delivers many useful indices and ratings including the Government's SAP Rating on a scale of 1–100, the NHER on a scale of 1–10, the Building Energy Performance Index (BEPI), the total CO_2 emissions produced by a house.

STARPOINT

STARPOINT, Home Energy Labelling, is an essential aid to achieving the needs and requirements set by the Government and the Housing Corporation, which was launched in 1990. It aims to assess existing energy efficient standards in housing, identify priority action areas and cost-effective improvements and test the extent to which these improvements can raise the rating and reduce carbon dioxide emissions and heating bills.

STARPOINT is computer software, and its products and consultancy services are based on the Government's SAP. It has been given to issue Home Energy Labels bearing the SAP rating, which illustrates the energy efficiency of the house on a scale of 1–100. The Star grading operates with ONE STAR being poor and FIVE STAR excellent (Lee et al., 2013).

In addition, a home energy label requires a home energy survey, which is simply the collection of the information necessary to estimate the cost of the energy needed to heat a house and hot water, based on a standard, national heating pattern and typical occupancy. The energy survey normally lasts less than 30 minutes. It collects information, such as dimensions of the property, how it is constructed, insulated and ventilated, and what type of heating system and controls are installed. A home energy label provides an energy efficiency rating to the property not to the lighting and appliances. The main aims of this scheme are:

- To improve the home energy rating of the house.
- To help reduce harmful CO_2 emissions in the environment.
- To save household money on house's heating bills.
- To improve the comfort conditions in the house and therefore make the house feel warmer.
- To reduce maintenance costs.
- To add to the value of the house.

ENVEST

Environmental Impact Estimating Design Software is the first UK software for estimating the life-cycle environmental impacts of a building from the early design stage. The first version is for office buildings and considers the environmental impacts of materials used during construction and the energy and other resources consumed over the building's life. ENVEST was developed and launched in early 2000 by BRE in conjunction with DETR (Sousa & Robinson, 2020). This is a software tool that helps specifiers determine the environmental impact of their construction decisions. BRE previously used it for internal consultancy. Data is provided by the tool and information about the size, fabric and service options for the proposed building are entered through input screens. By using minimal input data, ENVEST allows designers to instantly identify those aspects of the building, which have the greatest influence on the overall impact. All environmental impacts are measured on a single-point scale called 'Ecopoints' allowing the designer to compare different designs and specifications directly. 100 Ecopoints are equivalent to the environmental impact caused by one UK citizen in one year. It is extremely easy to use; the simplicity of the software belies the wealth of research data on materials, impacts and weightings that underpin its calculations.

In addition, the software has pre-selected default values set on typical office building specifications, which can also be edited by the user. Total life cycle environmental impact of the building in Ecopoints is also allowing designers to make direct comparisons between different designs and specifications. These are displayed both numerically and in graphic form. The graphs include the total life cycle environmental performance of the study building and environmental impacts comparison between the study building and a previously saved building. The key benefits for users of this software are:

- Optimises the building form for least environmental impact.
- Informs choice on main construction materials.
- Enables the environmental impacts of construction and building operation to be balanced over the life of the building.
- Gives comparisons for different buildings and specifications.
- Graphically illustrates the environmental credentials of various design options to clients.

The environmental impacts of construction encompass a wide range of issues, including climate change, mineral extraction, ozone depletion and waste generation. Assessing such different issues in combination requires subjective judgements about their relative importance. Moreover, ENVEST also has great potential to be used as a benchmarking tool. UK Ecopoints is derived by adding together the score for each issue, calculated by multiplying the normalised impact with its percentage weighting. The annual environmental impact caused by a typical UK citizen therefore creates 100 Ecopoints. More Ecopoints indicate higher environmental impact (Li et al., 2022).

The Toolkit – The Office Toolkit

The Toolkit is a guide for facilities and office managers for reducing costs and environmental impact. It examines and evaluates all the aspects of an office-based business, which have an environmental impact. BRE in partnership with PA Consulting Group prepared the Toolkit in late 1995. This tool is not just the direct impacts of energy, water, paper usage and waste management but also commuting, business travel and working from home. In all, it provides a complete environmental assessment for your business or office. The Toolkit consists of spreadsheets, which are supplied on a 3.5-inch disk and require Microsoft Excel or Lotus 1-2-3 software. It is designed to help busy facilities, buildings or office managers improve environmental performance and save money. Existing users of the Toolkit found that it takes only a few days (typically five days for a 150-person office) to get to the point where they were achieving significant results. Some large organisations have managers responsible for a group of smaller offices, such as local sales offices and bank branches, which can easily use the Toolkit as well. Therefore, it can be said that it is also applicable to the small businessman, with one or several offices.

The focus of the Toolkit is the office and everything that is based upon it. It includes commuting, business travel and home working as well as the more direct impacts of energy, water, paper usage and waste management. In addition, the Toolkit includes the environmental impacts of a building-based business such as the impacts caused by generating the power used. Within the office, small kitchens, laboratories, workshops and computer rooms are also included and this can also be applied to most warehouses and shops.

Furthermore, the Toolkit also covers health and safety topics that can be included within the spreadsheet if required. Initial benefits may be up to 20% if not previously taken any environmental initiatives, however, after a little more effort and time it should enable to extract more benefits and then to develop a process of continuous improvement, gaining 5–10% every year (Trotta, 2018). The key decision-makers in the business should support the aims of the initial review, and all staff should be aware of what is about to happen. It may have been that other people should be involved from several parts of the organisation, or even from several organisations. In order to maximise the gain and make it worthwhile, the following steps should be considered when using the Toolkit:

- If you are a tenant, then you need to involve the landlord. Only he can commit to significant changes in the building and how it is operated – for example, the building probably has a centrally controlled central heating system, and waste will be collected centrally.
- If you are the landlord, then you may want to get all the tenants involved. Everyone has something to gain, and you can use this approach to develop business relationships.
- If you run a branch office of a large organisation, then office supplies and energy may be purchased centrally.

- If you are responsible for the facilities or environmental management in an organisation with a number of sites, then you will want to involve local managers.

Use the Toolkit spreadsheet to fill in numbers, for each section of the review, as displayed on the example spreadsheet and enter guidance notes to help estimate numbers where the facts are not readily available. From these numbers, the Ecopoints (the environmental impact) and the running costs calculations will be performed automatically by the Toolkit programme. The key to a successful environmental programme is gaining momentum. Most of the early actions involve everyone in the office changing their behaviour; particularly in switching off lights and office equipment. A carefully planned communication programme is needed to ensure that everyone gets the message and gets involved. Once the major environmental impacts are discovered, convey these impacts to all staff by using a series of training seminars or a poster campaign or an internal newsletter (or combinations of these) to generate discussion and lots of ideas on how best to get the improvements achieved. At the end of the implementation period, the actions should be reviewed in order to know what has been achieved, where things could be improved and what to tackle during the next period. At this stage, it is a second cycle of the process of continuous improvement.

APACHE

The Applications Program for Air-Conditioning and Heat Engineers is a software tool for thermal design and energy simulation related to buildings, which was originally developed by Facet Ltd in 1981 and bought by Integrated Environmental Solutions (IES) Ltd. In design mode, APACHE covers the calculation of heating, cooling and latent room loads, the sizing of room units, internal comfort analysis and codes/standards checks (Aragon et al., 2022). In simulation mode, APACHE performs a dynamic thermal simulation using hourly weather data, as shown in Figure 9.2. Linked modules deal with the performance of HVAC plants and natural ventilation. APACHE is a component of the IES Virtual Environment, an integrated computing environment encompassing a wide range of tasks in building design such as listed below:

- Thermal design (heating, cooling and latent load calculations).
- Equipment sizing.
- Codes and standards check.
- Dynamic building thermal performance analysis.
- Systems and controls performance.
- Energy use.

APACHE covers a wide range of requirements from basic design to detailed simulations of HVAC systems. Apache-HVAC is used to create HVAC system schematic models, subsequent to creating a building model using the IES building modeller ModelIT, as shown in Figure 9.3. The building models together with the HVAC system schematics are used as the data source for APsim. APsim HVAC system simulation may be used to estimate the expected fuel consumption of the building's space heating and cooling systems.

It may also be used to examine in detail the expected performance of an HVAC system and controls under a range of operating conditions (Loga et al., 2016). The APACHE software suite incorporates Chartered Institution of Building Services Engineers (CIBSE) facilities for performing fabric analysis; heat loss and heat gain calculations and thermal simulation. Geometrical models may alternatively be entered using facilities within the Virtual Environment (Eyre & Baruah,

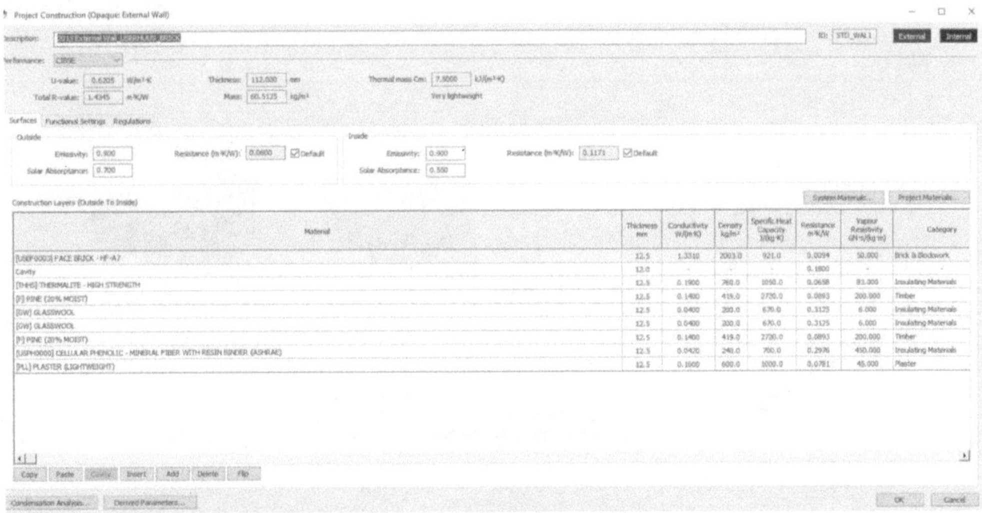

Figure 9.2 Assigned construction properties into the APACHE platform.

2015). Input of data relating to materials, occupancy, internal gains, climate, air movement and systems is managed via graphical interfaces and supported by extensive databases. APACHE presents a wide range of outputs in tabular and graphical form.

TAS

Thermal Analysis Software (TAS) is a suite of computer software products, which simulate the dynamic thermal performance of buildings and their systems. Environmental Design Solutions Limited (EDSL) was formed in 1989 to commercially develop TAS. The main module is TAS Building Designer, which performs dynamic building simulation with integrated natural and forced airflow. As a secondary module, it has 3D graphics-based geometry input that includes a CAD link. TAS Systems is a HVAC systems/controls simulator, which may be directly coupled with the building simulator. It performs automatic airflow and plant sizing and total energy demand. The third module, TAS Ambiens, is a robust and simple to use 2D CFD package which produces a cross section of microclimate variation in a space. TAS is a complete solution for the thermal simulation of new or existing buildings, allowing design professionals to compare alternative heating/cooling strategies and façade design for comfort, equipment sizing and energy demand (Ozarisoy, 2022). It is also a very responsive and accurate tool for concept development while being a powerful design tool for the optimisation of a building's environmental, energy and comfort performance. With TAS 3D Modeller, rendered 3D views that display comprehensive shading can be created and even calculate sunshine penetration through the interior of the building between spaces.

Construction materials and glazing types can be allocated from the comprehensive databases included in the software. These have their dynamic response automatically calculated in the model. The occupation of the building is represented by a calendar, with different day types having varying schedules of use. The final ingredient that brings the building model to life is the climate data used to drive the simulation. TAS has access to over 2,500 recorded weather sites worldwide (Ozarisoy & Elsharkawy, 2019). The data consists of hourly values for solar, temperature, humidity and wind speed and direction, as shown in Figure 9.4.

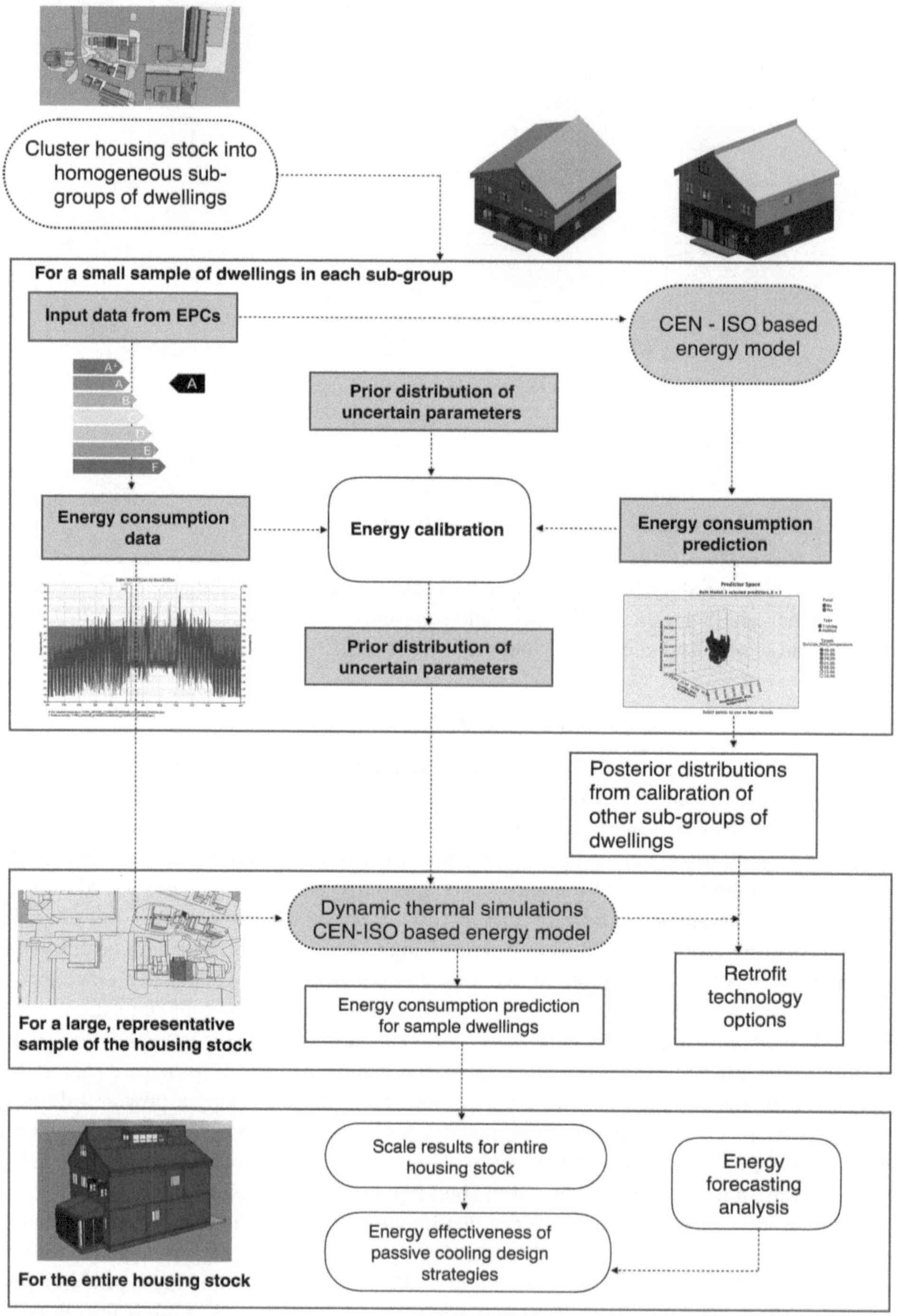

Figure 9.3 Building energy development for the simulation analysis.

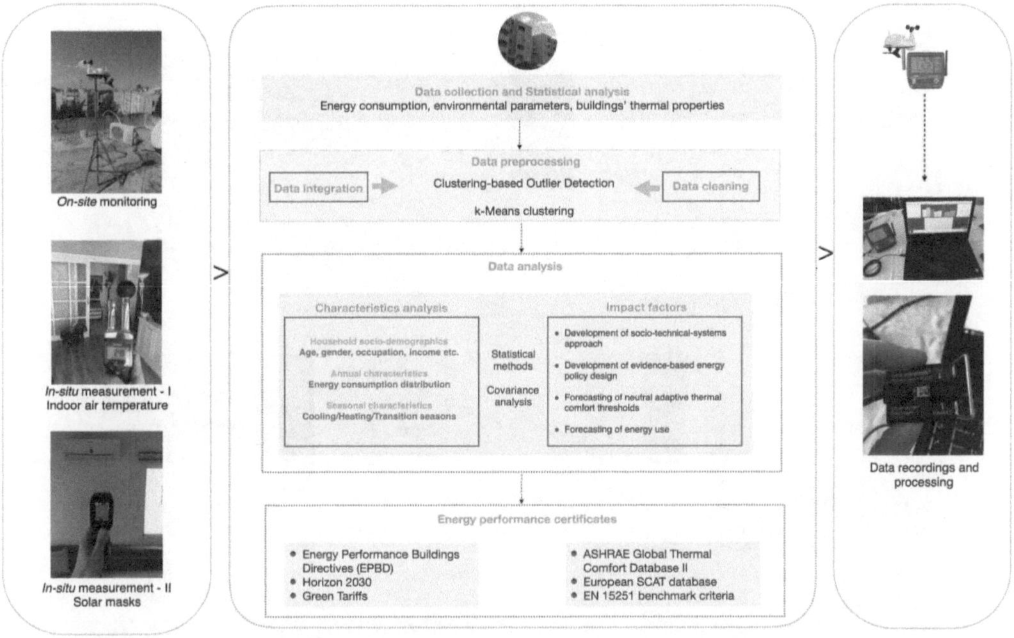

Figure 9.4 Environmental analysis parameters assigned to the simulation model.

TAS exports the various constructions and glazing types with their original TAS colours, and these can be edited within Lightscape to enhance the appearance of the building's interior finishes. The sky conditions can be specified, and the building can be located anywhere in the world for any date and time. From this, the lighting simulation is completed, and a photo-realistic model of the whole building is created. The model can be viewed from any position and used to generate fly-through animations. Daylighting analysis can be performed on the rendered model to determine the daylight levels on any of the internal surfaces. In addition, luminance lux levels can also be displayed numerically with contour lines or via colour-mapped images representing lux level banding, as shown in Figure 9.5.

Efficiency of Using Environmental Design Software

In this chapter, existing energy and environmental assessment methods for buildings in the UK have been reviewed and compared in order to assist people in choosing the right method for the building energy study and environmental impact assessment. As stated earlier in this chapter, tools and techniques discussed are the main energy rating and environmental impact analysis methods currently used in the country.

The following comparison, as shown in Table 9.1, has not been prepared to justify that there is only one method to be used, which would be the best method for the purpose, but to describe each method in more detail and therefore help users make decisions easier. As it can be seen, each method has its unique way of assessment, some may even cover issues that others may not, and therefore it is worthwhile to use more than one method. Architects, designers, environmental/energy consulting firms, authorities/clients, quantity surveyors and research universities presently use these methods.

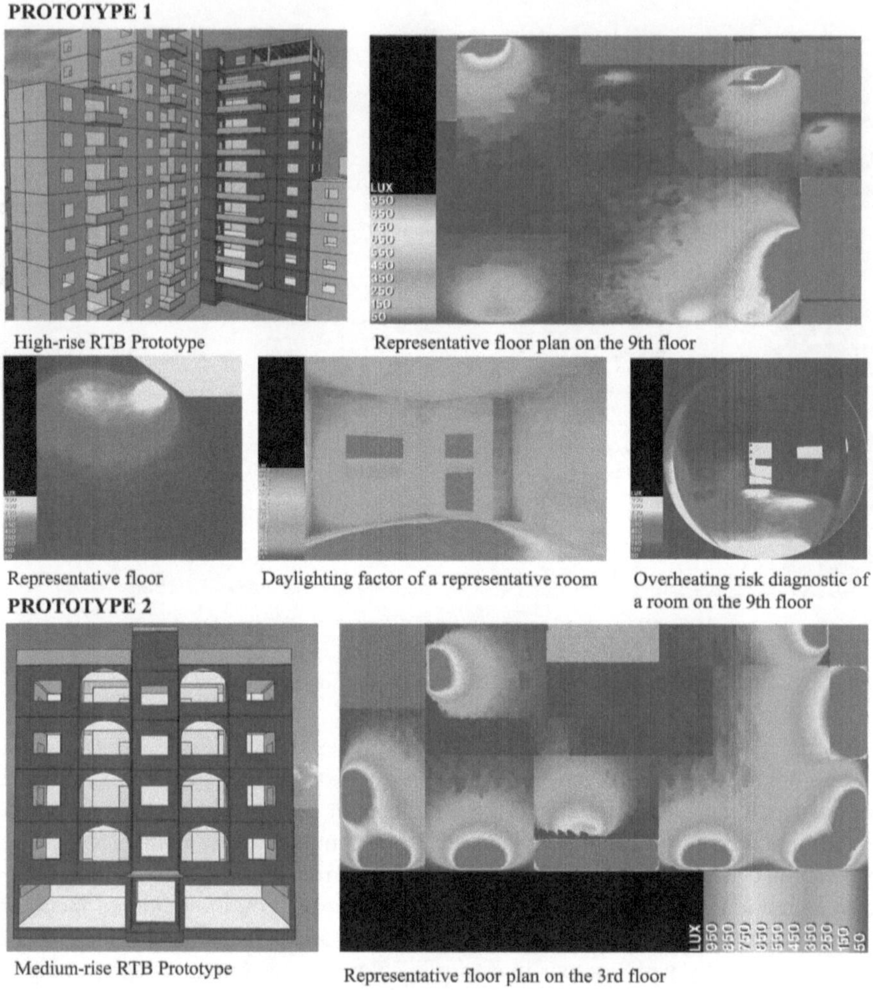

Figure 9.5 Daylighting analysis of a representative floor plan.

To summarise, it is suggested that using more than one method at a time would help achieving better results as some of these energy and environmental assessment methods cover whole range of issues considering energy rating and environmental impact assessment of a building.

Conclusions

As widely known, there is more than 20 years of research background that has produced broad understanding of the implications of the energy use and helped increasing the number of energy efficiency strategies and technologies with significant potential for energy savings in the UK. Building energy and environmental assessment methods described in this chapter are designed and developed to assess energy consumption of buildings and building designs against a range of environmental issues covering impacts on the environment.

Table 9.1 Comparison of existing energy and environmental assessment methods for buildings in the UK

Comparison issues	*BREDEM*	*SAP*	*BREEAM*	*NHER*	*STARPOINT*	*ENVEST*	*TOOLKIT*	*APACHE*	*TAS*
Produces an energy rating for a dwelling	No	Yes	Yes	Yes	Yes	Yes	Yes	Yes	Yes
Suitable for calculating SAP ratings	Yes	Yes	No	Yes	Yes	No	No	No	No
Provides a framework for the calculation of energy use in buildings	Yes	No	Yes	No	Yes	Yes	Yes	Yes	Yes
Estimates energy costs and potential savings in housing	Yes	Yes	Yes	Yes	Yes	Yes	Yes	Yes	Yes
When calculating energy costs, takes into account thermal insulation of the building envelope as well as the efficiency and control of the heating and hot water systems and the ventilation characteristics of the dwelling	No	Yes	No	Yes	No	Yes	No	Yes	Yes
Estimates the solar gain features of the building and the price of fuel used for space and water heating	No	Yes	No	Yes	Yes	No	No	Yes	Yes
Includes energy used for lighting and appliances	No	No	No	No	No	No	Yes	Yes	No
Consists of a worksheet that calculates the space and water heating energy requirement of a dwelling, and a rating that is based on fuel costs rather than primary energy use or CO_2 emissions	No	Yes	No	Yes	Yes	No	Yes	Yes	No
Provides simple means of reliably estimating the energy efficiency performance of dwellings	Yes	Yes	No	Yes	Yes	Yes	Yes	Yes	Yes
Enables the levels of energy efficiency in buildings to be compared	No	Yes	No	Yes	Yes	Yes	Yes	Yes	Yes
Additionally, taking into account CO_2 emissions resulting from energy use in dwellings; CFC and HCFC (hydro chlorofluorocarbon) emissions; natural resources and recycled materials; storage of recyclable materials; water economy; the ecological value of the site; ventilation; volatile organic pollutants of indoor origin; wood preservatives; man-made mineral fibres; asbestos and lead; lighting; smoke alarms; storage of hazardous substances	No	No	Yes	Yes	Yes	Yes	No	Yes	No

(*Continued*)

Table 9.1 (Continued)

Comparison issues	*BREDEM*	*SAP*	*BREEAM*	*NHER*	*STARPOINT*	*ENVEST*	*TOOLKIT*	*APACHE*	*TAS*
Provides guidance on ways of minimising the adverse effect of buildings on the global and local environments while promoting a healthy and comfortable indoor environment for inhabitants	No	No	Yes	No	Yes	Yes	Yes	Yes	Yes
Based on readily available and generally accepted information – establish criteria and standards going beyond those required by law and regulation	No	No	Yes	No	No	Yes	No	No	No
Covers issues such as global, local and indoor	No	No	Yes	No	Yes	Yes	Yes	Yes	Yes
Provides a common rating for the new and existing buildings and offers greater flexibility in benchmarking building portfolios	No	Yes	Yes	Yes	Yes	Yes	Yes	Yes	Yes
Encourages greater use of public transport and cycling, while discouraging the use of private cars for commuting	No	No	Yes	No	No	No	Yes	No	No
Provides the environmental assessment of materials as a major addition, green guide to specification and an environmental profiling system for building materials and components	No	No	Yes	No	No	Yes	No	Yes	No
Produces a simple scale rating for material's life expectancy and recyclability	No	No	Yes	No	No	Yes	No	Yes	No
Provides applicable ratings in both new and existing dwellings	Yes	Yes	Yes	Yes	Yes	Yes	Yes	Yes	Yes
Software based program	No	Yes	Yes	Yes	Yes	Yes	Yes	Yes	Yes

These methods are not only covering a large part in research background achieved, but also playing an essential role in increasing awareness of the importance of energy efficiency in buildings today and encouraging energy conscious building design throughout the country. In addition, some of these main assessment methods, such as SAP and NHER, have been added to the building regulations to improve energy efficiency in buildings and reduce carbon dioxide emissions caused by existing building stock. Nevertheless, these strategies and technologies have not been effectively transferred to the building design community, as recent studies suggested that the profile of the existing housing stock has remained with little change since 1996.

Against this background, my research advocates that more emphasis should be paid to encouraging property developers to design energy conscious buildings and property owners to implement energy efficiency improvements in existing buildings. Today, while being one of the most important issues, sustainable development is forever on the agenda and therefore this knowledge gap should be rapidly bridged. Moreover, the building regulations and associated policies could be more effective and efficient in this matter and the use of energy ratings should enable decision-makers to take energy efficiency into account on a rational basis when designing new dwellings or refurbishing existing ones. For building professionals, ratings can be used as a design tool to optimise energy efficient building design.

References

Adedoyin, F. F., & Zakari, A. (2020). Energy consumption, economic expansion, and CO_2 emission in the UK: the role of economic policy uncertainty. *Science of the Total Environment, 738*, 23–24. https://doi.org/10.1016/j.scitotenv.2020.140014.

Ahern, C., & Norton, B. (2020). A generalisable bottom-up methodology for deriving a residential stock model from large empirical databases. *Energy and Buildings, 215*. https://doi.org/10.1016/j.enbuild.2020.109886.

Aragon, V., James, P. A. B., & Gauthier, S. (2022). The influence of weather on heat demand profiles in UK social housing tower blocks. *Building and Environment, 219*, 35–37. https://doi.org/10.1016/j.buildenv.2022.109101.

Eyre, N., & Baruah, P. (2015). Uncertainties in future energy demand in UK residential heating. *Energy Policy, 87*, 641. https://doi.org/10.1016/j.enpol.2014.12.030.

Gupta, R., & Gregg, M. (2020). Domestic energy mapping to enable area-based whole house retrofits. *Energy and Buildings, 229*, 33–35. https://doi.org/10.1016/j.enbuild.2020.110514.

Hall, L. M. H., & Buckley, A. R. (2016). A review of energy systems models in the UK: prevalent usage and categorisation. *Applied Energy, 169*, 607. https://doi.org/10.1016/j.apenergy.2016.02.044.

Kane, T., Firth, S. K., & Lomas, K. J. (2014). How are UK homes heated? A city-wide, socio-technical survey and implications for energy modelling. *Energy and Buildings, 86*, 817. https://doi.org/10.1016/j.enbuild.2014.10.011.

Lee, T., Yao, R., & Coker, P. (2013). An analysis of UK policies for domestic energy reduction using an agent-based tool. *Energy Policy, 66*, 267. https://doi.org/10.1016/j.enpol.2013.11.004.

Li, X., Arbabi, H., Bennett, G., Oreszczyn, T., & Densley Tingley, D. (2022). Net zero by 2050: investigating carbon-budget compliant retrofit measures for the English housing stock. *Renewable and Sustainable Energy Reviews, 161*, 35–37. https://doi.org/10.1016/j.rser.2022.112384.

Loga, T., Stein, B., & Diefenbach, N. (2016). TABULA building typologies in 20 European countries—making energy-related features of residential building stocks comparable. *Energy and Buildings, 132*, 4. https://doi.org/10.1016/j.enbuild.2016.06.094.

Lomas, K. J., & Porritt, S. M. (2016). Overheating in buildings: lessons from research. *Building Research & Information, 45*(1–2), 1. https://doi.org/10.1080/09613218.2017.1256136.

Mckenna, R., Hofmann, L., Kleinebrahm, M., & Fichtner, W. (2018). A stochastic multi-energy simulation model for UK residential buildings. *Energy and Buildings, 168*, 470. https://doi.org/10.1016/j.enbuild.2018.02.051.

Natarajan, S., Padget, J., & Elliott, L. (2011). Modelling UK domestic energy and carbon emissions: an agent-based approach. *Energy and Buildings, 43*(10), 2602. https://doi.org/10.1016/j.enbuild.2011.05.013.

Oliveira, S., Marco, E., Gething, B., & Organ, S. (2016). Evolutionary, not revolutionary – logics of early design energy modelling adoption in UK architecture practice. *Architectural Engineering and Design Management, 13*(3), 168. https://doi.org/10.1080/17452007.2016.1267606.

Ozarisoy, B. (2022). Energy effectiveness of passive cooling design strategies to reduce the impact of long-term heatwaves on occupants' thermal comfort in Europe: Climate change and mitigation. *Journal of Cleaner Production, 330,* Article 129675. https://doi.org/10.1016/j.jclepro.2021.129675.

Ozarisoy, B., & Elsharkawy, H. (15 March 2019). Assessing overheating risk and thermal comfort in state-of-the-art prototype houses that combat exacerbated climate change in UK. *Energy and Buildings, 187*, 201–217, https://doi.org/10.1016/j.enbuild.2019.01.030.

Preston, S., Mazhar, M. U., & Bull, R. (2020). Citizen engagement for co-creating low carbon smart cities: practical lessons from Nottingham City council in the UK. *Energies, 13*(24), 121–123. https://doi.org/10.3390/en13246615.

Rosenow, J., Kern, F., & Rogge, K. (2017). The need for comprehensive and well targeted instrument mixes to stimulate energy transitions: the case of energy efficiency policy. *Energy Research & Social Science, 33*, 95. https://doi.org/10.1016/j.erss.2017.09.013.

Sandberg, N. H., Sartori, I., Heidrich, O., Dawson, R., Dascalaki, E., Dimitriou, S., Vimm, R.-T., Filippidou, F., Stegnar, G., Šijanec Zavrl, M., & Brattebø, H. (2016). Dynamic building stock modelling: application to 11 European countries to support the energy efficiency and retrofit ambitions of the EU. *Energy and Buildings, 132*, 26. https://doi.org/10.1016/j.enbuild.2016.05.100.

Simpson, K., Childs, P., & Whyte, J. (2021). *Sensitivity analysis of heating a typical UK dwelling and implications for retrofit design.* IOP Publishing. https://doi.org/10.1088/1742-6596/2042/1/012148.

Sousa, G., Jones, B. M., Mirzaei, P. A., & Robinson, D. (2017). A review and critique of UK housing stock energy models, modelling approaches and data sources. *Energy and Buildings, 151*, 66. https://doi.org/10.1016/j.enbuild.2017.06.043.

Sousa, G., & Robinson, D. (2020). Enhanced EnHub: dynamic simulation of housing stock energy systems. *Journal of Building Performance Simulation, 13*(5), 516. https://doi.org/10.1080/19401493.2020.1788641.

Strachan, N., Pye, S., & Kannan, R. (2008). The iterative contribution and relevance of modelling to UK energy policy. *Energy Policy, 37*(3), 850. https://doi.org/10.1016/j.enpol.2008.09.096.

Summerfield, A. J., Lowe, R. J., & Oreszczyn, T. (2010). Two models for benchmarking UK domestic delivered energy. *Building Research & Information, 38*(1), 12. https://doi.org/10.1080/09613210903399025.

Taylor, P. G., Upham, P., Mcdowall, W., & Christopherson, D. (2014). Energy model, boundary object and societal lens: 35 years of the MARKAL model in the UK. *Energy Research & Social Science, 4*, 32. https://doi.org/10.1016/j.erss.2014.08.007.

Trotta, G. (2018). Factors affecting energy-saving behaviours and energy efficiency investments in British households. *Energy Policy, 114*, 529. https://doi.org/10.1016/j.enpol.2017.12.042.

10
FEED FORWARD INTERVIEWS

Introduction

The survey questionnaire was conducted in order to explore the potential relationships between energy performance in both social housing estates and a number of factors, such as housing issues, occupants' energy consumption behaviour, occupancy patterns and socio-demographic characteristics. The questionnaire was split into four sections exploring the issues affecting social housing energy performance, such as housing conditions, energy use patterns and behaviour, attitudes of energy management applications, as well as occupants' socio-demographic characteristics. This chapter starts with a description of current conditions of the two social housing estates in a variety of aspects through presenting the general survey findings. The data analysis is reported with the compiled results of both case studies. Then, the correlation analysis was conducted towards the different variables. It is found that a number of energy consumption behaviours and socio-demographic characteristics potentially influence building energy performance. The research findings may help to form a set of evidence-based recommendations for policymakers to promote occupants' energy behaviour and improve retrofit delivery. It also helps develop recommendations for energy management application which provides behavioural intervention directly on the energy users' level. The detailed research findings from the questionnaire survey are demonstrated below according to the different sections and corresponding questions.

General Survey Findings

Basic Information and Housing Conditions

This section aims to draw an overview picture of the research sample with results by tackling a few key issues, such as their occupancies, tenancy status, energy bills and housing conditions. This information is fundamental for the data analysis.

Tenancy Status and Living Years

According to Figure 10.1, social renters form the majority of the participating households comprising 76% of overall tenancy status. Owner-occupied properties comprise 16% of the overall tenancy status and housing association renters only make up 9% of it.

DOI: 10.1201/9781003490364-13

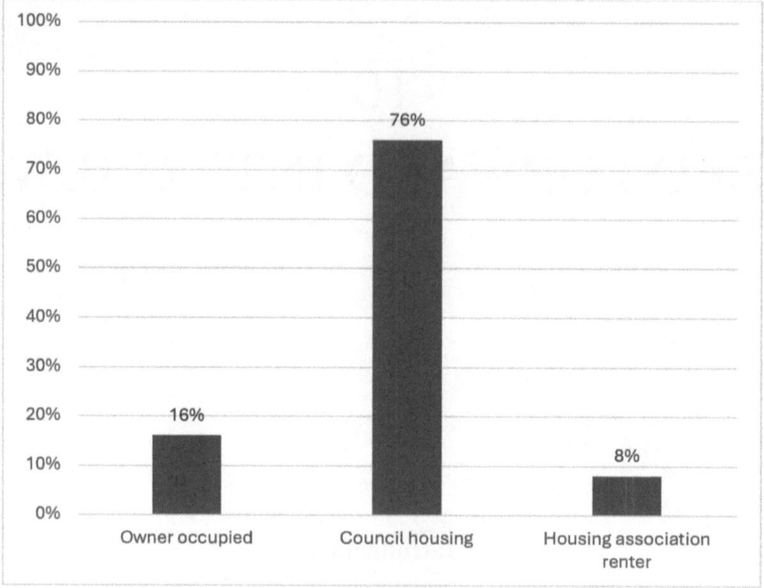

Tenancy status

Figure 10.1 Tenancy status.

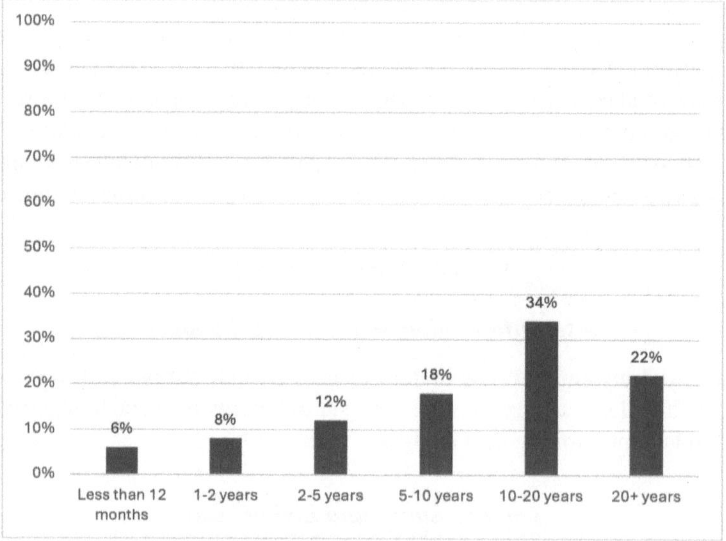

Tenancy status

Figure 10.2 Living years.

According to Figure 10.2, there are 6% of households living in their current property for 'less than 12 months', 8% of households living there for '1 up to 2 years', 12% of households living there for '2 to 5 years', 18% of households living there for '5 to 10 years', 34% of households are living there for '10 to 20 years', and 22% of households living there for 'more than 20 years'. The result shows an increasing trend of number of households along with the increase of the living years. Households living in their current properties for 'more than 10 years' form 56% of all respondents. De Castella (2013) stated that most people decide to rent a property only if they have to do it with relatively high mobility than the private rental sector. However, the distinct difference in the social rental sector indicates that financial issue is the most significant factor among social renters as they pay much lower rentals and are exempt from the costs of housing upgrades and retrofits.

Quarterly Electricity and Gas Bills

Occupants are also asked to provide their quarterly electricity and gas bills in the questionnaires. It is found that each household pays almost the same amount on electric and gas bills. In general, households' gas bills may be slightly higher due to high gas demands in the winter.

According to Figure 10.3, among the participants, 2% of them had their quarterly electricity bills between '£0–£49'; 20% of the households each paid their quarterly electricity bills between '£50–£99', '£100–£149' and '£150–£199'; 16% of the households' quarterly electricity bills were between '£200–£249'; 6% of them spent '£250–£299' on their quarterly electricity bills; 10% of them spent '£300–£349' on their quarterly electricity bills; and 6% of the households tended to spend 'more than £350' on their quarterly electricity bills. According to Figure 10.4, 22% of the participating households each paid their quarterly gas bills between '£50–£99' and '£100–£149';

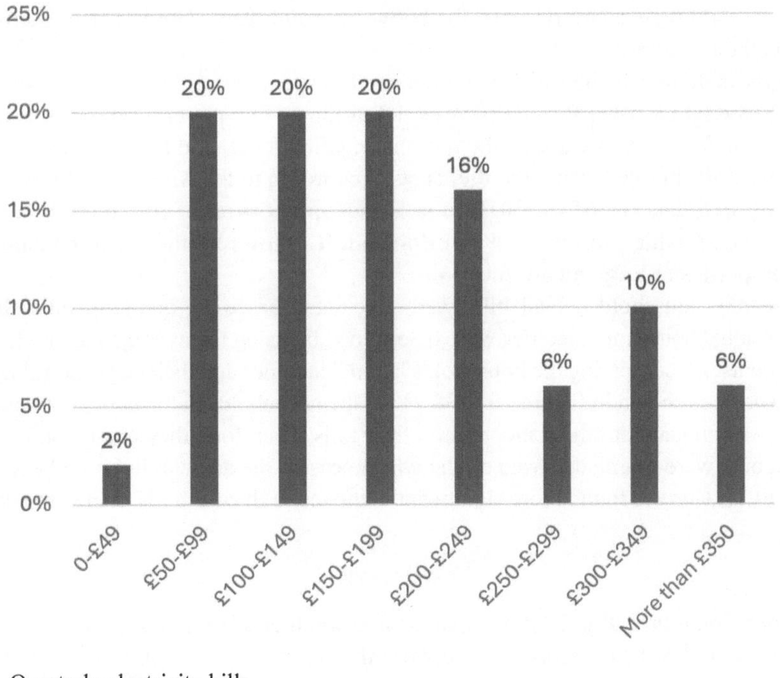

Figure 10.3 Quarterly electricity bills.

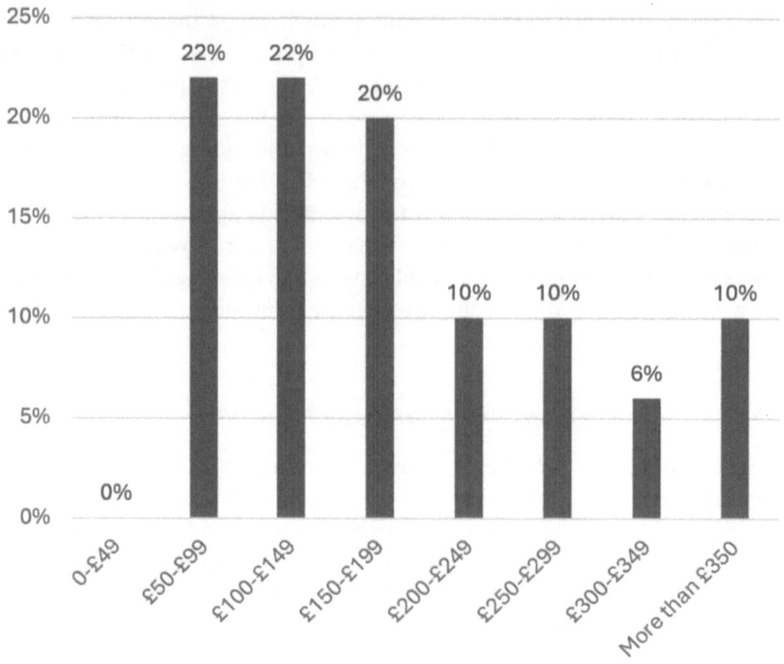

Figure 10.4 Quarterly gas bills.

20% of them paid '£150-£199' for their quarterly gas bills; 10% of the households paid '£200–£249' on their gas consumption; 10% of them paid their quarterly gas bills between '£250–£299'; 6% of the respondents paid their quarterly gas bills between '£300–£349'; and 10% of them paid 'more than £350' on their quarterly gas bills.

Fuel poverty is defined as households which need to spend more than 10% of their incomes on fuel. By 2013, the DECC (2013) re-defined its definition as fuel poverty households need to meet two criteria: their fuel costs are above the national median level, and their incomes are below the official poverty line after deducting the fuel costs. According to the Annual Fuel Poverty Statistics Report, 2018 (2016 data) (DBEIS, 2018), the highest proportion of fuel poverty appears in the private rental sector, which forms 19.4% in England. It follows by the social housing sector that 16% of the properties in England are fuel poor.

In the research, household annual bills were calculated based on quarterly bills shared by the respondents. Annual household incomes were determined by using the average income level indicated by the respondents. By comparing the households' annual incomes and their corresponding energy bills by each household, as shown in Figure 10.5. 38.7% of the households in the data sample need to spend more than 10% of annual household incomes on their bills. Therefore, they are at risk of fuel poverty. As the annual bills were estimated based on the winter season, the actual bills could be less. However, it shows that fuel poverty is found more significant in the research case study than at the national level.

Energy Advice

In the questionnaire, it investigated if occupants have obtained sufficient support on energy conservation from local authorities, energy suppliers or other sources. According to Figure 10.6, 40% of

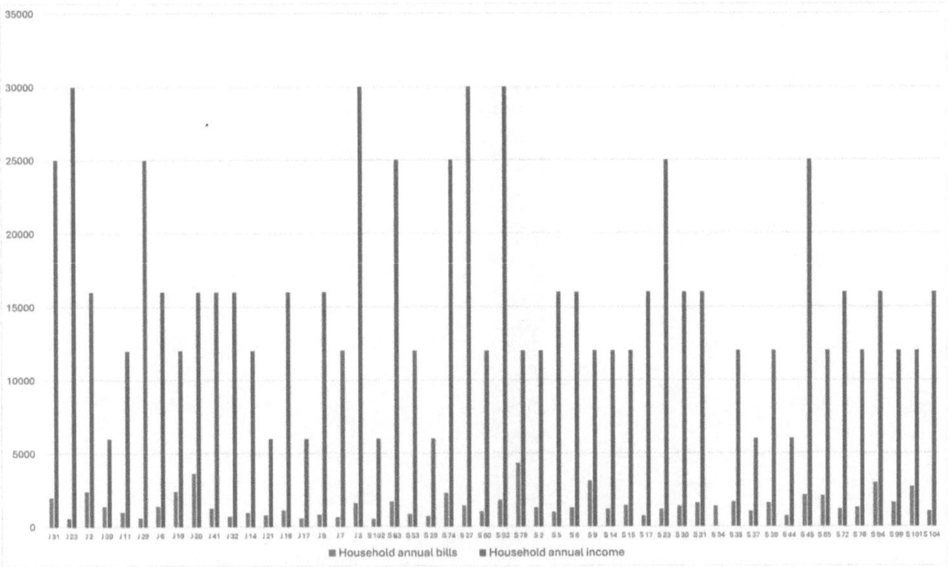

Figure 10.5 Household annual incomes and annual bills.

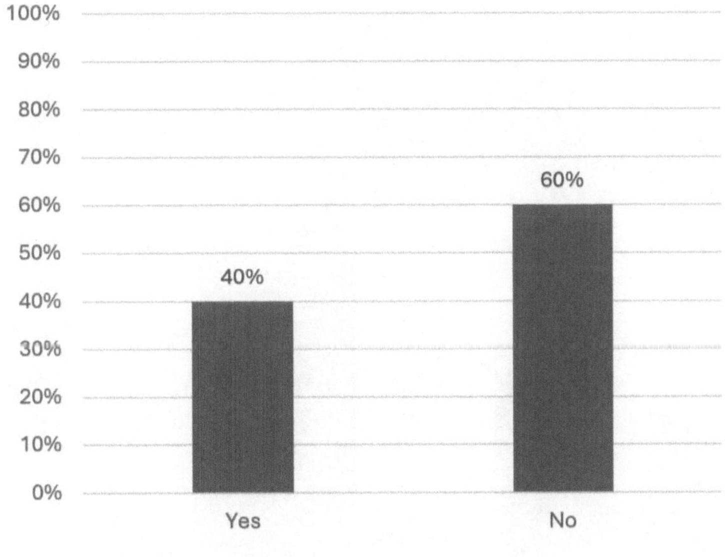

Figure 10.6 Have you received energy advice?

the households had previously received energy advice and 60% of them were not told how to save their energy from the energy companies or local authorities before.

The sources of energy advice received were indicated in Figure 10.7. Among the respondents who have received energy advice, 5% of them only received energy advice from 'the council'; another 5% of them received energy advice from 'both the council and the energy suppliers'; 5% of them received energy advice from 'the energy suppliers and other sources'; another 85% of

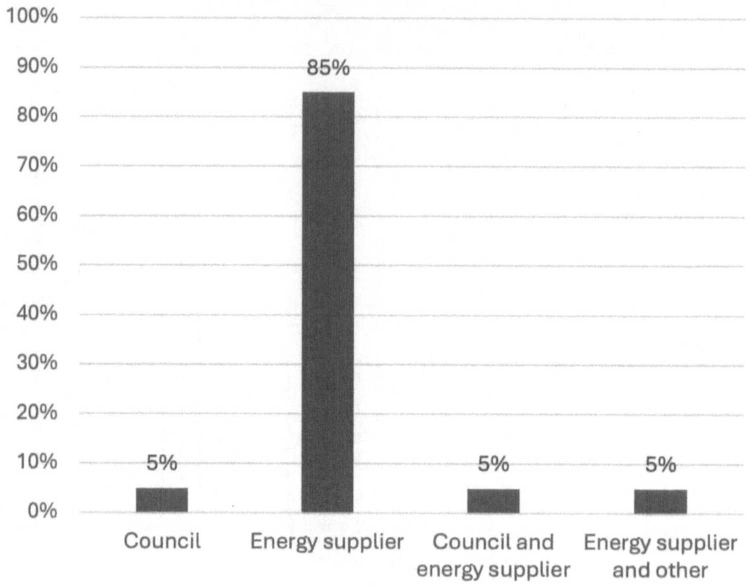

Figure 10.7 Where did you receive advice from?

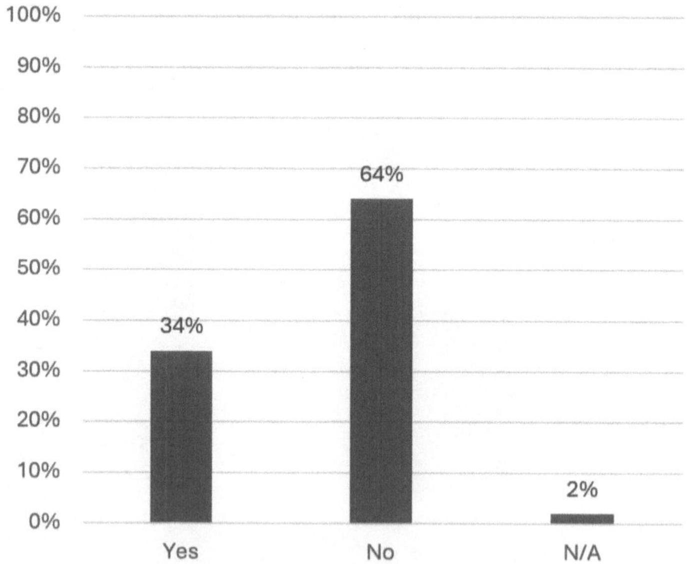

Figure 10.8 Have you changed suppliers/plans?

them expressed that they have received energy advice from 'the energy suppliers' only. It shows that energy company is currently taking the major responsibility on informing and educating occupants. However, local authorities, communities and other organisations need to work together to maximise the influence of energy conservation (Bogin et al., 2024).

According to Figure 10.8, 64% of the respondents had not considered changing energy suppliers or plans; 34% of the participants had previously changed their energy supplier/energy plans;

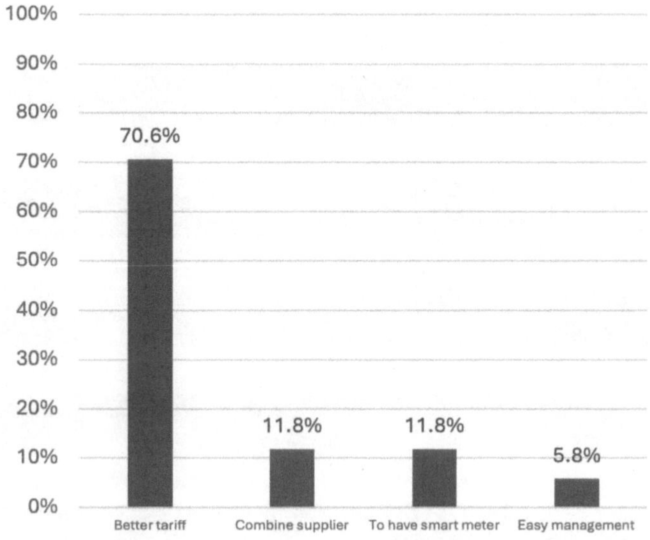

Figure 10.9 Why did you change it?

4% of them expressed that they wished to do so but had not started. It shows that the respondents who have previously changed energy suppliers or energy plans may have certain levels of energy awareness and may be willing to save energy. The reasons for changing energy tariffs and plans were also explored and are indicated in Figure 10.9. Among the respondents who have changed their energy plans or energy suppliers, 60% of them changed their energy plans or energy suppliers for 'better tariffs'; 34% of them did it for 'easy energy management' or 'installation of smart meters'; 6% of the occupants either 'plan to do it' or 'have tried but not successful'. The results indicate that financial saving is the dominant reason for occupants to make changes to current energy tariffs. Other reasons that could increase occupants' motivations to make changes also include meter upgrades and easy energy management (Colclough & Salaris, 2023). Occupants expressed that they would like to change their energy suppliers if the new supplier could offer them the installation of smart meters. Besides, they would like to do it if their gas and electricity supplies could be unified under the same suppliers to save hassles.

It was also believed that issues within the internal environment may have a significant impact on home energy performance. Hence, the issues that occurred inside the properties were also investigated in the questionnaires, such as cold, dampness, mould, draught and condensation. Participants were asked to report on how frequently they experienced those issues in their homes. Notably, a significant variance has been indicated in the results of the two cases under study, which is shown in Figure 10.10.

As shown in Figure 10.10, 'draught' and 'mould' are the most experienced housing issues as 66.6% of the respondents 'always or usually' experience them at their homes in Estate A; 'condensation' is the second most experienced housing issue at Estate A as 55.6% of the respondents expressed that they 'always or usually' experience it at their homes. On the other hand, 'cold' is the least problematic issue at Estate A as only 38.9% of the participants 'always or usually' feel 'cold' inside their dwellings; 'dampness' is also not too obvious at Estate A as 38.9% of the participants 'never or quite a few' feel it at their homes and only 33.3% of them will 'always' feel 'dampness' at homes.

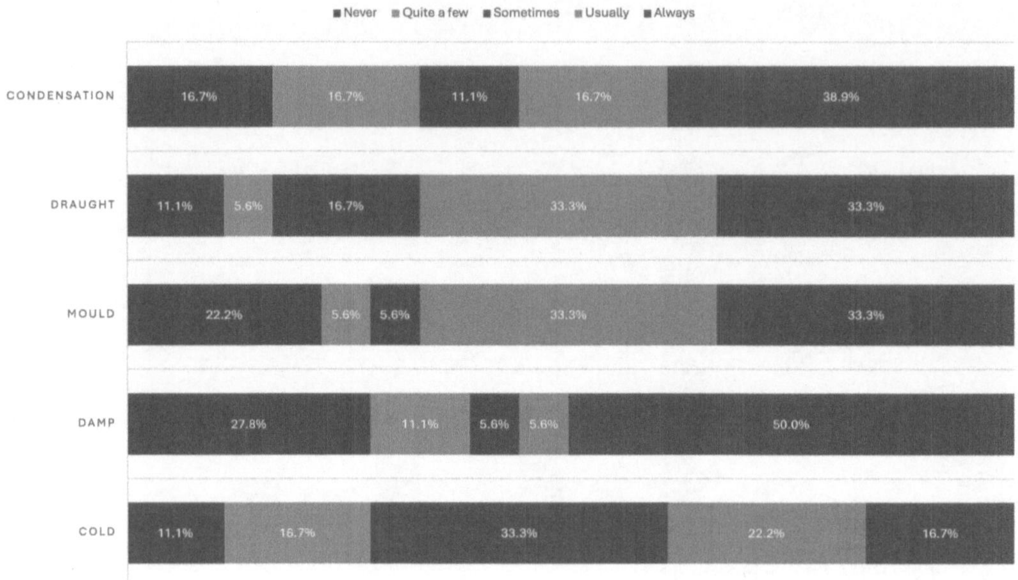

Figure 10.10 Housing conditions at Estate A.

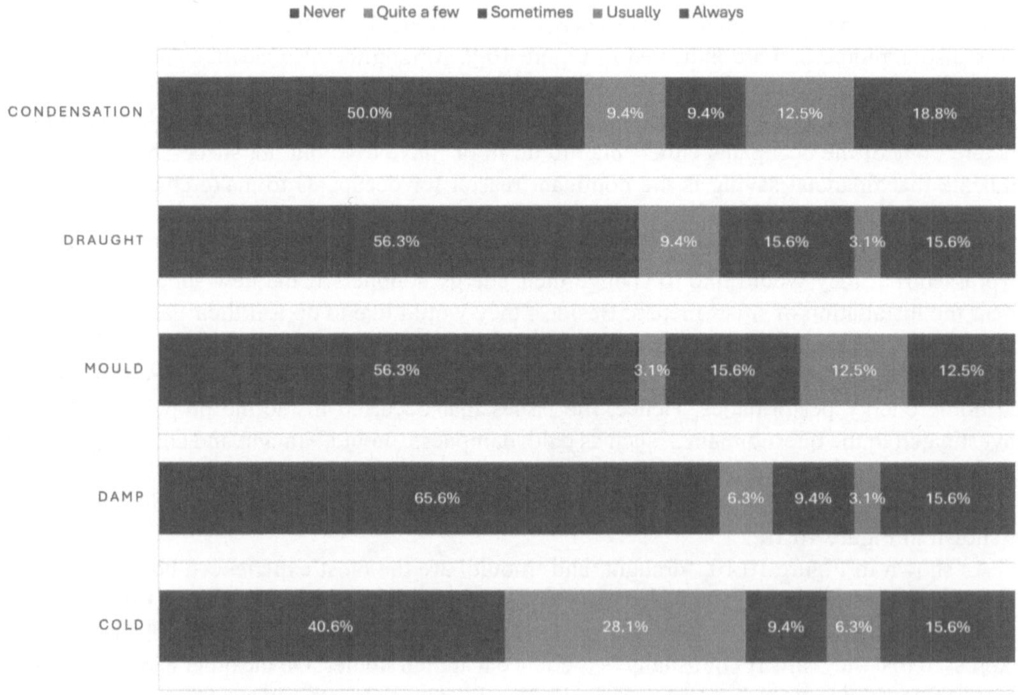

Figure 10.11 Housing conditions at Estate B.

On the other hand, as shown in Figure 10.11, 'condensation' is the most experienced housing issue at Estate B as 31.3% of the respondents expressed that they 'always or usually' experience it at their homes; the second problematic housing issue is 'mould' where 25.0% of the participants 'always' feel it at their homes; the percentages that respondents experience 'cold', 'dampness' and 'draught' are 21.9%, 18.7% and 18.7%. Hence, 'draught' and 'dampness' are the least problematic issues at Estate B.

Variations were indicated between the two social housing estates regarding the level of housing problems experienced: the average percentage of respondents who had 'never or quite a few' experienced all housing issues at Estate A is 28.92%. On the other hand, there were 65% of the respondents expressed that they had 'never or quite a few' experienced all housing issues at Estate B; the average percentage of respondents who had 'always or usually' experienced all housing issues at Estate A is 56.48% and 23.12% at Estate B. As a result, the housing conditions at Estate B is much better than at Estate A. That explains why the local authority decided to retrofit Estate A prior to Estate B. The distinct differences in housing conditions at both estates informed that the correlation analysis, which is related to 'housing problems', needs to be conducted separately.

Energy Consumption Behaviour and Energy Use Patterns

Occupants' Heating and Occupancy Patterns

In order to explore the implication of occupants' occupancy and heating patterns on home energy performance, the hours that occupants occupied their flats and the hours that heaters were turned on were asked in the questionnaire survey. As recording the occupancy and heating patterns requires a large amount of detailed information, the respondents could not remember the details in every room of their flats. As a result, the occupancy and heating patterns in the 'living room' as the most complete answer is used for the analysis in the future (Edalatnia & Das, 2024).

According to Figure 10.12, the occupants who spent '0–4 hrs' in the living room during the weekdays comprise 17.8% of the overall respondents; the ones spending '5–9 hrs' in the living room during the weekdays comprise 24.4% of the overall respondents; the ones spending '10–12 hrs' in the living rooms during the weekdays comprise 20.0% of overall respondents; the rest of 17.8% of respondents would spend 'more than 12 hrs' in their living rooms. On the other hand, their occupancy patterns in the living room during weekends were also explored. According to Figure 10.13, the occupants who spent '0–4 hrs' in the living room during the weekdays comprise 13.3% of the overall respondents; the ones who spent '5–9 hrs' in the living room during the weekdays comprise 26.7% of the overall respondents; the ones spent '10–12 hrs' in the living rooms during the weekdays comprise 37.8% of overall respondents; the rest of 22.2% of respondents would spend 'more than 12 hrs' in their living rooms. The results showed different occupancy patterns in the living room between weekdays and weekends. The occupancy is generally higher during the weekends than the weekdays. It may imply that the employed family members spend less time at their homes during weekdays but more time during weekends.

Apart from respondents' occupancy patterns, their heating patterns in the living room during weekdays and weekends were also explored. According to Figure 10.14, the occupants who would turn on the heaters for '0–4 hrs' in the living room during the weekdays comprise 30.5% of the overall respondents; the ones that kept the heating turned on for '5–9 hrs' in the living room during the weekdays comprise 39.1% of the overall respondents; the ones with heating turned on for ', 10–12 hrs' in the living rooms during the weekdays comprise 26.1% of overall respondents; the rest of 4.3% of respondents would turn on the heaters for 'more than 12 hrs' in their living rooms.

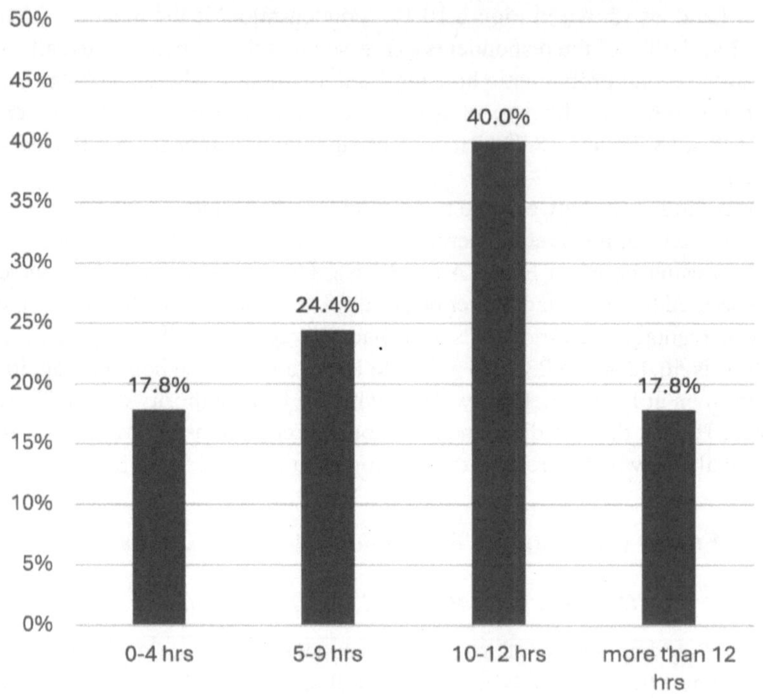

Figure 10.12 Hrs occupied in weekdays (living room).

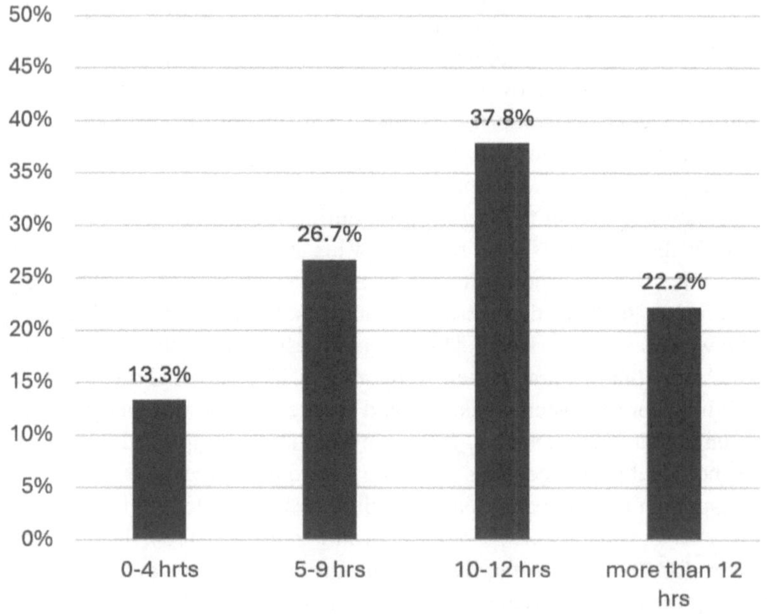

Figure 10.13 Hrs occupied in weekends (living room).

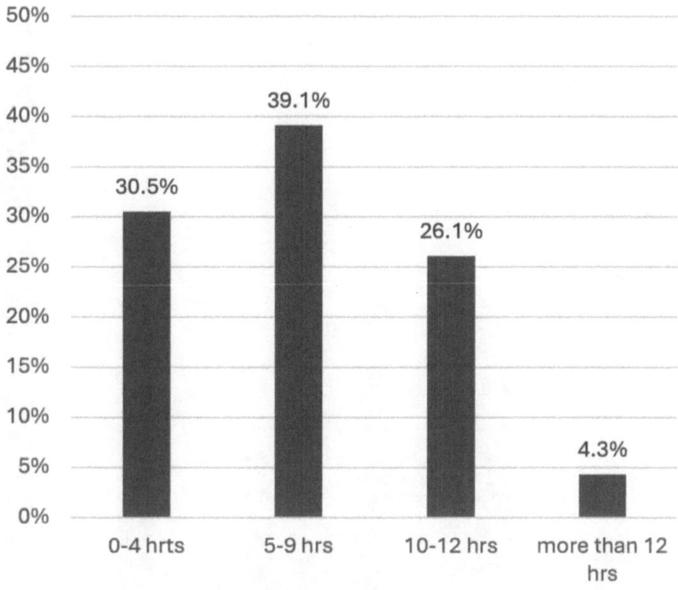

Figure 10.14 Hrs heaters on in weekdays (living room).

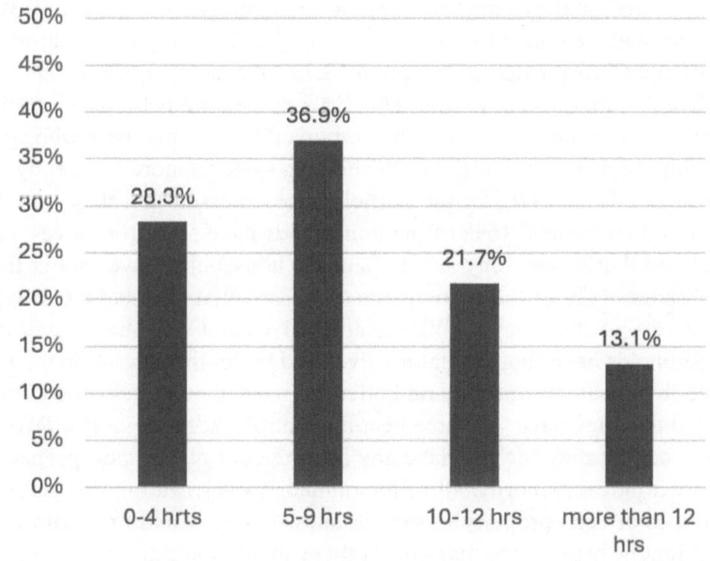

Figure 10.15 Hrs heaters on in weekends (living room).

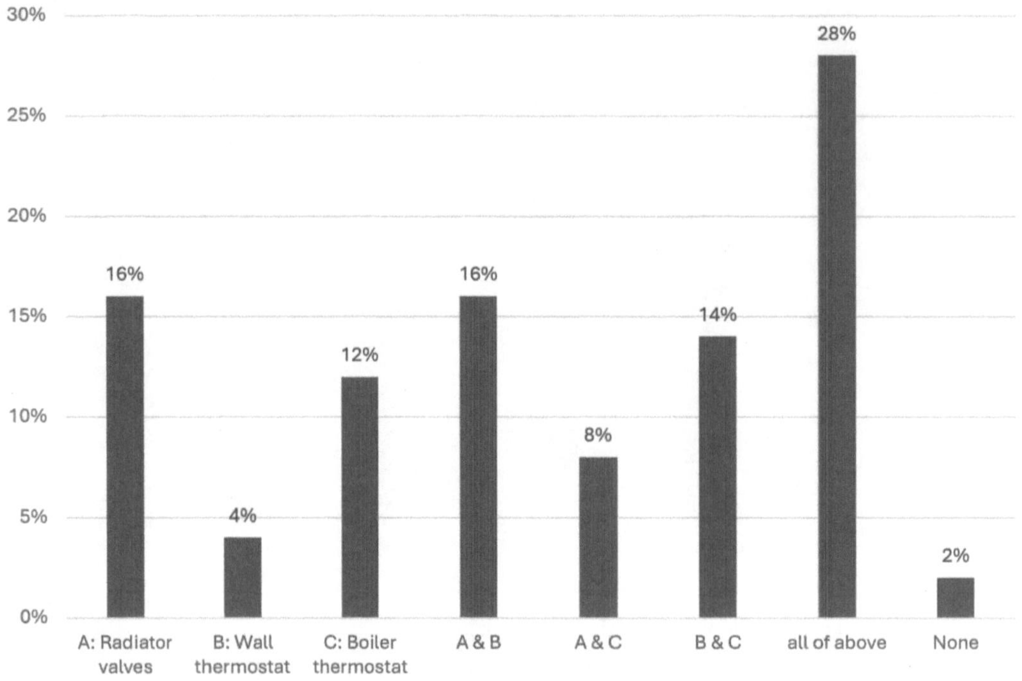

Figure 10.16 What heating controls do you have at home?

According to Figure 10.15, the occupants who would turn on the heaters for '0–4 hrs' in the living room during the weekdays comprise 28.3% of the overall respondents; the ones that kept the heating turned on for '5–9 hrs' in the living room during the weekdays comprise 36.9% of the overall respondents; the ones with heating turn on for '10–12 hrs' in the living rooms during the weekdays comprise 21.7% of overall respondents; the rest of 13.1% of respondents would turn on the heaters for 'more than 12 hrs' in their living rooms. The different heating patterns are indicated between weekdays and weekends. It may imply that the employed family members who spend more time at their homes during weekends will also use the heating systems more frequently.

According to Figure 10.16, 32% of the participants expressed that they only have one type of heating control in their homes: 16% of the households have 'radiator valves' only, 4% of the households have 'wall thermostat' only and 12% of the households have 'boiler thermostat' only at their homes. Besides, 38% of the participants expressed that they have two types of heating controls at homes: 16% of them have 'both radiator valves and wall thermostat' in their properties, 8% of the households have 'both radiator valves and boiler thermostat' in their properties and 14% of them have 'both wall thermostat and boiler thermostat' at homes. In addition, 28% of the households claimed that they have 'all three heating controls' at their homes. Two per cent of the participants expressed that they 'do not have any heating controls' in their properties. According to the discussions with local authority, all of the radiator valves, wall thermostat and boiler thermostat were equipped at each property of both estates. Except small proportion of them whose heating controls might be broken, the majority of them should have all of them in working conditions. The results indicate that respondents generally do not have sufficient knowledge regarding the functions of heating systems. Hence, the heating controls might be ignored when operating the homes (Ho-Kim & Jun-Moon, 2024).

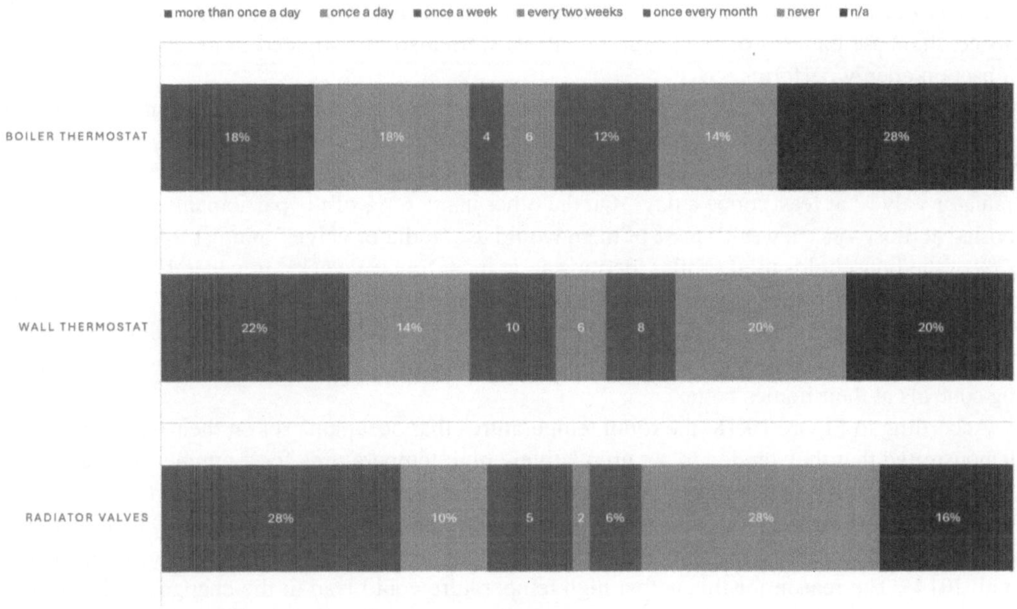

Figure 10.17 How often do you use your heating controls?

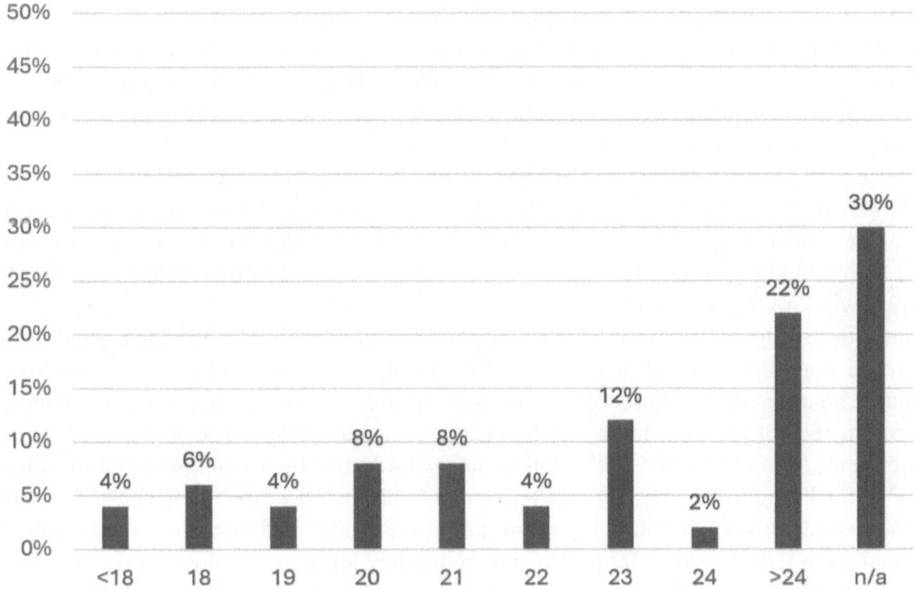

Figure 10.18 What temperature do you set thermostat?

It is noticed that the majority of the respondents knew that they have heating controls at home. Hence, their use patterns of the heating controls were also investigated in order to explore their impact on energy performance.

Occupants were asked how frequently they use each heating control at their homes. According to Figure 10.17, 36% of the respondents used their 'boiler thermostat' at least 'once a day'; 36% of the participants would use their 'wall thermostat' at least 'once a day'; and 38% of them used 'radiator valves' at least 'once a day'. On the other hand, 64% of the participants used 'wall thermostat' at most 'once a week'; 64% of them would use 'radiator valves' at most 'once a week'; and 62% of the households used 'boiler thermostat' at most 'once a week'. In general, around 36.67% of respondents used their heating controls at least once a day, which may imply that occupants understood the significance of the heating controls to a certain level. This indicated that occupants need to be further educated and advised in order to appreciate the functions and operations of heating controls at their homes better.

According to Figure 10.18, the room temperatures that occupants set on their wall thermostats demonstrated that they tended to set up relatively high temperatures for a number of reasons that were explored in the later sections. Forty-eight per cent of the respondents tended to set their wall thermostat 'more than 21°C', which may not be necessary. The respondents (24%) who have set their room temperature at more than 24°C could encounter the cardiovascular disease risk (Saeki et al, 2014). The reason for this is that high temperature could lead to the change in human body signs, such as blood pressure, blood viscosity and heart rate. The high indoor temperature in winter also enlarges the temperature differences between inside and outside. Hence, it increases the risk of cardiovascular diseases (Gasparrini et al., 2015). Recent reports (Gram-Hanssen, 2014) also stated that the main causes of high heat consumption are relatively high indoor temperatures, extensive infiltration due to poor building physical conditions and hot water over-consumption.

It is believed that opening the windows will effectively increase the air circulation within the properties and improve indoor air quality and comfort in a natural approach (Stazi, 2017). Besides, it may also help to solve the condensation, mould and dampness experienced inside. However, opening windows for longer hours may reduce the indoor temperature thus requiring more heating (Stazi, 2017). The frequency of opening the windows in the winter and the hours the windows are opened have also been investigated.

According to Figure 10.19, 28% of the participants expressed that they 'always' open their windows in the winter for a certain time; 24% of the occupants would 'usually' open their windows in the winter; 26% of them would 'sometimes' open their windows in the winter; 14% of them opened their windows 'quite a few times' in the winter while 8% of the occupants would 'never' open their windows in the winter.

According to Figure 10.20, 4% of the respondents would never open their windows in the living room when they were at home; 46% of the occupants tended to leave their windows open 'less than 2 hours' in the living room; 8% of them left their windows open for '2 to 4 hours' in the living room; 16% of the households would leave their windows open for '4 to 6 hours in average in winter in the living room; 6% of the households would leave their windows open for 'more than 6 hours' in the living room; and 20% of them could not remember how many hours they tended to leave their windows open in the living room during the winter. Although some occupants did not like opening the windows very frequently, the hours they left their windows open were relatively long. On the other hand, although some occupants may open their windows every day in the winter, they tend to leave their windows open for a very short time. These different patterns of opening the windows will significantly affect their energy consumption and issues experienced (Berneiser et al., 2023).

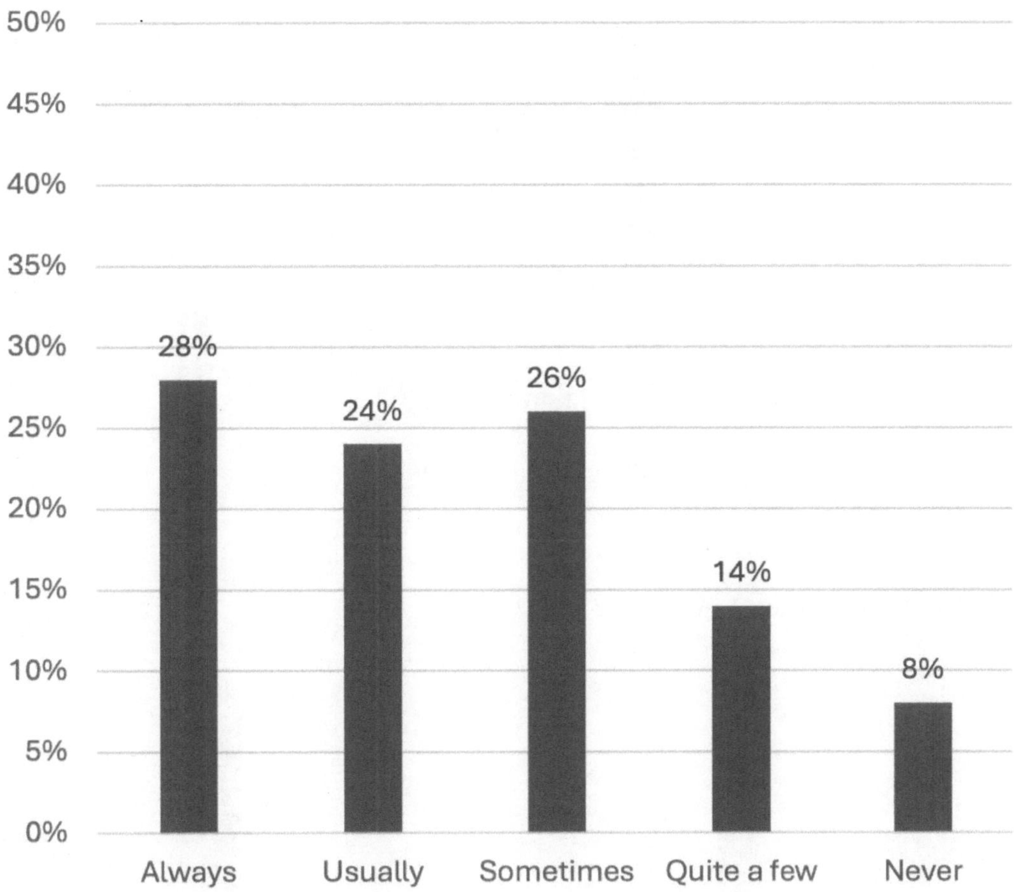

Figure 10.19 Frequency of opening the windows in the winter.

According to Figure 10.21, 38% of the participants expressed that they 'always' turned on the extractor fans when taking a shower; 2% of them would 'usually' turn on the extractor fan when taking a shower; 10% of the occupants would 'sometimes' turn on the extractor fans; 2% of them would turn on the extractor fan 'quite a few times' when taking a shower; and 48% of the participants 'never' turned on the extractor fans when taking a shower which may dramatically increase the possibilities of mould, condensation and dampness in their properties.

A trickle vent is a line of small slots in the window to allow air to filter into the indoor spaces even when the window is closed (Designing buildings, 2015). Occupants can control it when needed. 58% of the participants expressed that they were familiar with the purpose of the trickle vent. However, 42% of them did not know the purpose of it. Among the occupants who are familiar with the purpose of trickle vents, they are asked how the trickle vents in their homes are operated.

According to Figure 10.22, 40% of the participants 'always' kept the trickle vents open; 2% of them 'usually' kept it open; 10% of the occupants would keep it open 'sometimes'; 2% of them only kept it open for 'quite a few times'; 20% of participants 'never' kept the trickle vents open; and 26% of them 'did not answer' as they have no idea what the trickle vent is. Therefore, they either leave the trickle vents open or close them all the time.

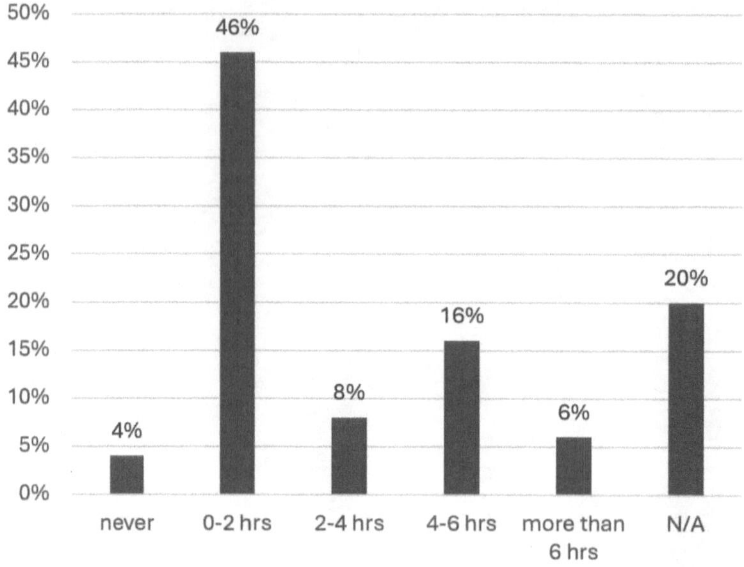

Figure 10.20 Hours windows opened in the winter?

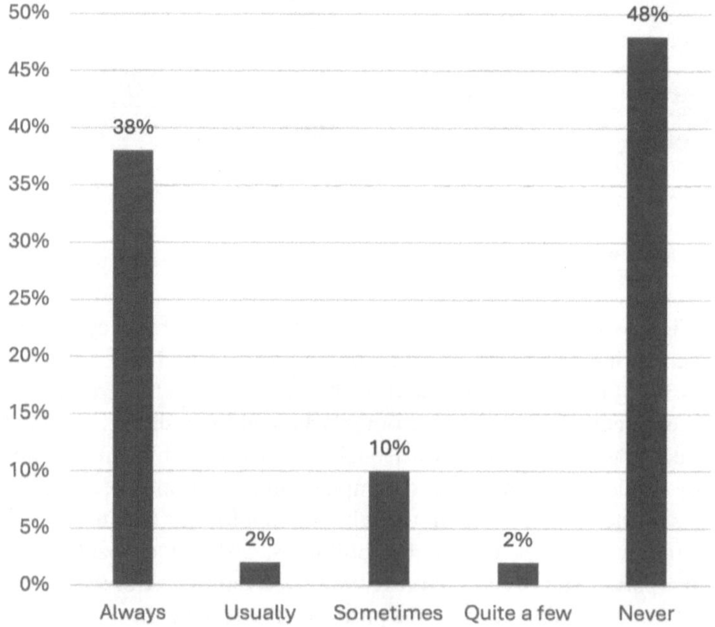

Figure 10.21 How often do you turn on the extractor fan?

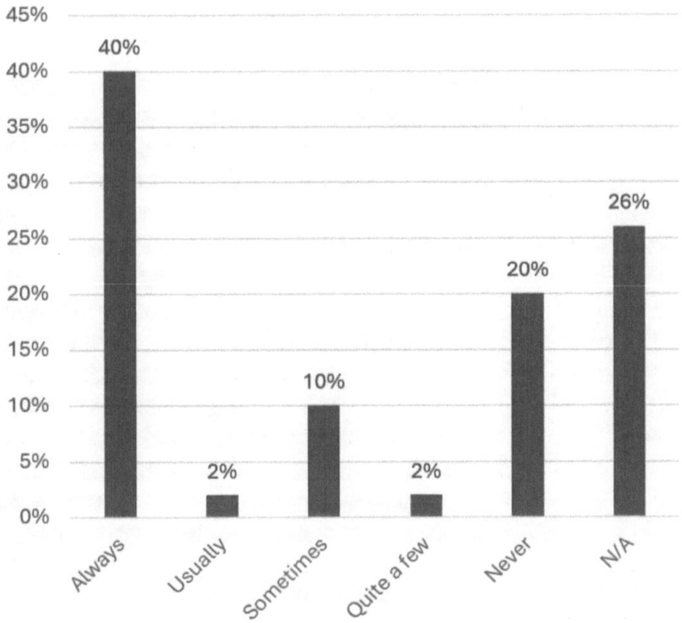

Figure 10.22 How often do you keep trickle vents open?

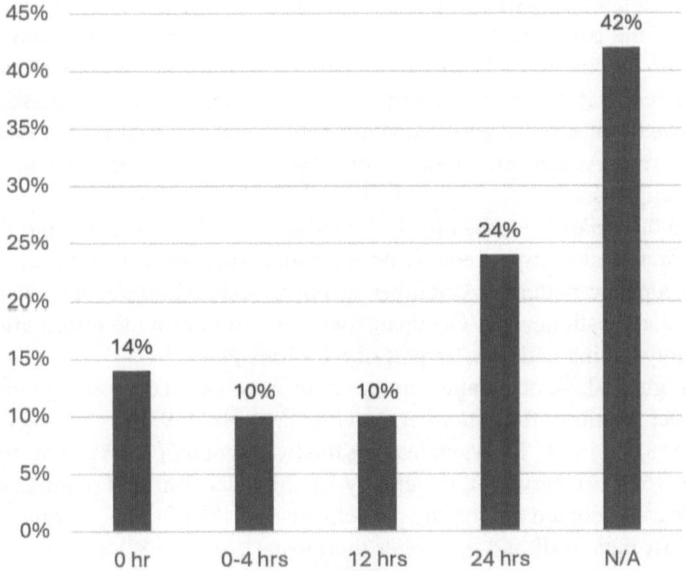

Figure 10.23 How long do you keep the trickle vents open?

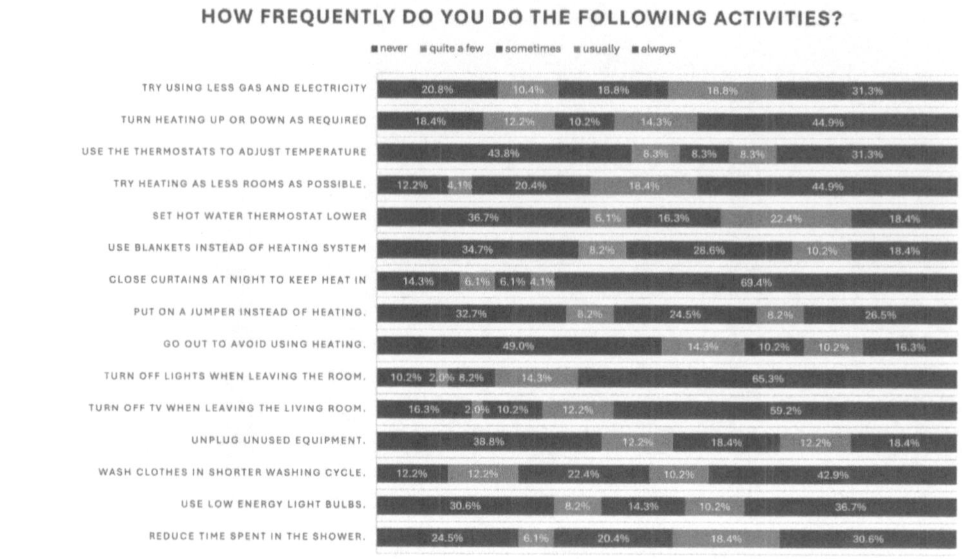

Figure 10.24 Occupants' preferred energy saving behaviours.

According to Figure 10.23, 14% of them have 'never' kept it open; 10% of them would keep the trickle vent open for 'less than 4 hours'; 10% of participants would leave it open only during the daytime for '12 hours'; 24% of occupants would always keep the trickle vent open all day '24r hours'; and 42% of the participants 'did not answer' as they did not deal with the trickle vents and consequently do not know if it is closed or open. Occupants' behaviours of using trickle vents need to be considered as it is used to maximise the indoor temperature while keeping the spaces ventilated. It showed that 82% of respondents would leave it closed all the time, leave it open all the time or ignore it. As a result, guidance on how to properly operate trickle vents should be provided to the occupants.

It is recognised that apart from the physical conditions of the properties and the uses of appliances, a series of factors also impact energy performance, such as occupants' energy consumption behaviour and occupancy patterns. A number of energy-related approaches were proposed and incorporated into the questionnaires. Occupants were asked to provide information on how often they do a few energy-saving activities as part of their lifestyle.

According to Figure 10.24, occupants 'always or usually' saved their energy through more conventional ways, such as 'turn off lights when leaving the room' (79.6%), 'close curtains at night to keep heat in' (73.5%), 'turn off TV when leaving the living room' (71.4%) and 'try heating as less rooms as possible' (63.3%). However, the energy-saving behaviour that requires more knowledge and skills was not well adopted among the participants: 52.1% of the occupants would 'never' or 'quite a few' 'adjust their wall and hot water thermostat', and 42.8% of them would 'never' or 'quite a few' 'set hot water thermostat lower'. Besides, people did not want to save energy by compromising their comfort, which is why 63.3% of the participants would 'never' or 'quite a few' 'go out avoid using heating', 40.9% of them would 'never' or 'quite a few' 'put on a jumper instead of heating' and 42.9% of them would 'never' or 'quite a few' 'use blankets instead of heating system'.

Occupants may over-consume energy due to a lack of knowledge and aspiring to achieve a more comfortable living environment. According to Figure 10.25, respondents' perception of energy use was explored. As a result, 46% of the respondents believed they use more energy than

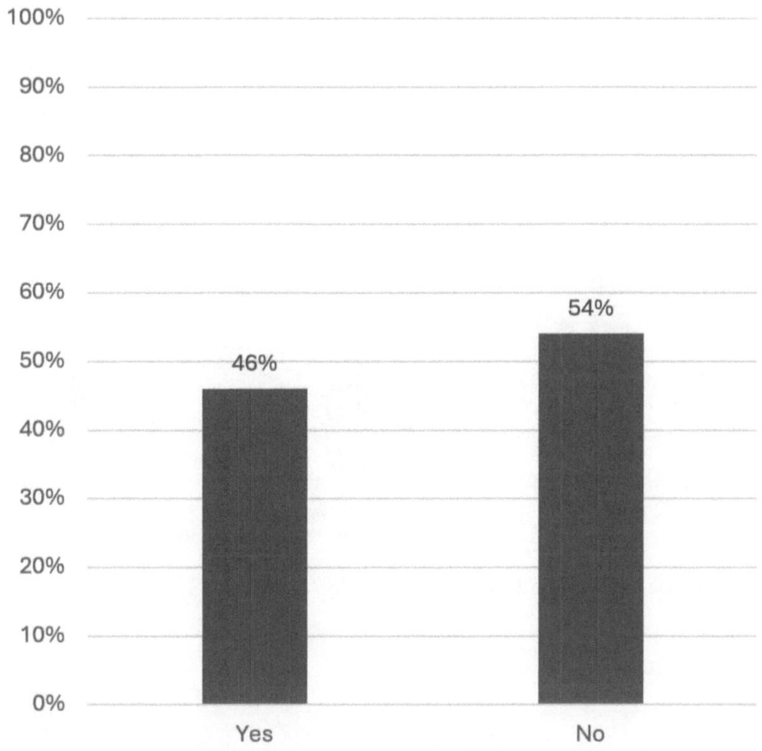

Figure 10.25 Do you use more energy than you should?

they should to have a comfortable home. On the other hand, 54% of the respondents do not think they used more energy than they should.

To further explore the reason behind this, the occupants were asked why they thought they used more energy than they should through an open-ended question. Their answers were categorised with different themes and shown in Figure 10.26. Among the respondents who thought they used more energy than they should, 36% of them claimed that the reason of energy over-consumption is due to the draught and cold experienced in their flats; 9% of them needed to provide warmer homes to their children; 27% of them did it for their own comfort and satisfaction; 5% of the respondents had to consume more energy due to illness, while 23% of the respondents did not answer this question. This shows that occupants may consume more energy for children's needs, personal comforts and illness (Abrahamse & Steg, 2009; Neuhoff et al., 2011). The poor housing conditions also led to energy over-consumption.

Attitudes Towards Saving Energy Through Energy Management Applications

According to Figure 10.27, 86% of the respondents asserted that they or their household members have smartphones; only 14% of the households did not have them.

According to Figure 10.28, 32% of the respondents expressed that they are 'very uncomfortable' to use application on smartphones; 4% of them felt 'not very uncomfortable' to use applications on their smartphones; 12% of them feel 'neutral' to use application on their smartphones; 12% of them felt 'somewhat comfortable'; and 40% of the respondents felt 'very comfortable'

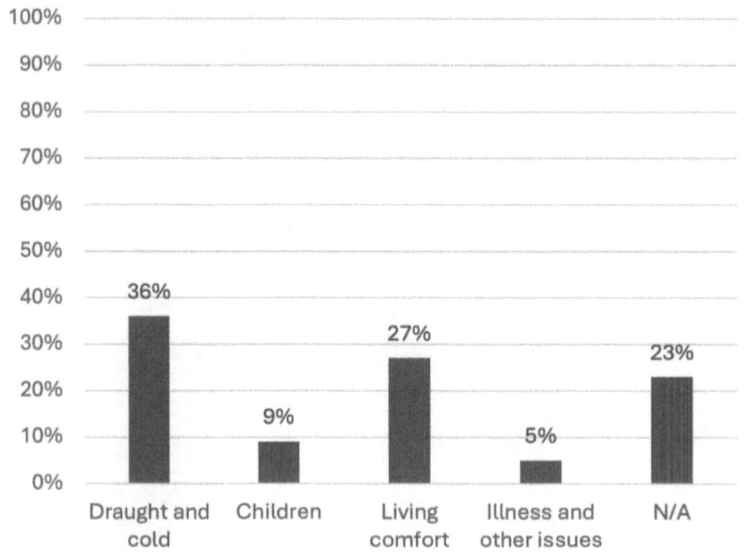

Figure 10.26 If you use more energy than you should, why?

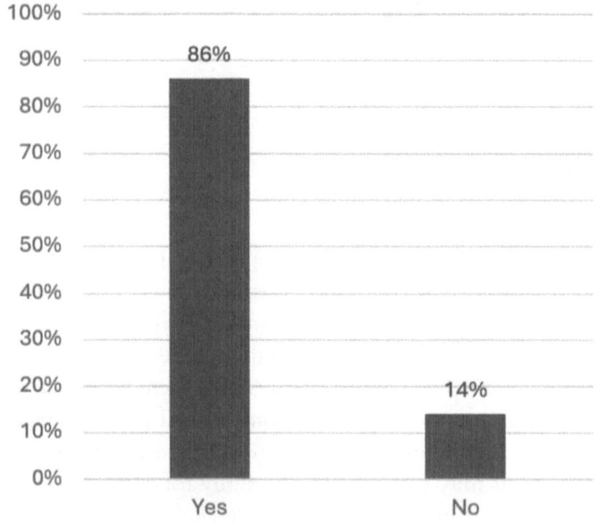

Figure 10.27 Do you have a smart phone?

to use application on their smartphones. It was noticed that there was a large proportion of the respondents did not feel very comfortable to use the application. It is because of the large proportion of elderlies and carer-needed persons living in the social housing estates. They would find it very difficult to adopt smart technologies and applications.

According to Figure 10.29, only 20% of the households have smart meters installed in their homes and 80% of the households do not have smart meters installed. It is noticed that although the roll-out of smart meters was set up by the government and committed to finish by 2020 by the

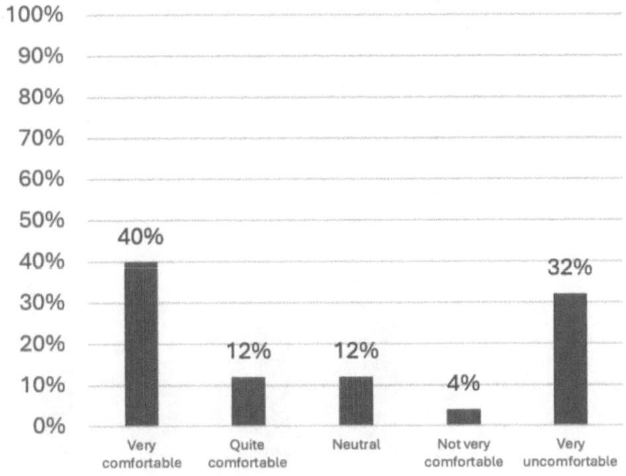

Figure 10.28 How comfortable do you like to use app on it?

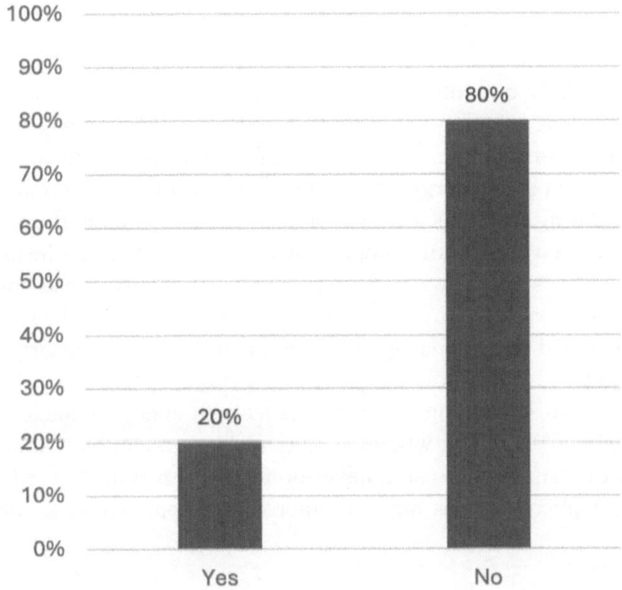

Figure 10.29 Do you have smart meter at home?

energy companies, it is still a long way to equip smart metering devices in all homes, especially social housing and housing association estates. Blackman (2017) stated the pace of smart meter implementation and its implications of fuel poverty for social housing are behind the schedule.

According to Figure 10.30, among the ten respondents who have smart meters installed at their homes, 10% of them 'always' read smart meters and adjust energy use accordingly; 30% of them 'usually' did it; 30% of them will do it 'sometimes'; 10% of the respondents only read the smart meter and adjusted energy use 'quite a few' times; and 20% of them never did it although the smart meters are at homes. It shows that 70% of the respondents who have smart meters installed at their

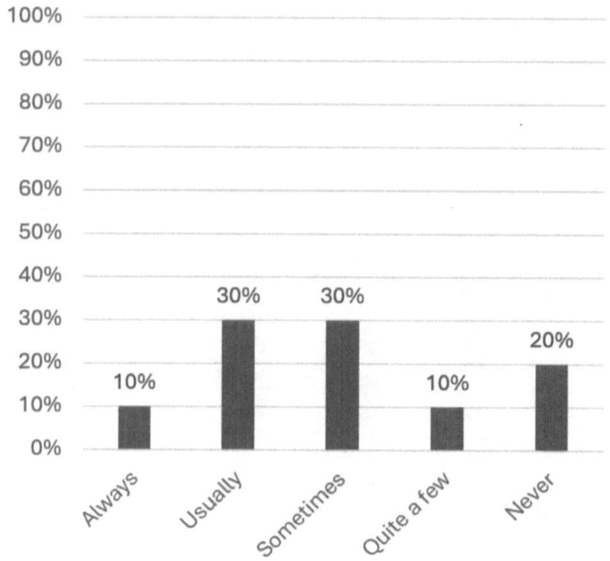

Figure 10.30 How often do you read smart meters?

homes will 'always, usually or sometimes' adjust their energy uses according. It is much more than the frequency of using 'heating controls' and 'extractor fans' among all of the respondents regardless of smart meter installation. Therefore, the result shows the positive influences of smart meters on improving occupants' energy behaviour. The low implementation rate of smart meters was further explored during the focus group interview and discussed in the following section.

It has been found that the number of occupants who have smart meters installed to monitor their energy use accordingly was very low. The smart control of energy consumption through energy management applications was also investigated. According to Figure 10.31, 10% of the respondents indicated that they had energy management applications on their smartphones while 90% of the respondents did not.

According to Figure 10.32, among the respondents with energy management applications on their smartphones, only 20% of them would 'sometimes' give attention to it and adjust energy use accordingly, the rest of them (80%) would never do that. The results showed that the adoption of energy management applications has not been thoroughly approached to the energy end user's (occupant) level.

Preferred Energy Saving Features in an Energy Application

This section aims to explore which features on the energy management applications may mostly attract people's attention and raise their engagement. A series of potential application aspects for energy saving were incorporated according to the literature review and current energy management applications in the market.

According to Figure 10.33, occupants rated their preferences of a number of application features, which they felt would help them to reduce home energy consumption in the case study of social housing estates. Fifty respondents all have answered this question. Although many of them have not experienced those application features, they all expressed their opinion about them.

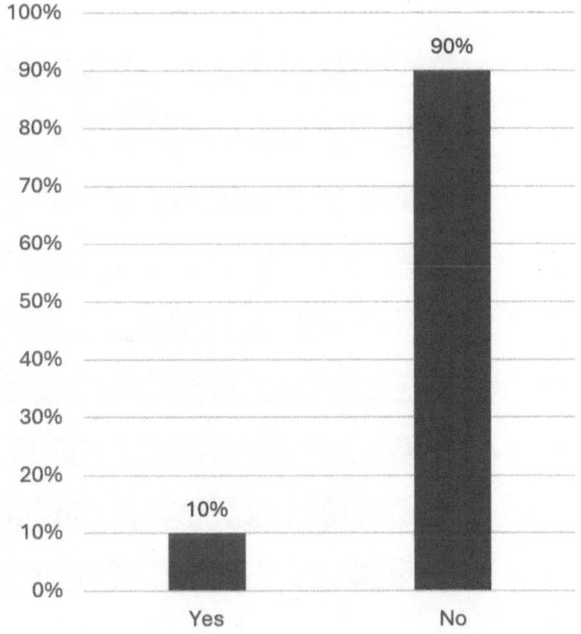

Figure 10.31 Do you have energy management app on your phone?

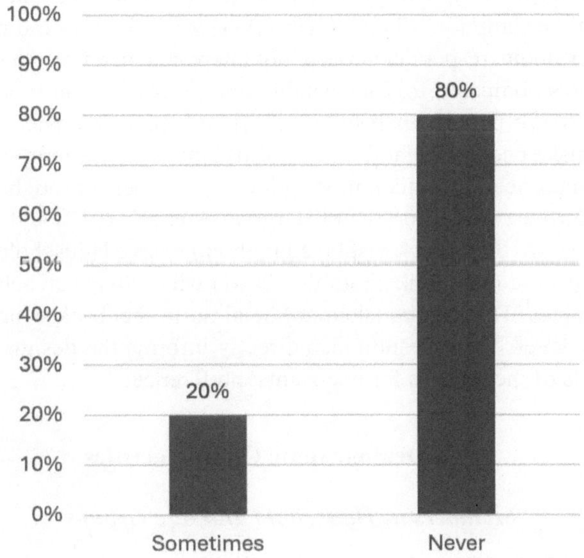

Figure 10.32 How often do you use it?

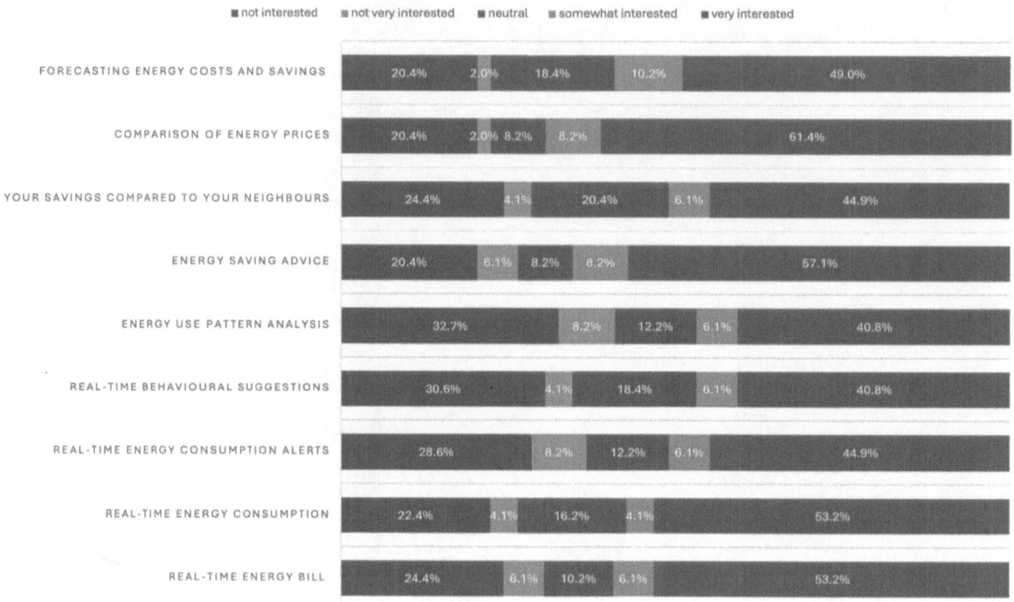

Figure 10.33 Preferred smart application aspects.

Respondents had expressed they are 'very interested' in some application features, such as 'comparison of energy prices' (61.4%), 'energy saving advice' (57.1%), 'real-time energy bill' (53.2%) and 'real-time energy consumption' (53.2%). However, some features did not raise interest from the participants. For example, respondents were 'not interested' in a few other application features, such as 'energy savings compared to your neighbours' (24.4%), 'real-time behavioural suggestions' (30.6%), 'energy use pattern analysis' (32.7%) and 'real-time energy consumption alerts' (28.6%). In order to make occupants familiar with those innovative energy-saving aspects, energy suppliers and the council need to initiate more pilots within their boroughs. In the research produced by Ehrhardt-Martinez et al. (2010) and Hargreaves et al. (2010)'s, they both indicated that households with comparative feedback displayed in the Intensive Household Density (IHDs) tend to use less energy as people may think about the reason why others can achieve low energy consumption than themselves. This can be taken as social norm feedback which is normally carried out at the community level. This question also directly informs the design of the focus group to probe more into details of the reasons for occupants' preferences.

Socio-Demographic Characteristics

Members in Household and Age Groups

According to Figure 10.34, 15 participating households are occupied by one person, making up 30% of overall households; 24 households are occupied by two or three persons, making up 48% of overall households; nine participating households have four or five people living in the properties, which corresponds to 18% of overall households; and two of the households are occupied by six or seven people, each making up 4% of the overall households.

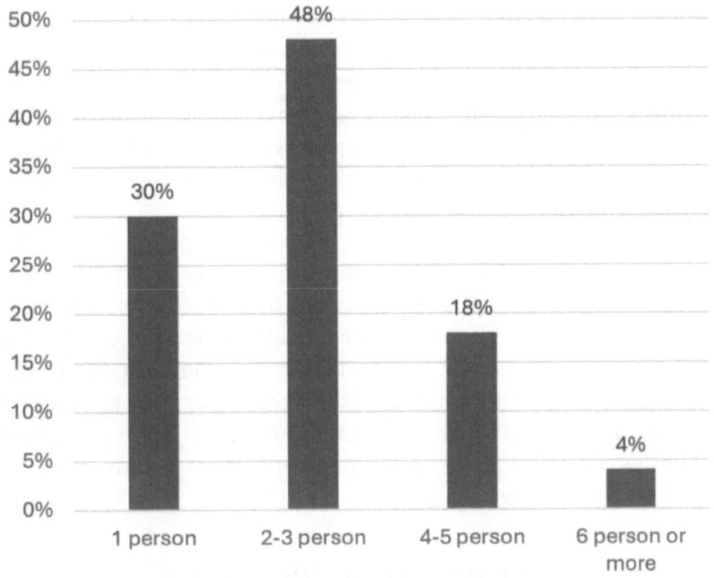

Figure 10.34 Number of family members.

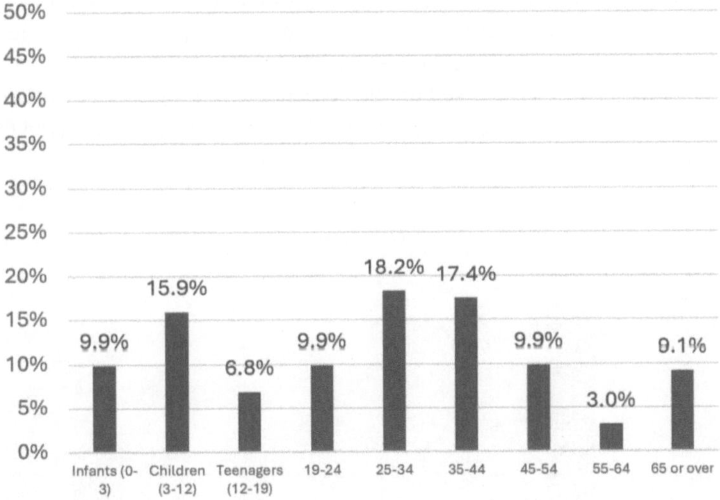

Figure 10.35 Age band of family members.

The age band of each family member was also explored in the questionnaire. According to Figure 10.35, there are 13 'infants (0–3 yrs)', making up 9.9% of overall family members in participating households; 21 'children (3–12 yrs)', making up 15.9% of overall members; nine 'teenagers (12–19 yrs)', which corresponds to 6.8% of overall family members; 13 adults between '19 and 24 yrs', which corresponds to 9.9% of overall family members in participating households; 24 adults between '25 and 34 yrs', making up 18.2% of overall family members; 23 adults between '35 and 44 yrs', making up 17.4% of overall family members in participating households; 13 adults between '45 and 54 yrs', making up 9.9% of the overall family members; four adults between '55 and 64 yrs', making up 3.0% of overall family members; and 12 adults with '65 yrs or over', which corresponds to 9.1% of the family members in the participating households. From the result, the dominant age groups in participating households are 'infants and children' (25.8%), 'adult (25–34)' (18.2%) and 'adult (35–44)' (17.4%), while 'adult (65 or over)' comprised 9% of the total household residents.

At the national level, properties occupied by one person comprise 42.3% of the overall social housing sector, properties occupied by two to three people comprise 37.7% of the overall sector, the ones with four to five people comprise 16.6% of the overall social housing sector, and properties with six or more persons comprise 3.4% of the overall sector (DCLG, 2017). In this research, the composition of one person properties is a little lower, the composition of two to three person properties is slightly higher, and the compositions of other categories are more or less the same as in the DCLG report. It showed that the research sample generally represents the condition of social housing estates in terms of number of occupants. The representativeness of the data sample is further discussed in the following section.

Representativeness of the Research

Household Economic Status

Occupants' economic status was investigated in the questionnaire survey as it may closely relate to energy-saving awareness and the degree of knowledge and perceptions. Therefore, it may impact occupants' energy use patterns and affect home energy consumption.

The economic status statistics are based on the percentage of households with different options for employability. According to Figure 10.36, households with full-time employed family members

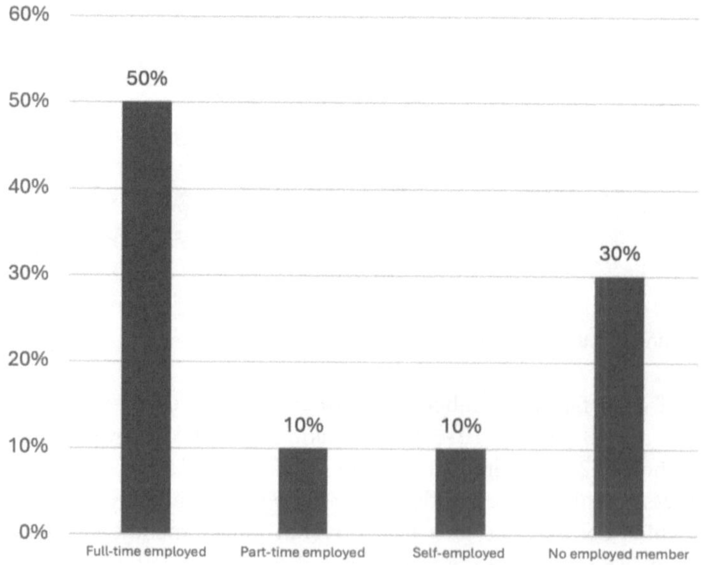

Figure 10.36 Economic status.

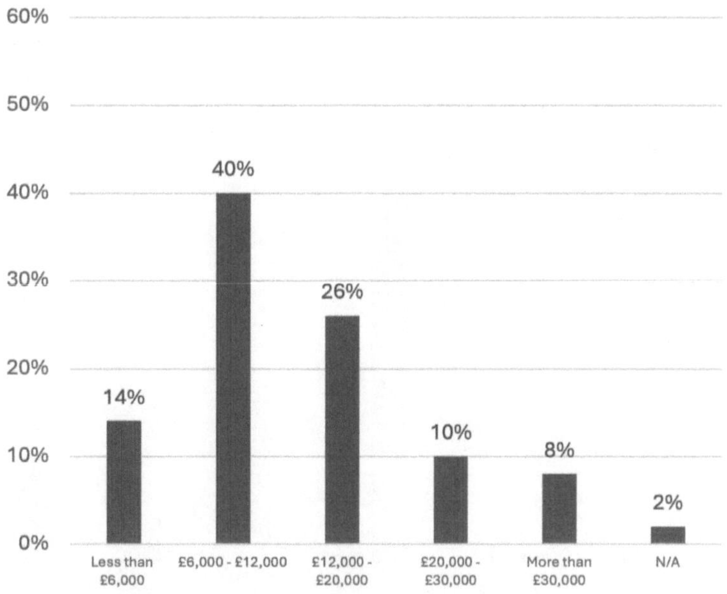

Figure 10.37 Total household income level.

comprised 50% of overall participating households; 10% of the households had part-time employed family members only; 10% of them had self-employed family members only, and 30% of them indicated that all of their family members were not able to work. According to Figure 10.37, the majority (54%) of the households earned less than £12,000 per year, which is considered lower than the UK's minimum wage rate for full-year work. This also increases the possibility of fuel poverty. Among these households, 17.5|% of them earned less than £6,000 total annual income.

The mapping of participants' ethnic backgrounds helps to explore if it is correlated with energy performance in the later stage. According to Figure 10.38, 20 respondents are Black or Black British, comprising 40% of all respondents; 15 respondents are white British, making up 30% of overall participants; 6 respondents are Pakistani or Bangladeshi, making up 12% of overall participants; 5 of them are from other ethnicity background, including Arab, Turkish and Jamaican, which corresponds to 10% of respondents; two respondents are Asian or Asian British, making up 4% of all respondents; one respondent is mixed ethnicity, comprising 2% of the overall participants; and one respondent is Indian, comprising 2% of the data sample. It shows that 'White' and 'Black or Black British' are the dominating ethnic groups in this case study.

Level of Education and Health Condition

Occupants' education level may influence perceiving the knowledge and environmental awareness. According to Figure 10.39, 30% of the respondents had no degree; 20% of them held secondary (GCSE) qualifications; 10% of the participants held A or AS level qualifications; 6% of them held diploma, teaching or nursing; 2% held Level 5 certificate; 24% of them expressed that they had obtained degree with honours; and 8% of them held Master's or PhD degree qualifications.

Finally, participants' health conditions were investigated as they may potentially impact home energy consumption in the case study of social housing estates. According to Figure 10.40, 22%

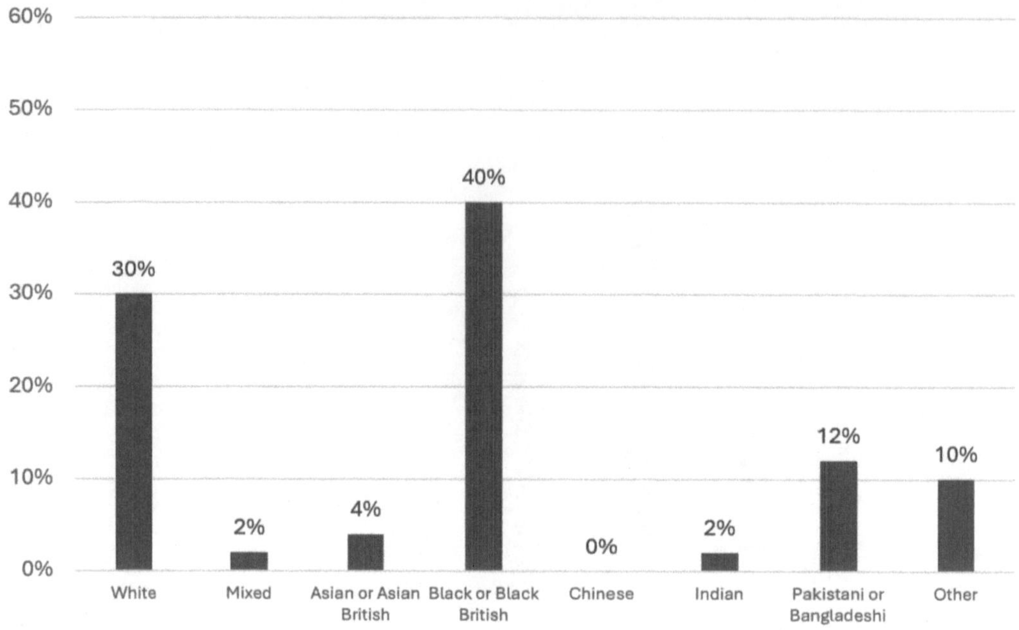

Figure 10.38 Ethnic background.

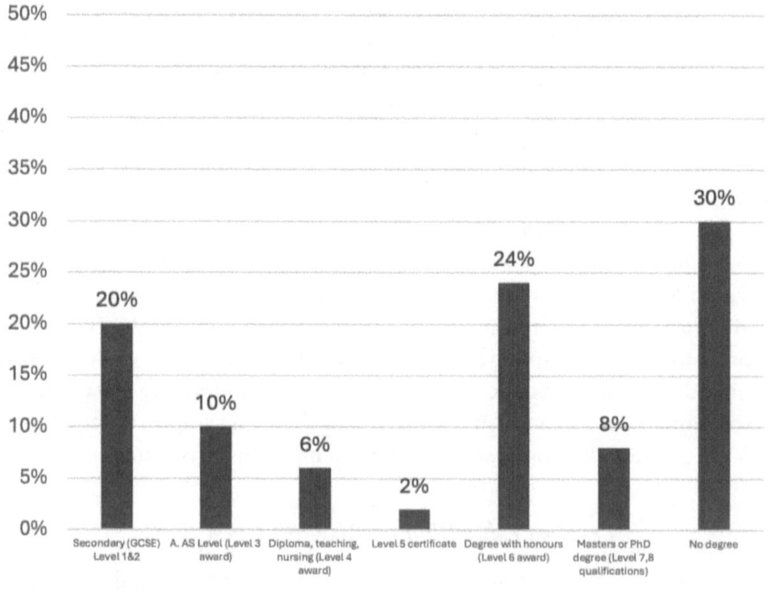

Figure 10.39 Level of education.

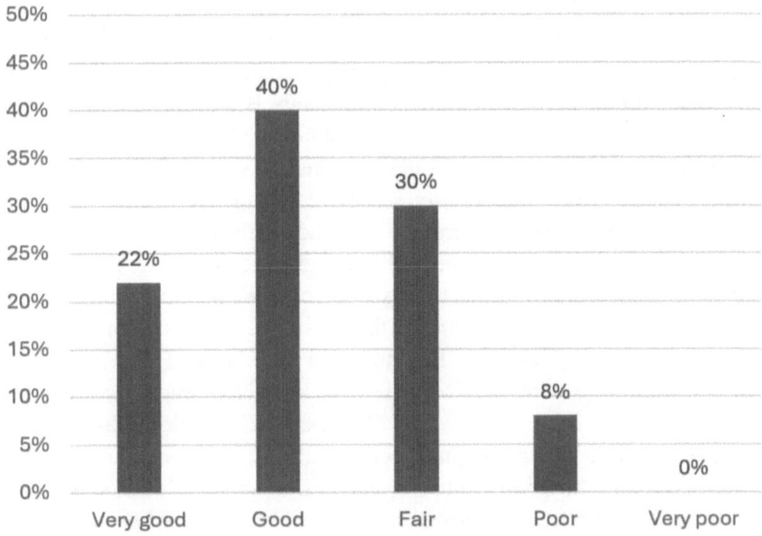

Figure 10.40 Health condition.

of the respondents believed they were very good in health; 40% felt that they were good in their health; 30% of the respondents felt fair; 8% of the participants felt that they were in poor health condition; and nobody felt very poor in health. In general, 62% of the participants felt either 'very good' or 'good' concerning their health conditions, while 38% rated their health as 'fair' or 'poor'.

Summary of the General Findings

The previous section demonstrated the general survey findings of both case studies in the aspects of housing characteristics, occupants' energy consumption behaviour, occupancy patterns and their socio-demographic information. It draws the current conditions of two particular social housing estates in one of the London boroughs and how occupants operate their homes. The important findings are summarised below.

As stated from the above sections, the majority of the households are currently renting the properties either through the council (76%) or housing association (8%). Although local authority allows occupants to purchase these properties by offering relatively cheap prices, only a small proportion of them (16%) bought these properties. However, it does not mean that occupants do not have a long-term living plan in social housing as 52% of respondents expressed that they have been living in their homes for more than five years. With its distinct characters, the mobility of renters in these two social housings is low. According to DBEIS (2018), the average quarterly electricity bill applying standard electricity tariffs supplied by home energy supplier was £154.75 per quarter in 2017. The average quarter gas bill was £157.5 in 2017. In the research, the majority of the households (68%) tend to spend their electricity and gas within a reasonable range (£0–£99 and £100–£199) in the winter. Occupants also tend to spend the same on both electricity and gas bills. However, 30% paid their quarterly electricity and gas bills in between £200 and £400. Some of them (2%) even paid more than £400 for their quarterly bills, which is unrealistically high. The reason for that will need to be further investigated. Around half of the respondents (46%) expressed that they have previously received energy advice and this is mainly from energy companies (90%). The council also provided energy support (10%) to the local occupants. Only one-third of them have changed their energy suppliers/energy plans before for financial savings and other promotions such as the installation of smart meters. In addition, a huge variation of housing conditions was identified between Estate A and Estate B and occupants were more satisfied with their living conditions at Estate B rather than Estate A. This also helps to explain the reason that the council decided to carry out retrofit interventions at Estate A first. It also informed that the further correlation analysis was conducted separately regarding these variables.

Although 34% of the respondents expressed that they only have one heating control at their homes among radiator valves, wall and boiler thermostats; according to the local authority, most of them have all heating controls equipped unless some of these heating controls are not in service. It showed that occupants may not be aware of the existence of these heating controls, and thus ignore using them in their daily routines. Besides, less than 20% of them will use their heating controls at least 'once a day', which may contribute to inefficient energy consumption. Fifty-two per cent of the respondents will 'always' or 'usually' open their windows during the winter, which could help to relieve 'mould', 'dampness' and 'condensation'. Forty-six per cent of them will open their windows for less than 2 hours in winter for each time but some of them (6%) will keep their windows open for more than 6 hours or always keep it closed (24%). These behaviours may lead to energy over-consumptions and more problematic housing issues as using heating controls more frequently may effectively reduce heating demands. Lack of ventilation may cause mould and condensation in the flats and consequently impact on indoor environment. It is worth mentioning that 46% of the respondents will never open extractor fans at home when taking showers. Local authorities need to make sure extractor fans are installed at every home. Additionally, trickle vents play an important role in indoor living environments where 40% of respondents tend to 'always' leave it open and 20% of them tend to 'never' open it. Another 20% of the respondents did not know what the trickle vent is, thus never turning it on or off accordingly. Furthermore, occupants tend to save their energy through conventional approaches without compromising their living comfort: 46% of respondents thought they use more energy than they should for a variety of reasons such as children's needs, illness and comfort.

In both social housing estates, there are 14% of the respondents do not have smartphones and only 20% of the households have smart meters installed. Smart technology is deeply involved in people's lives and significant in helping occupants to achieve energy efficiency through interactions. Besides, among the ones with smartphones, only 10% of them have energy management applications installed on them and only 2% of the occupants will actually use the application.

The results indicate that occupants' energy conservation awareness is relatively low and need to be further educated. Therefore, the interventions need to be adopted through a variety of ways to improve occupants' energy consumption behaviours and increase their energy awareness. By focusing on the government's level, making more effective policies and schemes may stimulate society's incentives for energy saving and consequently improve behaviours. The energy consumption behaviours could also be improved by increasing interactions between occupants and energy management systems, such as IHDs and energy management applications. In addition, more advanced energy-saving approaches need to be introduced to the occupants for more efficient energy uses such as using thermostats, increasing ventilation and installing energy monitoring and management devices. On the other hand, the support given needs to be increased due to the low take-up rate of smart meters and low energy-saving awareness. Energy suppliers are currently the major provider of energy advice and guidance. Additional, support needs to be provided by local authorities, communities and other organisations.

A significant number of participating households (30%) have only one family member where the majority of them are single elderly. Thirty per cent of the households do not have any employed family members, which leads to low household income and fuel poverty where 14% of households earn less than £6,000 annual income. The majority (54%) of them identify their annual income levels between £6,000 and £12,000. Besides, the education level in these housing estates is relatively low: half of the respondents have 'no degree' or only 'GCSE level 1 and 2' and 32% of the respondents have a degree qualification or above. Their educational background may be a factor affecting their income levels and may also impact the way that they operate their homes.

Correlation Results and Analysis

Sample Size and the Validity of the Results

As a result, there were 18 questionnaires returned from Estate A and 32 questionnaires returned from Estate B. Among those returned questionnaires, one questionnaire reflects unrealistic energy consumptions that were consumed by a single occupant with high energy knowledge and sustainable awareness. As this does not follow the general pattern of the whole data set, it was taken off from the sample. Therefore, the data set in the research is 49. However, some analyses might be based on a smaller number of variables due to the missing answers. The number of variables will be detailed in the following sections if they are less than 49. Before running the correlation analysis, the validity of the data sample is discussed. To examine if it is an appropriate data set, comprehensive studies adopting correlation analysis using statistical tests are presented below.

Technically, the correlation coefficient can be calculated from 'n = 2', where 'n' means the number of data set. However, with the increasing number of data set, more valuable data will be collected with more accurate prediction during the analysis of correlation coefficient. According to Weaver and Koopman (2014), when 'n < 3' in a data set, 'p value' and 'coefficient intervals' are not computed; when 'n < 4' in a data set, 'coefficient intervals' is not computed; when 'n < 10' in a data set, normal approximation is poor. Hence, a data set with more than 10 is recommended. Besides, Bonett and Wright (2000) recommended that in order to run valid Pearson, Kendall and Spearman correlation tests, the sample set need to be equal or superior to 25. Field (2009) also suggests that the sample size for running correlation tests needs to be at least 30 for accurate predictions. The same point of view was also suggested by Pajula and Tohka (2016) in their study. They found that 20 subjects in a data set may be converged close to a large data set that comprises 120 subjects. A notable improvement was seen if the data set was increased from 20 to 30 or more. This indicates that a data set of 49, as in the current study, is capable to produce sufficient and accurate predictions for the study.

Correlation Between Occupants' 'Energy Consumption Behaviour, Socio-Demographic Characteristics and Energy Consumption

As previously stated, various kinds of aspects were investigated through the questionnaire survey. It is believed that those aspects potentially influence home energy consumption, such as housing issues, occupants' energy consumption behaviour, occupancy patterns and their socio-demographic information. Thus, all of the collected data were put into correlation analysis in SPSS to explore those influences. According to the nature of data acquired, analysis of variances (ANOVA) was also employed as the most appropriate test to further explore the potential relationships between unordered categorical data (mixed discrete data) and ordered data (Leon & Zhu, 2008). As a result, a significant number of interesting findings were identified and demonstrated below.

Energy Advice and the Energy Consumption Behaviour

The research first investigated whether the occupants who have previously received advice on how to reduce their energy bills will perform better than the others on using heating controls, ventilation and energy-saving behaviour.

According to Table 10.1, significant and moderate correlations were identified between 'Have you received advice on how to reduce your energy bills?' and a number of energy-related variables at both 0.05 and 0.01 levels. As a result, respondents who expressed that they 'have previously received energy advice' tend to use 'wall thermostat' (0.257*, $p < 0.05$) and 'boiler thermostat' (0.334*, $p < 0.05$) at homes in winter more frequently; people who 'have previously received energy advice' also tend to 'try using less gas and electricity' (0.339**, $p < 0.01$), 'turn heating up or down as required' (0.273*, $p < 0.05$), 'set hot water thermostat lower' (0.326*, $p < 0.05$), 'unplug unused equipment' (0.401**, $p < 0.01$) and 'use low energy light bulbs' (0.260*, $p < 0.05$) more frequently.

The correlation analysis can easily identify correlations among numerical data and ordinal data that have ordered series or ranking sequences among categories, such as different 'frequency' or 'satisfaction degrees'. However, if the analysed information belongs to binary or nominal data which are not in ordered series, an ANOVA test needs to be adopted in order to help interpret the established correlations. According to Table 10.2, significant differences of 'frequency of using boiler thermostat', 'try using less gas and electricity', 'turn your heating up or down', 'set how water thermostat lower', 'unplug unused equipment' and 'use low energy light bulbs' have also been identified with different categories of 'Have you received advice on how to reduce bills?'. The 'means plot' diagram Figure 10.41 clearly presents the significant differences between a number of variables and different categories of the chosen factor: 'Have you received advice on how to reduce your energy bills?'.

According to Figures 10.41 and 10.42, occupants who have received advice on how to reduce energy bills tend to use wall thermostats and boiler thermostats more frequently. On the other hand, the occupants who have not received energy advice tend to use wall thermostats and boiler thermostats less frequently.

According to Figures 10.43 and 10.44, occupants who have received advice on how to reduce energy bills tend to follow the energy conservation actions more frequently, such as 'try using less gas and electricity' and 'turn heating up or down as required'. On the other hand, the occupants

Table 10.1 Correlation between 'Have you received energy advice?' and the 'energy-related questions'

Q4.1 Have you received advice on how to reduce bills?	*Correlation coefficient*	*Sig. (two-tailed)*	*N*
q10.2. How often do you use your wall thermostat? (in winter)	0.257*	0.048	49
q10.3. How often do you use your boiler thermostat? (in winter)	0.334*	0.011	49
q16.1. Try using less gas and electricity	0.339**	0.010	48
q16.2. Turn your heating up or down as required	0.273*	0.038	49
q16.5. Set the hot water thermostat lower	0.326*	0.013	49
q16.12. Unplug unused equipment	0.401**	0.002	49
q16.14. Use low-energy light bulbs	0.260*	0.048	49

* Correlation is significant at the 0.05 level (two-tailed).
** Correlation is significant at the 0.01 level (two-tailed).

Table 10.2 ANOVA test between 'Have you received energy advice' and the 'energy-related questions'

Dependent variable	*F*	*Sig.*
q10.2. How often do you use your heating controls? (in winter) Wall thermostat	4.377	0.042
q10.3. How often do you use your heating controls? (in winter) Boiler thermostat	7.478	0.009
q16.1. Try using less gas and electricity	8.446	0.006
q16.2. Turn your heating up or down as required	5.336	0.025
q16.5. Set hot water thermostat lower	6.905	0.012
q16.12. Unplug unused equipment	11.023	0.002
q16.14. Use low-energy light bulbs	4.912	0.032

Table 10.3 Correlation between 'Have you changed energy suppliers/plans?' and the 'energy-related questions'

q5.1. Have you changed your energy supplier/energy plan?	*Correlation coefficient*	*Sig. (two-tailed)*	*N*
q16.5. Set hot water thermostat lower	0.328*	0.023	48
q16.9. Go out to avoid using heating	0.397**	0.005	48

* Correlation is significant at the 0.05 level (two-tailed).
** Correlation is significant at the 0.01 level (two-tailed).

Table 10.4 ANOVA test between 'Have you changed energy plans/tariffs' and the 'energy-related questions'

Dependent variable	*F*	*Sig.*
q16.5. Set hot water thermostat lower	5.552	0.023
q16.9. Go out to avoid using heating	8.631	0.005

who have not received energy advice tend to follow energy conservation actions less frequently. The rest of the identified energy conservation actions also follow the same trends. The positive influence of energy information and advice was also highlighted in recent studies (Gupta & Gregg, 2016) that communications with occupants regarding the way of operating the homes are crucial for the improvement of home energy performance.

Change of Energy Suppliers/Plans and Energy Consumption Behaviour

The research also investigated whether the occupants who have 'previously changed their energy suppliers or energy plans' will perform better than the others on using heating controls, ventilation and energy consumption behaviour. The analysis was based on 48 variables due to missing answers.

According to Table 10.3, significant and moderate correlations was indicated between 'Have you changed your energy supplier/energy plan?' and a few energy related variables at both 0.05 and 0.01 levels. As a result, occupants who 'have previously changed their energy suppliers/plans' tend to 'set hot water thermostat lower' (0.328*, $p < 0.05$) more frequently. Besides, people who 'have changed energy supplier/plan' tend to 'go out to avoid using heating' (0.397**, $p < 0.01$) more frequently. The analytical result indicated that people who 'have previously changed energy suppliers/plans may have stronger awareness on energy saving (Seyedrezaei et al., 2023).

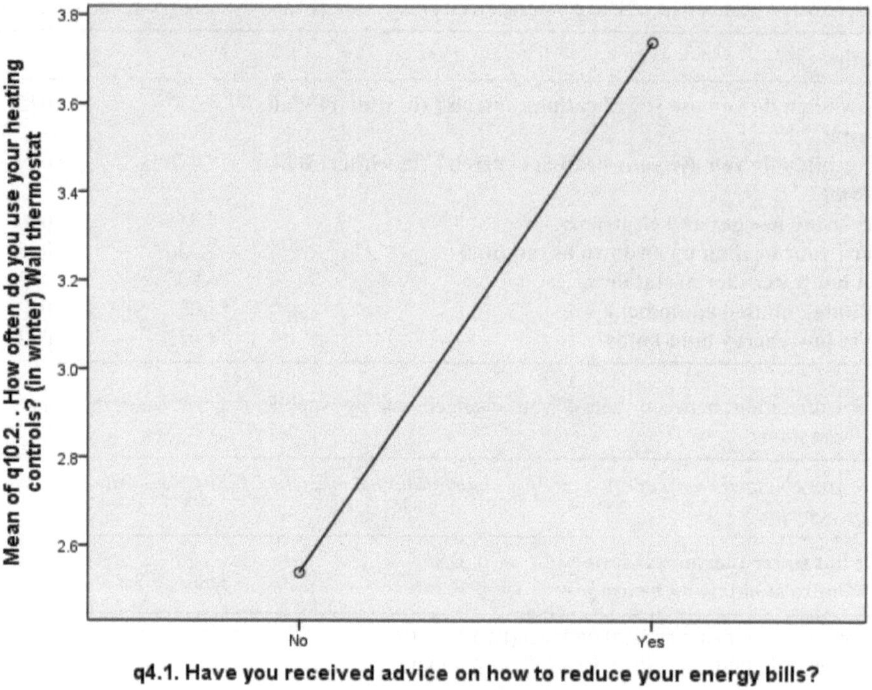

Figure 10.41 ANOVA test: energy advice and use of wall thermostat.

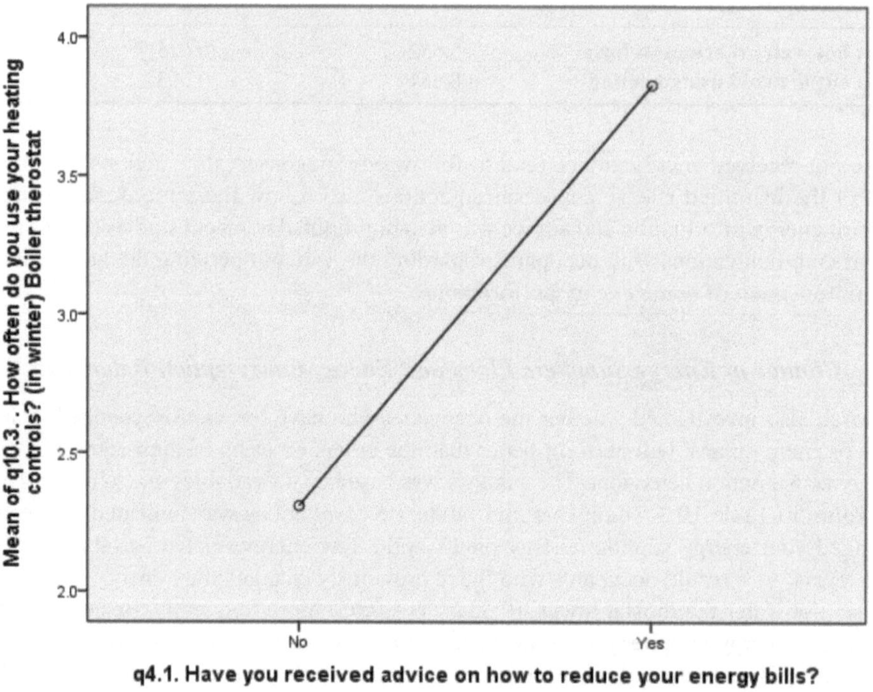

Figure 10.42 ANOVA test: energy advice and use of boiler thermostat.

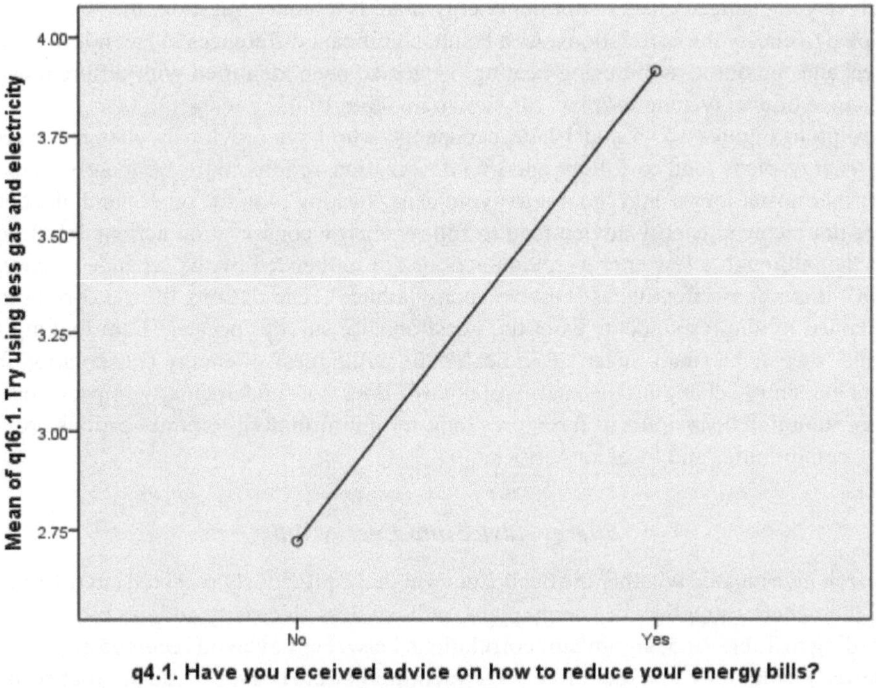

Figure 10.43 Energy advice and try use less gas and electricity.

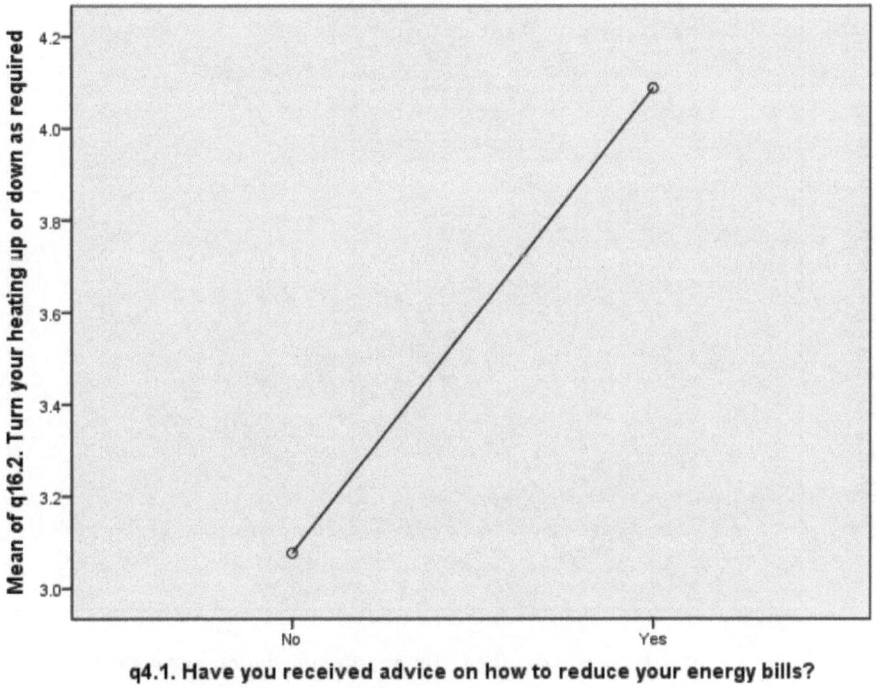

Figure 10.44 Energy advice and adjust heating as required.

As 'Have you changed energy supplier/energy plan' is a binary question, the ANOVA test was also adopted to clarify the correlations. As a result, significant differences in 'set hot water thermostat lower' and 'go out to avoid using heating' have also been identified with different categories of 'the change of energy supplier/plan', as shown in Table 10.4.

According to Figures 10.45 and 10.46, occupants' who have previously changed energy suppliers or energy plans tend to follow energy conservation actions more frequently, such as 'set hot water thermostat lower' and 'go out to avoid using heating'. On the other hand, the occupants who have not received energy advice tend to follow energy conservation actions less frequently. It shows that although a few energy-saving actions are influenced by the 'change of energy supplier/plan', it is not as effective as 'receive energy advice'. The reasons for the correlation were also explained by the respondents from the questionnaire survey 'because I am looking for less to pay' and 'received a smart meter'. It indicated the willingness of energy conservation from the occupants but simply changing the energy plan/tariff does not fundamentally improve occupant's energy consumption behaviours as it requires long-term informative communications provided by suppliers, communities and local authorities.

Energy Advice and Energy Bills

The research investigated whether the occupants who have previously received energy advice, and changed their energy suppliers or energy plans, will pay less electricity and gas bills than others.

According to Table 10.5, significant correlations between 'have you received energy advice?' and 'quarterly electricity and gas bills' were identified at 0.01 level. It shows that respondents

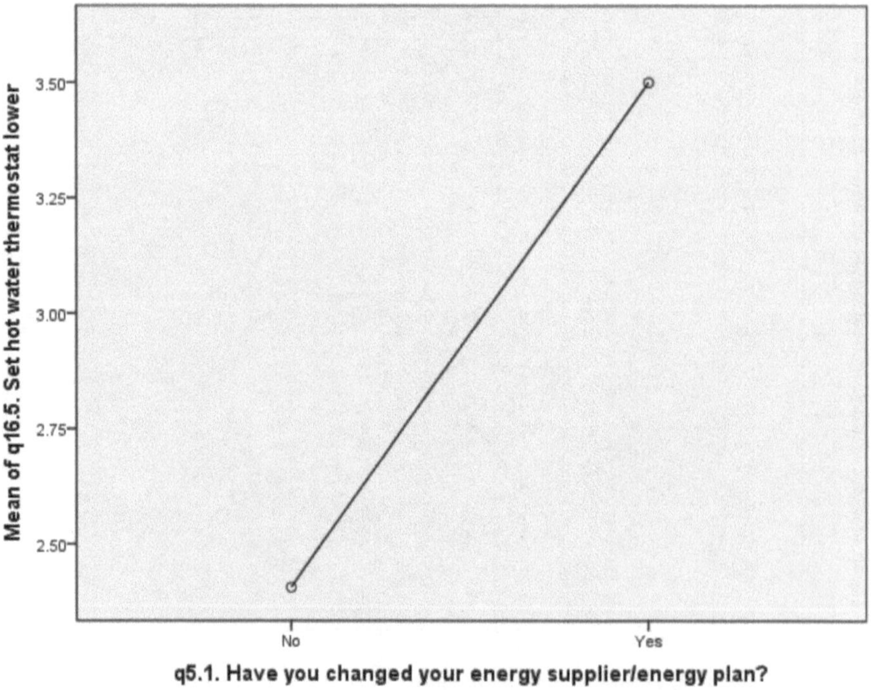

Figure 10.45 Change of supplier/plan and set thermostat lower.

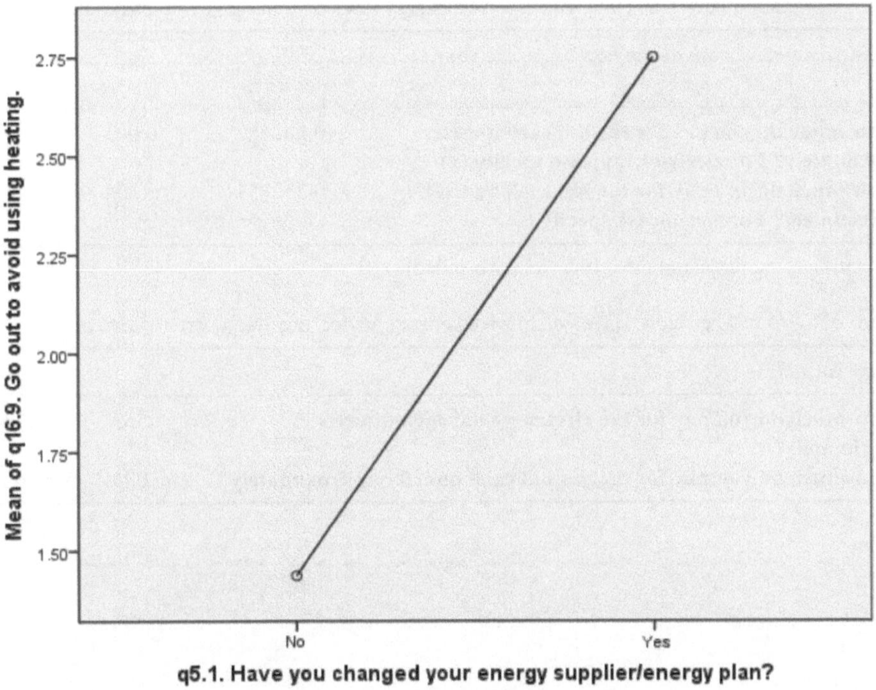

Figure 10.46 Change of supplier/plan and go out to avoid using heating.

who 'have previously received energy advice' tend to spend less on the 'quarterly electricity bills' (−0.340**, $p < 0.01$) and 'quarterly gas bills' (−0.352**, $p < 0.01$). On the other hand, correlations between 'have you changed energy tariffs/plans' and 'quarterly electricity and gas bills' were also analysed but nothing was indicated. As a result, it implies that the way of operating the homes more effectively will significantly impact home energy performance (Ozarisoy & Altan, 2023a).

ANOVA analysis was also conducted to further interpret the correlations between numerical and categorical data. According to Table 10.6, significant differences in 'quarterly electricity bills' and 'quarterly gas bills' have also been identified with different categories of 'energy advice'.

According to Figures 10.47 and 10.48, the relationships between different answers of 'energy advice' and the average of 'quarterly electricity and gas bills' were demonstrated. It showed that occupants who received energy advice tended to spend much less on 'quarterly electricity and gas bills' than the ones who had not received energy advice. This is supported by Simpson et al. (2016) that occupants' energy consumption behaviours need to be improved through education.

Energy Bills and the Energy Consumption Behaviour

The research also explored whether the occupants who used more heating controls, ventilation and energy consumption behaviour, will spend less on their electricity and gas bills than others. The analysis of 'number of hours that window opened', 'frequencies and houses trickle vents opened', 'try using less gas and electricity' and 'use the thermostats to adjust temperature' were based on less variables (see Tables 10.7 and 10.8) due to missing answers.

Table 10.5 Correlation between 'Have you received energy advice', and 'quarterly electricity and gas bills'

q4.1. Have you received advice on how to reduce your energy bills?	*Correlation coefficient*	*Sig. (two-tailed)*	*N*
q3.1. How much do you pay for the bill each quarter approximately? For electricity, please specify (£)	−0.340**	0.005	49
q3.2. How much do you pay for the bill each quarter approximately? For gas, please specify (£)	−0.352**	0.004	49

** Correlation is significant at the 0.01 level (two-tailed).

Table 10.6 ANOVA test between 'Have you received energy advice' and the 'quarterly electricity and gas bills'

Dependent variable	*F*	*Sig.*
Q3.1 How much do you pay for the electricity bill each quarter approximately?	7.568	0.008
Q3.2 How much do you pay for the gas bill each quarter approximately?	6.084	0.017

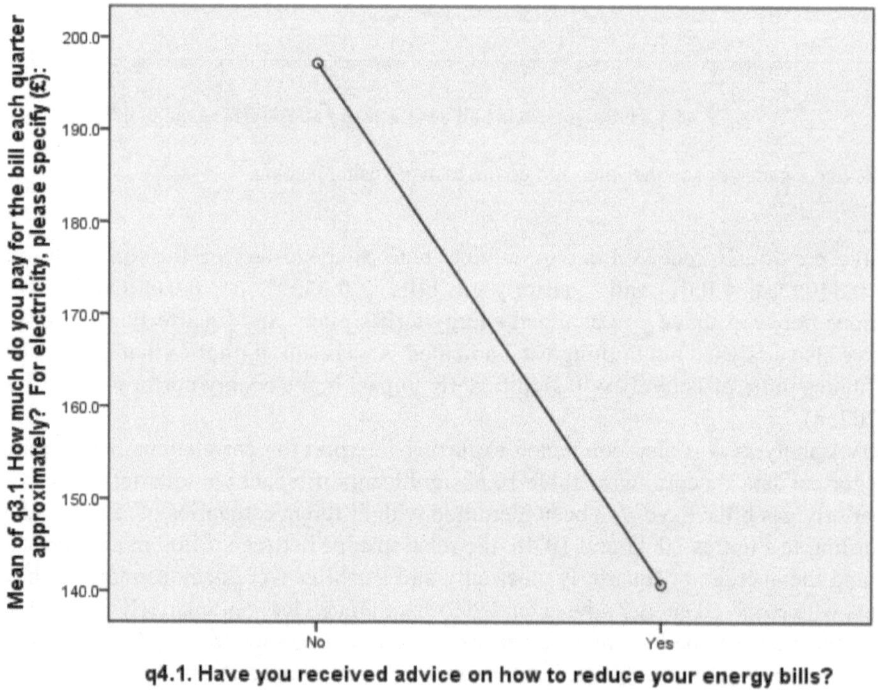

Figure 10.47 ANOVA: 'energy advice' and 'electricity bills'.

According to Table 10.7, significant and moderate correlations were indicated between occupants' 'quarterly electricity bills' and a number of energy-related variables at both the 0.01 and 0.05 levels. As a result, occupants who spend more on 'electricity bills' tend to 'use wall thermostat' (−0.372**, $p < 0.01$) less frequently in winter and set their 'wall thermostat' (0.224*, $p < 0.05$) higher. The importance of using thermostats was highlighted by Herring (2006) that

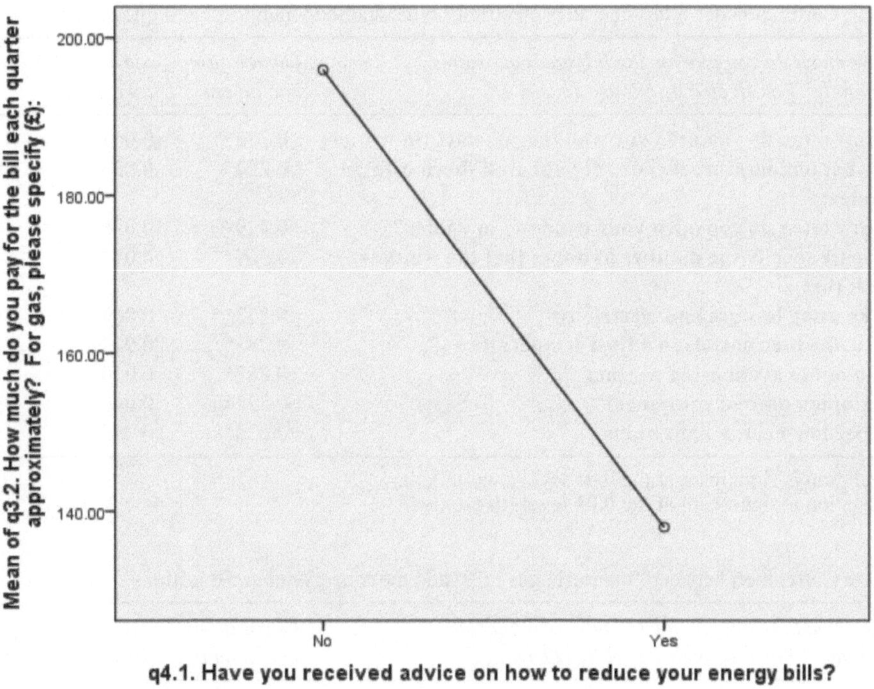

Figure 10.48 ANOVA: 'energy advice' and 'gas bills'.

energy demand could be reduced by lowering heating levels through turning down thermostat levels. According to ASHRAE Standard 55-2013, approximate acceptable indoor air temperature is between 67 and 82°F (19.4–27.8°C) depending on location and relative humidity (Ozarisoy & Altan, 2021). The Public Health England (Bone, 2014) suggests that occupants should keep their indoor temperature at least 18°C and 21°C for the recommendations. Ahn and Park (2016) demonstrated the correlation between occupants' active action to control heating/cooling systems and energy performance. Scholars who also support it include Yun and Steemers (2011) and Guerra-Santin (2011).

Besides, respondents who spend more on 'electricity bills' tend to 'open windows' (−0.229*, $p < 0.05$) less frequently. However, the 'hours that windows opened' (0.269*, $p < 0.05$) tend to be longer during the winter. Home energy consumption does not only relate to how often the windows are opened but also to the degree of openings (Fabi et al., 2012). The correlation between the hours that the window opened in the living room ($p < 0.05$) and energy performance was identified in Guerra-Santin (2011)'s research. Fabi et al. (2012) also stated that opening the windows could improve the indoor climate by increasing the air exchange rate and mitigating housing problems, such as 'mould', 'dampness' and 'condensation'. Hence, it decreases the heating demand. Furthermore, respondents who pay more 'electricity bills' tend to 'try using less gas and electricity' (−0.373**, $p < 0.01$), 'use the thermostats to adjust temperature' (−0.283*, $p < 0.05$), 'go out to avoid using heating' (−0.287*, $p < 0.05$), 'unplug unused equipment' (−0.222*, $p < 0.05$) and 'use low energy bulbs' (−0.328**, $p < 0.01$) less frequently.

According to Table 10.8, significant correlations were also found between occupants' 'quarterly gas bills' and a number of energy-related variables at both 0.05 and 0.01 levels. As a result,

Table 10.7 Correlation between 'quarterly electricity bills' and the 'energy-related questions'

q3.1. How much do you pay for the bill each quarter approximately? For electricity, please specify (£):	*Correlation coefficient*	*Sig. (two-tailed)*	*N*
q10.2. How often do you use your wall thermostat? (in winter)	−0.372**	0.001	49
q11. At what temperature do you set your wall thermostat? (in winter)	0.224*	0.036	49
q12.1. How often do you open your windows in winter?	−0.229*	0.038	49
q12.2. Please specify the number of hours that the window opened: (hrs)	0.269*	0.017	43
q16.1. Try using less gas and electricity	−0.373**	0.001	48
q16.3. Use the thermostats to adjust temperature	−0.283*	0.012	48
q16.9. Go out to avoid using heating	−0.287*	0.010	49
q16.12. Unplug unused equipment	−0.222*	0.043	49
q16.14. Use low-energy light bulbs	−0.328**	0.003	49

* Correlation is significant at the 0.05 level (two-tailed).
** Correlation is significant at the 0.01 level (two-tailed).

Table 10.8 Correlation between 'quarterly gas bills' and the 'energy-related questions'

q3.2. How much do you pay for the bill each quarter approximately? For gas, please specify (£):	*Correlation coefficient*	*Sig. (two-tailed)*	*N*
q10.2. How often do you use your wall thermostat? (in winter)	−0.314**	0.004	49
q11. At what temperature do you set your wall thermostat? (in winter)	0.468**	0.001	49
q12.1. How often do you open your windows in winter?	−0.353**	0.001	49
q13.1 How often do you turn on the extractor fan when taking the shower?	−0.354**	0.002	49
q15.1. How often do you keep the trickle vents open when it is cold?	0.318*	0.016	37
q15.2. Please specify the number of hours	0.332*	0.022	37
q16.1. Try using less gas and electricity	−0.371**	0.001	48
q16.12. Unplug unused equipment	−0.317**	0.004	49
q16.14. Use low-energy light bulbs	−0.494**	0.000	49
q8.1.2. Living room Hrs heaters on (weekdays)	0.256*	0.018	46
q8.1.2'. Living room Hrs heaters on (weekends)	0.306**	0.004	46

* Correlation is significant at the 0.05 level (two-tailed).
** Correlation is significant at the 0.01 level (two-tailed).

respondents who pay more 'quarterly gas bills' tend to use 'wall thermostat' (−0.314**, $p < 0.01$) and 'open windows' (−0.353**, $p < 0.01$) in winter less frequently. Occupants tend to spend more on 'quarterly gas bills' with higher 'temperature set on wall thermostat' (0.468**, $p < 0.01$). Besides, occupants who 'turn on the extractor fan' (−0.354**, $p < 0.01$) more frequently tend to pay less on 'quarterly gas bills'. The occupant who pays more 'quarterly gas bills' would 'leave the trickle vents open' (0.318*, $p < 0.05$) more frequently 'with longer hours' (0.332*, $p < 0.05$). The reason for this is that the hours that trickle vents opened are much longer than windows and extractor vents according to the survey findings. Thus, it decreases the indoor temperature and increases the heating demand. In addition, respondents who expressed that they pay more 'quarterly gas bills' tend to less frequently 'try using less gas and electricity' (−0.371**, $p < 0.01$), 'unplug unused equipment' (−0.317**, $p < 0.01$) and 'use low energy light bulbs' (−0.494**, $p < 0.01$).

Moreover, households who pay more 'quarterly electricity bills' tend to have longer 'heating patterns' during 'weekdays' (0.231*, $p < 0.05$) and 'weekends' (0.230*, $p < 0.05$). The same finding was also indicated between 'quarterly gas bills' and 'heating patterns' during 'weekdays' (0.256*, $p < 0.05$) and 'weekends' (0.306**, $p < 0.01$). This indicates that longer heating patterns may result in more energy consumption and consequently increase energy bills (Ozarisoy & Altan, 2023b).

Energy Bills and Socio-Demographic Characteristics

The research explored whether households with different socio-demographic characteristics, such as family size, tenancy status or economic status, will impact their quarterly bills.

According to Table 10.9, significant and moderate correlations were indicated between occupants' 'quarterly electricity bills' and a number of socio-demographic variables at both 0.05 and 0.01 levels. As a result, occupants with 'owner-occupied' status tend to spend more on their 'quarterly electricity bills' and 'social renters and 'housing association renters' tend to spend relatively less on it (−0.356**, $p < 0.01$). Based on the questionnaire survey, the average annual income for social renters is £13,097.6 and £21,375.0 for private renters. Higher incomes may imply more energy consumption. Guerra-Santin (2011) also argued that the high 'energy-intensity' of using large appliances and spaces at home was more apparent to homeowners with high incomes. An opposite view was held by Kavousian et al. (2013) that there was no correlation found between income level and energy performance based on their research.

The 'quarterly electricity bills' are found to be significantly or moderately correlated by the household profiles, such as the households with more 'children (3–12 yrs)' (0.302**, $p < 0.01$), 'adult (25–34 yrs)' (0.291*, $p < 0.05$) and 'student' (0.255*, $p < 0.05$) tend to spend more on 'quarterly electricity bills'. On the other hand, households with more 'retired' (−0.247*, $p < 0.05$) people tend to spend less on their 'quarter electricity bills'. Similar findings were also indicated by Kavousian et al. (2013) that individuals over 55 recorded lower energy consumption. It is also supported by Guerra-Santin (2011) that the presence of elderlies has a strong negative impact ($p < 0.001$) on energy performance. The 'number of occupants' (0.306**, $p < 0.01$) increases while the household 'quarter electricity bills' increases. Conan (1981) also identified a correlation between winter energy consumption and a number of occupants based on the questionnaire survey. The correlations

Table 10.9 Correlation between 'quarterly electricity bills' and the 'socio-demographic characteristics'.

q3.1. How much do you pay for the bill each quarter approximately? For electricity, please specify (£):	*Correlation coefficient*	*Sig. (two-tailed)*	*N*
q1. Tenancy status	−0.356**	0.003	49
q29.2. Members of household children (3–12 years)	0.302**	0.009	49
q29.5. Members of household 25–34 years	0.291*	0.012	49
q31.5. Economic status of each family member Retired	−0.247*	0.040	49
q31.6. Economic status of each family member Student	0.255*	0.029	49
q35. How is your health in general?	0.239*	0.034	49
qII. Number of occupants	0.306**	0.005	49

* Correlation is significant at the 0.05 level (two-tailed).
** Correlation is significant at the 0.01 level (two-tailed).

Table 10.10 Correlation between 'quarterly gas bills' and the 'socio-demographic characteristics'

q3.2. How much do you pay for the bill each quarter approximately? For gas, please specify (£):	*Correlation coefficient*	*Sig. (two-tailed)*	*N*
q1. Tenancy status	−0.275*	0.020	49
q29.3. Members of household Teenagers (12–19 years)	0.396**	0.001	49
q31.6. Economic status of each family member Student	0.244*	0.037	49

* Correlation is significant at the 0.05 level (two-tailed).
** Correlation is significant at the 0.01 level (two-tailed).

Table 10.11 ANOVA test between 'quarterly bills' and 'socio-demographic characteristics'

Dependent variable	*F*	*Sig.*
q3.1. How much do you pay for the bill each quarter approximately? For electricity, please specify (£):		
q1. Tenancy status	4.555	0.016
q35. How is your health in general?	3.298	0.044
q3.2. How much do you pay for the bill each quarter approximately? For gas, please specify (£):		
q1. Tenancy status	3.343	0.028

between household profiles and energy consumption were also identified by Guerra-Santin (2011) that the presence of both children and elderlies hugely influences heating and ventilation patterns. Larger families also tended to use large electricity appliances more frequently and consequently consume more energy. In addition, occupants with better 'health conditions' (0.239*, $p < 0.05$) tend to spend less on 'quarterly electricity bills'. Nicholls and Strengers (2017) emphasised the correlation between occupants' ill, health and the high cost of utility bills through interviews. Moreover, poor health condition increases the occupancy pattern at home for recovery.

Apart from the correlations between 'quarterly electricity bills' and socio-demographic characteristics, the correlations between 'quarterly gas bills' and the same variables were also conducted and demonstrated in Table 10.10.

According to Table 10.10, significant and moderate correlations were indicated between occupants' 'quarterly gas bills' and a number of socio-demographic variables at both 0.05 and 0.01 levels. As a result, occupants with 'owner-occupied' tenancy status tend to spend more on their 'quarterly gas bills' (−0.275*, $p < 0.05$). Similar findings were also identified by Elsharkawy (2013) that moderate correlations ($r = 0.465$, $p<0.005$) between 'social tenancy' and 'average monthly gas bills' were found in the Community Energy Saving Programme (CESP) survey. The households with more 'teenagers (12–19 yrs)' (0.396**, $p < 0.01$) and 'student' (0.244*, $p < 0.01$) tend to spend more on their 'quarterly gas bills'.

As 'tenancy status' and 'health condition' are unordered categorical data, the ANOVA test was also conducted to further explore the correlations. According to Table 10.11, significant differences in 'quarterly electricity bills' and 'quarterly gas bills' have also been identified with different categories of 'tenancy status' and 'health conditions'.

According to Figure 10.49, 'owner-occupied' households tend to spend highest amount of 'quarterly electricity bills' followed by 'social renters' and 'housing association renters'. Occupants in 'good' health condition tend to spend less on 'quarterly electricity bills' and it gradually increases along with 'fair' and 'poor' health conditions (see Figure 10.50).

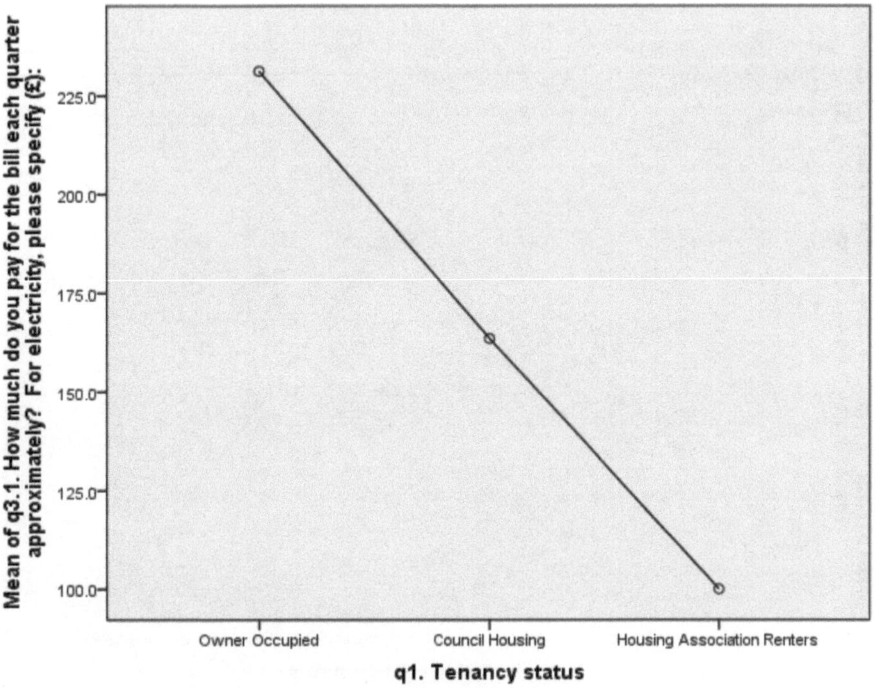

Figure 10.49 ANOVA: ‘electricity bills’ and ‘tenancy status’.

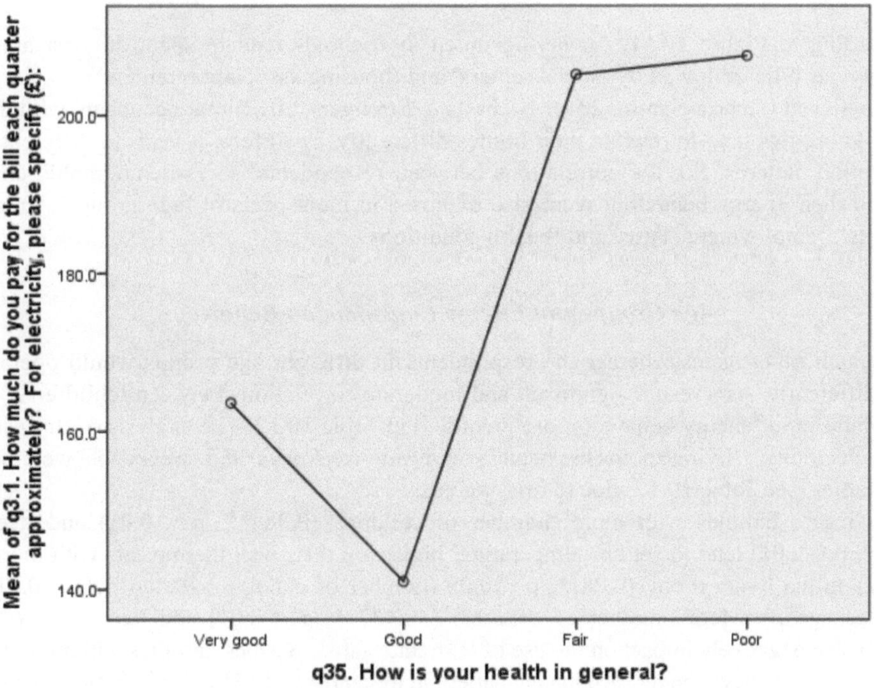

Figure 10.50 ANOVA: ‘electricity bills’ and ‘health conditions’.

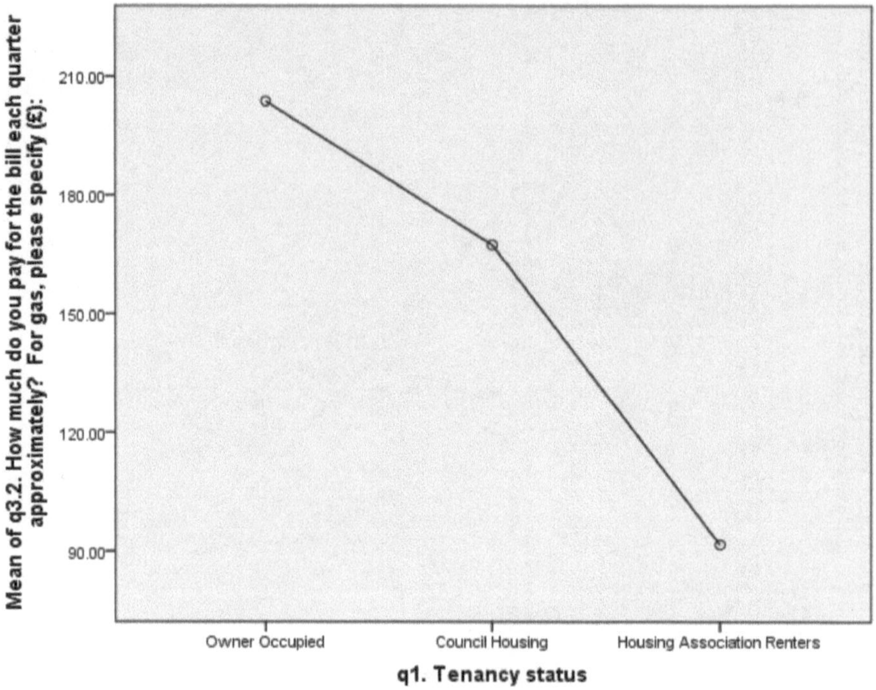

Figure 10.51 ANOVA test: 'gas bills' and 'tenancy status'.

According to Figure 10.51, 'owner-occupied' households tend to spend highest amount of 'quarterly gas bills' followed by 'social renters' and 'housing association renters'.

It is believed (Guerra-Santin, 2011; Nicholls & Strengers, 2017) that occupants with different household profiles tend to operate their homes differently. Therefore, it leads to different energy consumption patterns. So, the correlations between respondents' socio-demographic characteristics and their energy behaviour were also explored in the aspects of 'age groups', 'number of occupants', 'employment status' and 'health conditions'.

Age Groups and Energy Consumption Behaviour

The research investigates whether the respondents in different age groups would operate their homes differently. As a result, significant and moderate correlations between 'children', 'teenagers', 'infants' and 'energy behaviour' are identified in Table 10.12. The analysis of 'try using less gas and electricity', 'living rooms hrs heaters on during weekdays' and 'weekends' were based on less variables (see Table 10.12) due to missing answers.

As a result, families with more 'number of children' (0.384**, $p < 0.01$) and 'teenagers' (0.411**, $p < 0.01$) tend to set up 'temperature' higher on their wall thermostats with more 'heating hours in the living room' (0.281*, $p < 0.05$ (number of children)) (0.406**, $p < 0.01$ (number of teenagers)). More 'number of children' (−0.407**, $p < 0.01$) and 'teenagers' (−0.312*, $p < 0.05$) also negatively impact on the use of 'extractor fans'. Besides, families with more 'number of children' tend to have longer 'occupancy pattern in living room' (0.351**, $p < 0.01$), 'try using less gas and electricity' (−0.294*, $p < 0.05$) and 'try heating as less room as possible' (−0.314*, $p < 0.05$)

Table 10.12 Correlation between 'children', 'teenagers', 'infants' and 'energy consumption behaviour'

q29.2. Members of household children (3–12 years)	*Correlation coefficient*	*Sig. (two-tailed)*	*N*
q11. At what temperature do you set your wall thermostat? (in winter)	0.384**	0.006	49
q13.1. How often do you turn on the extractor fan when you take the shower?	−0.407**	0.004	49
q8.1.2. Living room Hrs heaters on (weekdays)	0.281*	0.023	46
q16.1. Try using less gas and electricity	−0.294*	0.042	48
q16.4. Try heating as less room as possible	−0.314*	0.033	49
q8.1.2. Living room Hrs occupied (weekdays)	0.351**	0.005	45
q29.3. Members of household Teenagers (12–19 years)	Correlation coefficient	Sig. (two-tailed)	N
q11. At what temperature do you set your wall thermostat? (in winter)	0.411**	0.003	49
q13.1. How often do you turn on the extractor fan when you take the shower?	−0.312*	0.029	49
q8.1.2'. Living room Hrs heaters on (weekends)	0.406**	0.001	46
q16.8. Put on a jumper instead of heating	−0.287*	0.049	49
q29.1. Members of household infants (0–3 years)	Correlation coefficient	Sig. (two-tailed)	N
q8.1.2. Living room Hrs heaters on (weekdays)	0.301*	0.016	46
q8.1.2'. Living room Hrs heaters on (weekends)	0.372**	0.002	46

* Correlation is significant at the 0.05 level (two-tailed).
** Correlation is significant at the 0.01 level (two-tailed).

less frequently. In addition, households with more 'number of teenagers' tend to 'put on a jumper instead of heating' (−0.287*, $p < 0.05$) less frequently. Families with more 'number of infants' tend to have longer 'heating hours in living room' during 'weekdays' (0.301*, $p < 0.05$) and 'weekends' (0.372**, $p < 0.01$). The results imply that infant, children and teenagers will impact the length of heating hours at home as respondents claimed that 'we use more energy when kids come back from college' and 'because we have small children and need to keep it warm for them'. As a result, households with more children could not follow energy-saving actions properly. This is supported by Brounen et al. (2012) that households with children generally consume more energy than the ones without children due to higher comfort levels required and greater use of entertainment appliances.

Apart from exploring how young people operate their homes, other age groups and their associated energy consumption behaviour are also identified in Table 10.13.

As a result, significant and moderate correlations between 'higher age groups' and 'energy behaviour' were identified. Families with more 'numbers between 55 and 64 yrs' tend to feel more difficult to understand utility bills (−0.275*, $p < 0.05$). They also less frequently use 'wall thermostat' (−0.291*, $p < 0.05$) and 'boiler thermostat' (−0.373*, $p < 0.05$). On the other hand, families with more 'members between 55 and 64 yrs' tend to 'put on a jumper instead of heating' (0.377*, $p < 0.05$) more frequently. Besides, the households with more 'members at 65 yrs or over' tend to use 'boiler thermostat' (−0.289*, $p < 0.05$) less frequently but 'try using less gas and electricity' (0.391**, $p < 0.01$) and 'try heating as less room as possible' (0.336*, $p < 0.05$) more frequently. It implies that older occupants may try to save energy through basic approaches in order to use as little energy as possible.

Table 10.13 Correlation between 'elderlies' and 'energy consumption behaviour'

q29.8. Members of household 55–64 years	*Correlation coefficient*	*Sig. (two-tailed)*	*N*
q17. Rate your ability to read the utility bill	−0.275*	0.037	49
q10.2. How often do you use your heating controls? (in winter) Wall thermostat	−0.291*	0.043	49
q10.3. How often do you use your heating controls? (in winter) Boiler thermostat	−0.373*	0.028	49
q16.8. Put on a jumper instead of heating	0.377*	0.022	49
q29.9. Members of household 65 years or over	**Correlation coefficient**	**Sig. (two-tailed)**	**N**
q10.3. How often do you use your heating controls? (in winter) Boiler thermostat	−0.289*	0.041	49
q16.1. Try using less gas and electricity	0.391**	0.008	49
q16.4. Try heating as less rooms as possible	0.336*	0.026	49

* Correlation is significant at the 0.05 level (two-tailed).
** Correlation is significant at the 0.01 level (two-tailed).

Table 10.14 Correlation between 'adults' and 'energy consumption behaviour'

q29.4. Members of household 19–24 years	*Correlation coefficient*	*Sig. (two-tailed)*	*N*
q16.16. Avoid using energy at peak time	−0.297*	0.039	49
q29.6. Members of household 35–44 years	**Correlation coefficient**	**Sig. (two-tailed)**	**N**
q13.1. How often do you turn on the extractor fan when you take the shower?	−0.349*	0.014	49
q11. At what temperature do you set your wall thermostat? (in winter)	0.308*	0.033	49
q16.11. Turn off TV when leave the living room	−0.290*	0.043	49
q29.7. Members of household 45–54 years	**Correlation coefficient**	**Sig. (two-tailed)**	**N**
q16.2. Turn your heating up or down as required	−0.297*	0.038	49
q16.6. Use blankets instead of a heating system	−0.283*	0.049	49

* Correlation is significant at the 0.05 level (two-tailed).

According to Table 10.14, moderate correlations were also identified between 'adults' and a number of energy consumption behaviours. Families with more 'adults between 19 and 24 yrs' tend to 'avoid using energy at peak times' (−0.297*, $p < 0.05$) less frequently. Families with more 'adults between 35 and 44 yrs' tend to set up 'temperature on wall thermostats' (0.308*, $p < 0.05$) higher and use 'extractor fans' (−0.349*, $p < 0.05$) and 'turn off the TV when leaving the living room' (−0.290*, $p < 0.05$) less frequently. Besides, households with more 'adults between 45 and 54 yrs' tend to 'turn heating up or down as required' (−0.297*, $p < 0.05$) and 'use blankets instead of heating system' (−0.283*, $p < 0.05$) less frequently. The results imply that those age groups did not follow energy-saving actions very well and tended to sacrifice energy efficiency for personal comfort (Ozarisoy, 2022). The reasons for this were explored during the focus group interview.

Table 10.15 Correlation between 'number of occupants' and 'energy consumption behaviour'

qII. Number of occupants	*Correlation coefficient*	*Sig. (two-tailed)*	*N*
q11. At what temperature do you set your wall thermostat? (in winter)	0.325*	0.023	49
q13.1. How often do you turn on the extractor fan when you take the shower?	−0.368**	0.009	49
q10.2. How often do you use your heating controls? (in winter) Wall thermostat	−0.288*	0.044	49
q16.1. Try using less gas and electricity	−0.362*	0.011	49
q16.4. Try heating as less room as possible	−0.314*	0.028	49
q16.11. Turn off TV when leave the living room	−0.337*	0.026	49
q16.16. Avoid using energy at peak times	−0.408**	0.002	49
q8.1.2. Living room Hrs occupied (weekdays)	0.323**	0.005	45
q8.1.2'. Living room Hrs occupied (weekends)	0.280*	0.016	45
q8.1.2. Living room Hrs heaters on (weekdays)	0.285*	0.013	46

* Correlation is significant at the 0.05 level (two-tailed).
** Correlation is significant at the 0.01 level (two-tailed).

Number of Occupants and Energy Consumption Behaviour

The study previously found that large families tend to consume more energy than small families. The study also investigates whether different family sizes determine the way occupants operate their homes. The analysis of 'living rooms, hours heaters on during weekdays' and 'weekends' were based on fewer variables (see Table 10.15) due to missing answers.

According to Table 10.15, significant and moderate correlations were identified between 'number of occupants' and occupants' 'energy consumption behaviour'. As a result, households with more 'number of occupants' tend to set up 'temperature on wall thermostat' (0.325*, $p < 0.05$) higher with longer 'occupancy' (0.323**, $p < 0.01$) and 'heating' (0.285*, $p < 0.05$) patterns. Besides, larger families tend to use 'extractor fans' (−0.368**, $p < 0.01$) and 'wall thermostats' (−0.288*, $p < 0.05$) less frequently. They also tend to 'try using less gas and electricity' (−0.362*, $p < 0.05$), try heating as less room as possible' (−0.314*, $p < 0.05$), 'turn off TV when leaving the living room' (−0.337*, $p < 0.05$), and 'avoid using energy at peak times' (−0.308**, $p < 0.01$) less frequently. The results imply that larger families with a more complex composition may not operate their homes effectively as each member's thermal comfort needs to be met. The greater number of occupants also increases the possibility of longer occupancy patterns and consequently increases the hours that heating systems are used during winter. The same findings were also identified by several scholars (Yun & Steemers, 2011; Abrahamse & Steg, 2009).

Employment Status and Energy Consumption Behaviour

Another socio-demographic characteristic that could potentially impact occupants' energy behaviour is the occupants' 'employment status'. The correlations between them were also explored. The analysis of 'living room, hours heaters on in weekdays' was based on 45 variables due to missing answers.

According to Table 10.16, significant and moderate correlations between 'employment status' and 'energy consumption behaviour' were indicated. As a result, households with more

Table 10.16 Correlation between 'employment status' and 'energy consumption behaviour'

q31.1. Economic status of each family member Full-time employed	*Correlation coefficient*	*Sig. (two-tailed)*	*N*
q16.2. Turn your heating up or down as required	−0.285*	0.048	49
q16.9. Go out to avoid using heating	−0.306*	0.033	49
q16.16. Avoid using energy at peak time	−0.388**	0.006	49
q31.7. Economic status of each family member Unemployed	**Correlation coefficient**	**Sig. (two-tailed)**	**N**
q8.1.2. Living room Hrs occupied (weekdays)	0.309*	0.015	45
q31.5. Economic status of each family member Retired	**Correlation coefficient**	**Sig. (two-tailed)**	**N**
q16.10. Turn off lights when you leave the room	0.350*	0.019	49

* Correlation is significant at the 0.05 level (two-tailed).
** Correlation is significant at the 0.01 level (two-tailed).

'full-time employed member' tend to 'turn heating up or down as required' (−0.285*, $p < 0.05$), 'go out to avoid using heating' (−0.306*, $p < 0.05$) and 'avoid using energy at peak time' (−0.388**, $p < 0.01$) less frequently. Besides, households with more 'unemployed members' tend to have longer 'occupancy patterns in living room' (0.309*, $p < 0.05$). Households with more 'retired members' tend to 'turn off lights when leaving the room' (0.350*, $p < 0.05$) more frequently. The results imply occupants with different employment statuses will determine their occupancy patterns at homes and consequently impact their capability to manage their energy use (Ozarisoy & Altan, 2022).

Health Condition and Energy Consumption Behaviour

The research also investigates if occupants with different 'health conditions' will operate their homes differently. The analysis of 'main bedroom hours heaters on in weekdays' was based on 46 variables due to missing answers.

According to Table 10.17, moderate correlations were indicated between occupants' 'health conditions' and their energy consumption behaviours. As a result, occupants' with poorer 'health conditions' tend to set 'temperature on wall thermostat' (0.275*, $p < 0.05$) higher with more hours with 'heaters on' (0.304*, $p < 0.05$), and 'open the windows' (−0.288*, $p < 0.05$) less frequently. The results imply that occupants with poor health conditions may require extra heating and consequently consume more energy than the ones in good health conditions.

Household Income Levels and Energy Consumption Behaviour

The research also explored what kind of behavioural variables may be affected by occupants' income levels. The analysis was based on fewer variables (see Table 10.18) due to missing answers.

According to Table 10.18, significant and moderate correlations were indicated between the 'household income level' and a number of energy consumption behaviours at both the 0.01 and 0.05 levels. As a result, households with higher 'income levels' tend to have longer 'heating patterns' in the living room (0.379*, $p < 0.05$) and main bedroom (0.309*, $p < 0.05$) during weekdays. Therefore, they also tend to turn on the 'thermostat' (0.336*, $p < 0.05$) for longer hours during weekdays. Besides, households with higher 'income levels' also tend to follow energy-saving

Table 10.17 Correlation between 'health condition' and 'energy consumption behaviour'

q35. How is your health in general?	*Correlation coefficient*	*Sig. (two-tailed)*	*N*
q11. At what temperature do you set your wall thermostat? (in winter)	0.275*	0.043	49
q12.1. How often do you open your windows in winter?	−0.288*	0.041	49
q8.2.2. Main bedroom Hrs heaters on (weekdays)	0.304*	0.025	46

* Correlation is significant at the 0.05 level (two-tailed).

Table 10.18 Correlation between 'total household income level' and the 'energy use patterns'

q33. Total household income level (annual):	*Correlation coefficient*	*Sig. (two-tailed)*	*N*
q8.1.2. Living room Hrs heaters on (weekdays)	0.379*	0.026	45
q8.2.2. Main bedroom Hrs heaters on (weekdays)	0.309*	0.048	45
q8.2.3. Hours thermostat on (weekdays)	0.336*	0.037	39
q16.4. Try heating as less room as possible	−0.310*	0.032	48
q16.9. Go out to avoid using heating	−0.353*	0.014	48
q16.13. Wash clothes in shorter washing cycle.	−0.349*	0.015	48
q16.15. Reduce time spend in the shower	−0.451**	0.001	48

* Correlation is significant at the 0.05 level (two-tailed).
** Correlation is significant at the 0.01 level (two-tailed).

actions less frequently, such as 'try heating as less room as possible' (−0.310*, $p < 0.05$), 'go out to avoid using heating' (−0.353*, $p < 0.05$), 'wash clothes in shorter washing cycle' (−0.349*, $p < 0.05$) and 'reduce time spend in the shower' (−0.451**, $p < 0.01$). It shows that higher income levels positively impact heating patterns and consequently increase home energy consumption. A few recent studies (Sugiura et al., 2013; Chen et al., 2013) also support this research finding. Sukarno et al. (2017) also stated that occupants with higher income levels tend to have larger electricity appliances and are less willing to compromise their personal comforts for less energy consumption.

Housing Issues, Energy Bills and Energy Consumption Behaviours

As stated in General survey findings, the housing issues that occupants are experiencing in Estate A and Estate B showed huge variations. Thus, the correlation analysis was carried out separately for the questions about 'housing issues'. The research tried to explore whether different degrees of problematic housing issues will affect energy bills and the way people operate their homes. The analysis of 'hours extractor fans opened' was based on 15 variables due to missing variables. The results are shown in Table 10.19.

According to Table 10.19, significant correlations were indicated between the 'housing issues' at Estate A and a number of energy use variables at both the 0.01 and 0.05 levels. As a result, correlations were found between 'Dampness' and 'quarterly electricity bills' (0.562*, $p < 0.05$) and 'quarterly gas bills' (0.626**, $p < 0.01$). The electricity and gas bills tend to increase while the 'Dampness' issue at homes increases. Besides, correlations were also found between 'Mould' (−0.545*, $p < 0.05$), 'Condensation' (−0.528*, $p < 0.05$) and the 'hours of opening the extractor fans'. According to the result, the hours of opening the extractor fans increase while the 'Mould'

Table 10.19 Correlation between 'quarterly bills', 'energy consumption behaviour' and 'housing issues' at Estate A

q7.2. How often do you experience the issues below in your home? (1 is never; 5 is always) Dampness	*Correlation coefficient*	*Sig. (two-tailed)*	*N*
q3.1. How much do you pay for the bill each quarter approximately? For electricity, please specify (£)	0.562*	0.015	18
q3.2. How much do you pay for the bill each quarter approximately? For gas, please specify (£)	0.626**	0.005	18
q7.3. How often do you experience the issues below in your home? (1 is never; 5 is always) Mould	**Correlation coefficient**	**Sig. (two-tailed)**	**N**
q13.2. Please specify the number of hours (hrs) opening extractor fan	−0.545*	0.036	15
q7.5. How often do you experience the issues below in your home? (1 is never; 5 is always) Condensation	**Correlation coefficient**	**Sig. (two-tailed)**	**N**
q13.2. Please specify the number of hours (hrs) opening extractor fan	−0.528*	0.043	15

* Correlation is significant at the 0.05 level (two-tailed).
** Correlation is significant at the 0.01 level (two-tailed).

Table 10.20 Correlation between 'quarterly bills', 'energy consumption behaviour' and 'housing issues' at Estate B

q7.1. How often do you experience the issues below in your home? (1 is never; 5 is always) Cold	*Correlation coefficient*	*Sig. (two-tailed)*	*N*
q3.1. How much do you pay for the bill each quarter approximately? For electricity, please specify (£)	0.289*	0.038	32
q7.5. How often do you experience the issues below in your home? (1 is never; 5 is always) Condensation	**Correlation coefficient**	**Sig. (two-tailed)**	**N**
q3.1. How much do you pay for the bill each quarter approximately? For electricity, please specify (£)	0.295*	0.035	32
q10.2. How often do you use your heating controls? (in winter) Wall thermostat	−0.441**	0.003	32

* Correlation is significant at the 0.05 level (two-tailed).
** Correlation is significant at the 0.01 level (two-tailed).

and 'Condensation' issues at their homes decrease. 'Dampness', 'mould' and 'condensation' increase the possibilities of low energy efficiency as they will directly impact indoor air quality and potentially influence human health and wellbeing (World Health Organization, 2009). Shortt and Rugkasa (2007) also argued that the 'dampness', 'mould' and 'condensation' could decrease home energy efficiency and increase the chances of fuel poverty.

According to Table 10.20, significant and moderate correlations were also found between the 'housing issues' at Estate B and a number of energy use variables at both the 0.01 and 0.05 levels. As a result, the 'Cold' (0.289*, $p < 0.05$) and 'Condensation' (0.295*, $p < 0.05$) issues increases while the 'quarterly electricity bills' increases. Besides, the 'frequency of using wall thermostat in winter' (−0.441**, $p < 0.01$) increases while the 'Condensation' issue decreases.

Flat Orientation and Other Variables

The orientations of flats are determined by the location of windows. According to Figure 10.52, there are four flats on each floor which are facing 'southwest and northwest', 'northwest and northeast', 'northeast and southeast' and 'southeast and southwest'. According to Figure 10.53, there are five flats located on each floor, which are facing 'southwest and northwest', 'northwest and northeast', 'northeast and southeast', 'southeast and southwest' and 'southwest only'.

It is believed that different orientations of flats will significantly impact heat gains and indoor thermal comfort (Li et al., 2002). Therefore, it affects the way that occupants operate their homes. In order to test if the energy consumption and occupants' energy consumption behaviour in the case studies are affected by orientations, the correlations and ANOVA tests were conducted.

According to Table 10.21, significant correlations were identified between 'orientation of flats' and a number of variables at both 0.05 and 0.01 levels, such as the 'orientation of flats' and 'quarterly electricity bills' (0.305*, $p < 0.05$), the frequency of experiencing 'cold' at home (0.436**, $p < 0.01$), the frequency of experiencing 'mould' at home (0.241*, $p < 0.41$), temperature set up on wall thermostat (0.337*, $p < 0.05$) and the frequency of opening the windows (−0.388**, $p < 0.01$). As the options of orientation are categorical data which are not in order, the relationships between different orientations and the abovementioned variables need to be further explored by using the ANOVA test.

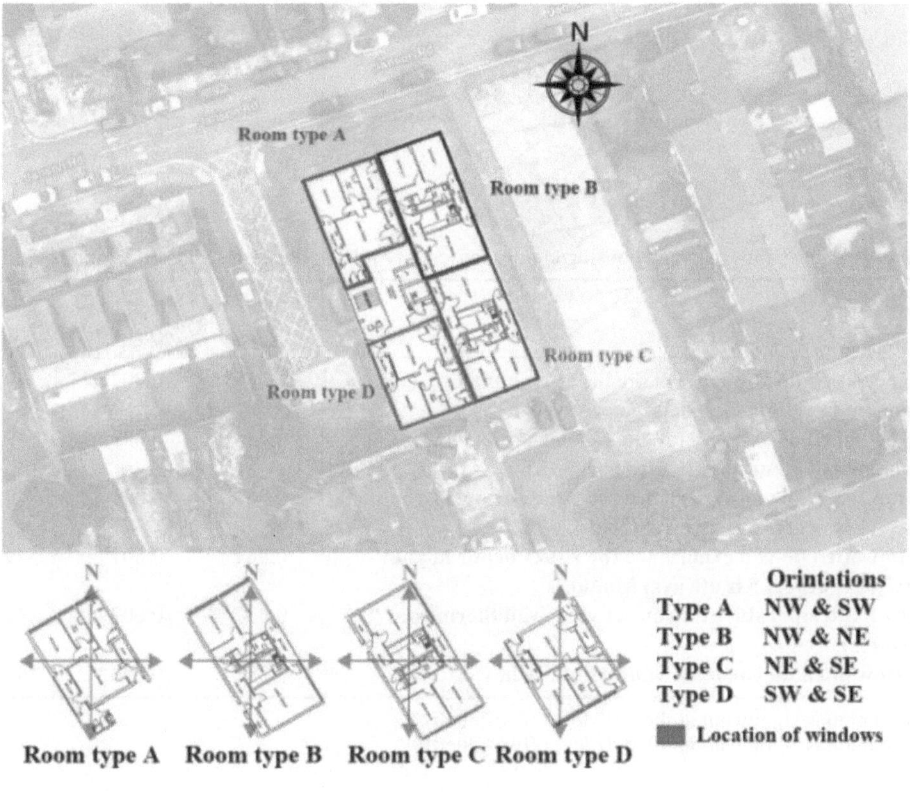

Figure 10.52 Flat orientations at Estate A.

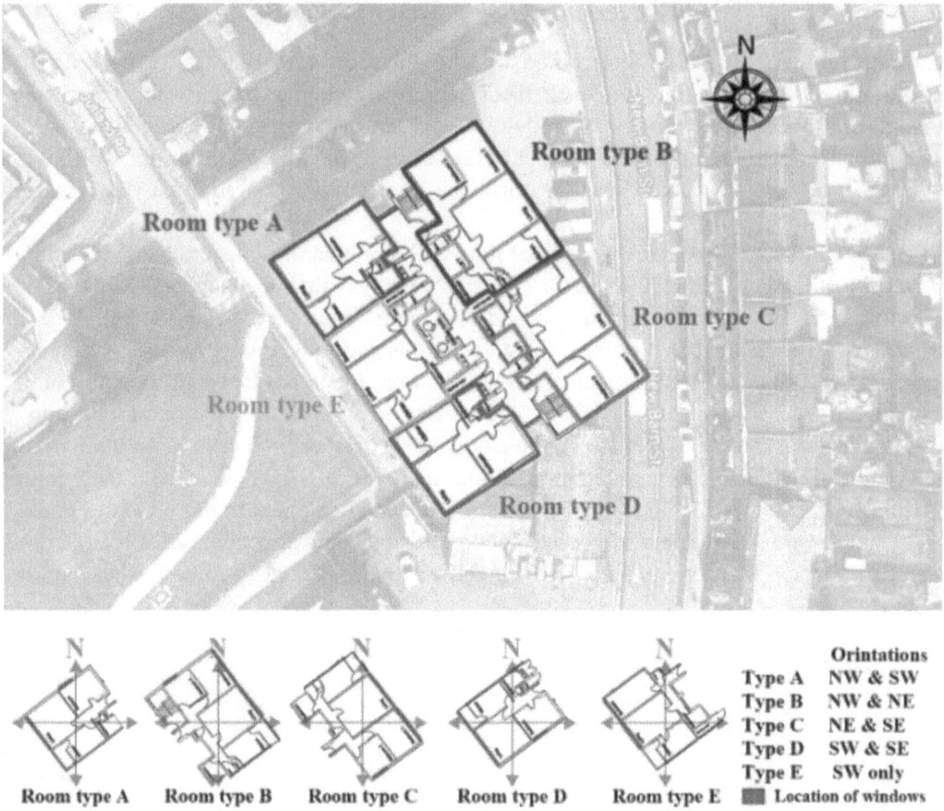

Figure 10.53 Flat orientations at Estate B.

Table 10.21 Correlations between housing conditions, energy related questions and 'orientations'

Orientation of flats	*Correlation coefficient*	*Sig. (two-tailed)*	*N*
q3.1. How much do you pay for the bill each quarter approximately? For electricity, please specify (£)	0.439**	0.002	49
q3.2. How much do you pay for the bill each quarter approximately? For gas, please specify (£)	0.436**	0.002	49
q7.1. How often do you experience the issues below in your home? (1 is never; 5 is always) Cold	0.369**	0.009	49
q7.3. How often do you experience the issues below in your home? (1 is never; 5 is always) Mould	0.461**	0.001	49
q11. At what temperature do you set your wall thermostat? (in winter)	0.437**	0.002	49
q12.1. How often do you open your windows in winter?	−0.388**	0.006	49

* Correlation is significant at the 0.05 level (two-tailed).
** Correlation is significant at the 0.01 level (two-tailed).

Table 10.22 ANOVA test between housing conditions, energy-related questions and 'orientations'

Dependent variable	*F*	*Sig.*
q3.1. How much do you pay for the bill each quarter approximately? For electricity, please specify (£)	3.746	0.010
q3.2. How much do you pay for the bill each quarter approximately? For gas, please specify (£)	2.976	0.029
q7.1. How often do you experience the issues below in your home? (1 is never; 5 is always) Cold	3.079	0.025
q7.3. How often do you experience the issues below in your home? (1 is never; 5 is always) Mould	3.190	0.022
q11. At what temperature do you set your wall thermostat? (in winter)	4.287	0.005
q12.1. How often do you open your windows in winter?	4.193	0.006

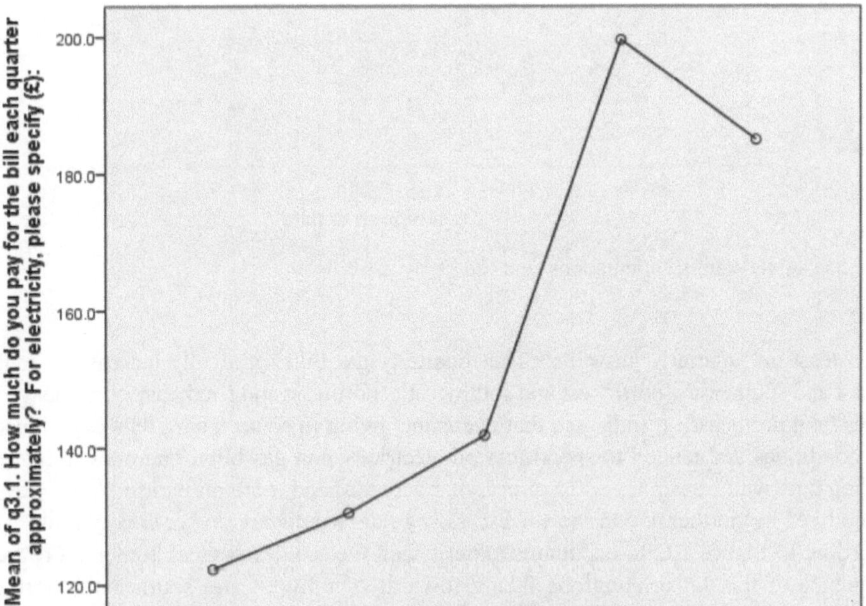

Figure 10.54 ANOVA test: 'orientations' and 'quarterly electricity bills'.

According to Table 10.22, significant differences in 'quarterly electricity bills', 'cold', 'mould', 'temperature set on wall thermostat' and 'frequency of opening the windows' were identified with different categories of 'orientation of flats'.

According to Figure 10.54, occupants who live in the flats with 'southwest and southeast' facing tended to spend the least on 'quarterly electricity bills'. The 'quarterly electricity bills' gradually increase in a sequence of 'southwest and southeast', 'northeast and southeast', 'northwest and southwest', 'southwest only' and 'northwest and northeast'. On the other hand, similar results were obtained from ANOVA test between 'orientation of flats' and 'quarterly gas bills'. According to Figure 10.55, occupants who live in the flats with 'southwest and southeast' facing tended to also

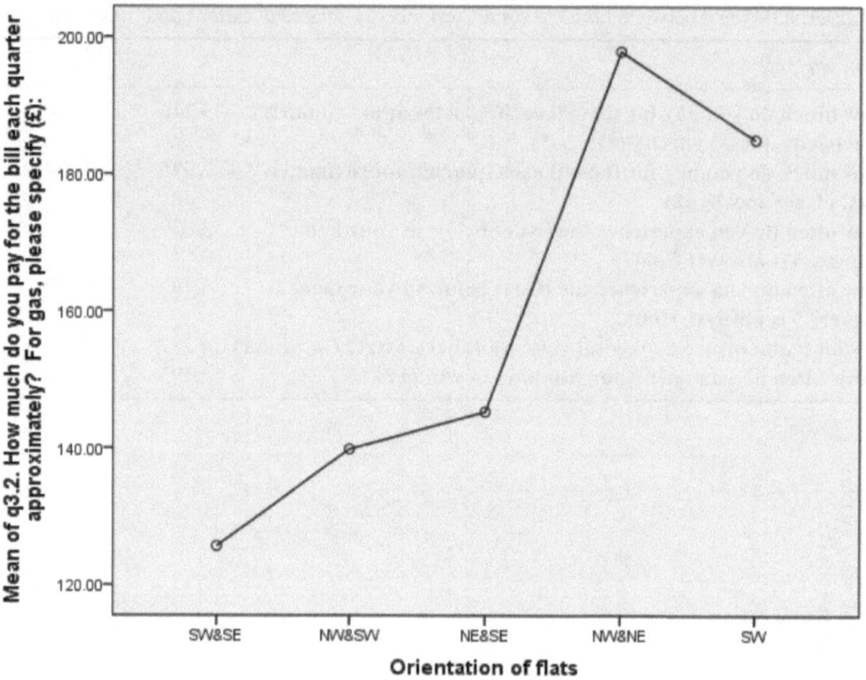

Figure 10.55 ANOVA test: 'orientations' and 'quarterly gas bills'.

spend the least on 'quarterly gas bills'. The 'quarterly gas bills' gradually increase in sequence of 'southwest and southeast', 'northwest and southwest', 'northeast and southeast', 'northeast only' and 'northwest and northeast'. It indicated that occupants living in south facing flats experienced better thermal conditions and tended to spend less on electricity and gas bills. The ones living in east or west facing flats with certain levels of direct solar access spend more on their bills. However, occupants who lived in 'northeast' and 'northwest' facing flats would pay higher energy bills.

According to Figure 10.56, occupants experienced the 'cold' issues at homes more frequently in a sequence of the flat orientations facing towards 'southwest and southeast', 'northeast and southeast', 'northwest and southwest', 'southwest only' and 'northwest and northeast'. According to Figure 10.57, occupants had experienced the 'mould' issues at homes more frequently in a sequence of flat orientations facing towards 'southwest and southeast', 'northwest and southwest', 'northeast and southeast', 'southwest only' and 'northwest and northeast'. Similarities were indicated between the two different housing issues and their relationships with the orientations of flats. However, the 'cold' issues were more obvious for the flats facing 'northwest and southwest' and the ones facing 'northeast and southeast'. In contrast, the 'mould' issues were more obvious in the flats facing 'northeast and southeast' than the ones facing 'northwest and southwest'.

According to Figure 10.58, the temperatures set on wall thermostats by the occupants increase in the sequence of the flat orientations facing 'northwest and southwest', 'southwest and southeast', 'northeast and southeast', 'southwest only' and 'northwest and northeast'. It implies that flats with more south-facing windows normally tend to receive more solar radiation thus occupants' heating requirements may be less than the ones living in the north-facing flats. According to Figure 10.59, the 'frequency of opening the windows' decreased in sequence with the flat orientations' towards

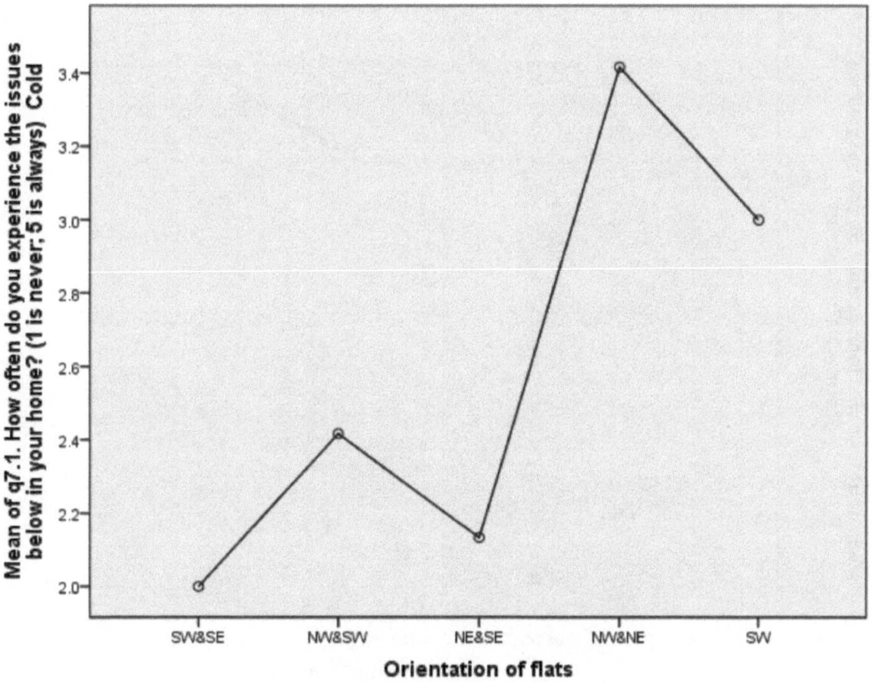

Figure 10.56 ANOVA test: ‘orientations’ and ‘cold’.

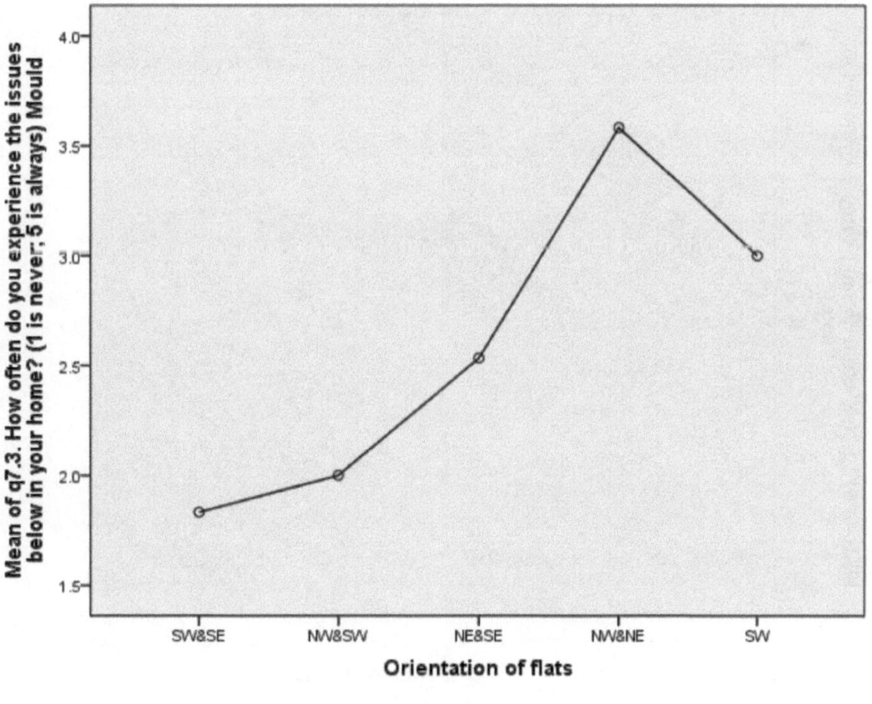

Figure 10.57 ANOVA test: ‘orientations’ and ‘mould’.

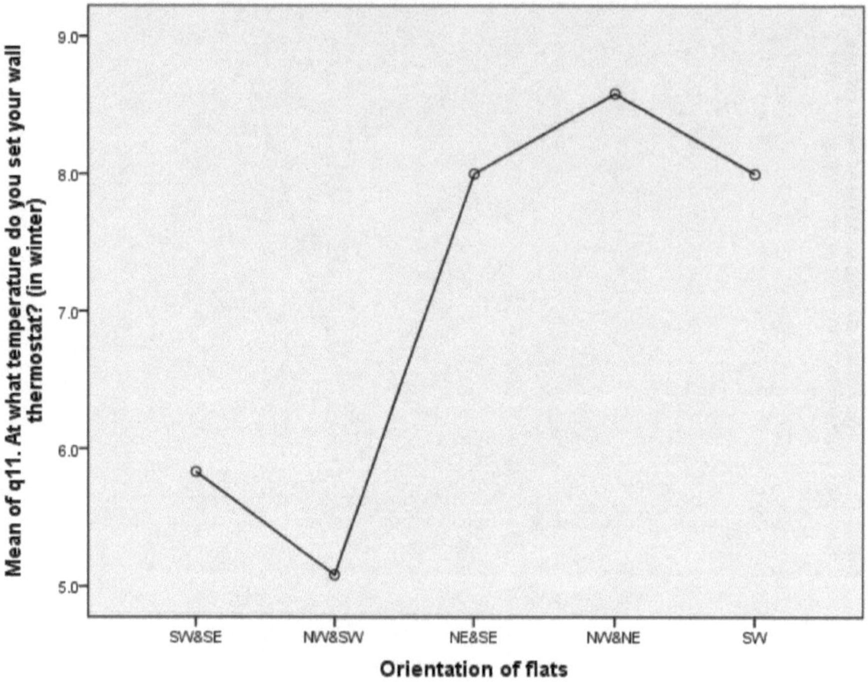

Figure 10.58 ANOVA test: 'orientations' and 'temperature'.

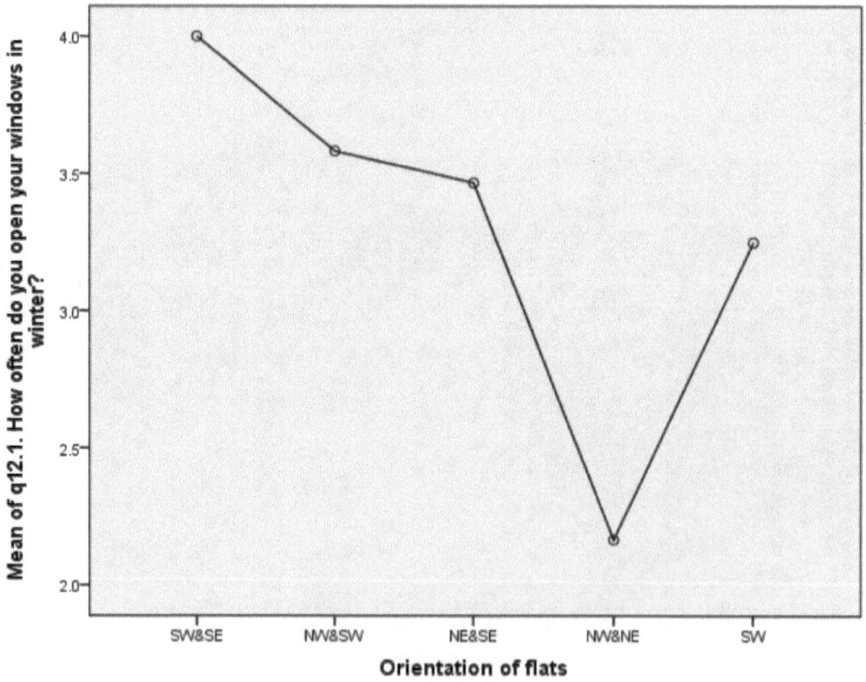

Figure 10.59 ANOVA test: 'orientations' and 'use of windows'.

'southwest and southeast', 'northwest and southwest', 'northeast and southeast', 'southwest only' and 'northwest and northeast'. In general, it showed that occupants living in south-facing flats tend to open their windows more than the ones living in north-facing flats.

Significant differences between different flat orientations with a number of variables were identified in the ANOVA test. Due to the flat orientation factors, south-facing flats in the UK receive more heat gain and less heat loss with abundant sufficient solar radiation. As a result, occupants tend to open the windows and set the lower temperature on the wall thermostat more frequently, and consume less heating and electricity. Kontoleon and Zenginis (2017) also indicated that the south-oriented buildings were characterised with the most heat gains and the north-oriented buildings were characterised with the most heat losses in the UK. The same point of view was also supported by Abanda and Byers (2016) through modelling the building performance of different orientations that building orientation significantly contributes to building energy performance. As a result, the flats with more south-facing windows tend to have less 'cold' and 'mould' issues than the ones with north-facing windows. Flats with different orientations demonstrate different levels of housing problems. It implies that the retrofit interventions may need to be specifically tailored towards different building facades rather than consistently applied to the whole building.

Conclusions

A significant of findings are identified to describe the current condition of home energy performance in the aspects of occupants' energy consumption behaviour, occupancy patterns and socio-demographic characteristics from the last section. The correlation analysis indicates that tenancy status significantly impacts home energy performance. Fuel poverty is also obvious in the research case study as 38.7% of the households are at risk of it when correlating their annual incomes and energy bills from the questionnaires.

It is also found that household profiles influence home energy performance in different aspects. Age groups as one of the dominating factors significantly impact the use of heating systems, ventilation components and energy-saving activities. Besides, employment status could potentially impact the occupancy patterns at homes and consequently impact occupants' flexibility in energy management. In addition, occupants in poor health conditions could also require extra heating and ignore the importance of air ventilation. Moreover, the household profiles also influence both the occupancy and heating patterns at homes and consequently impact home energy consumption. For instance, the occupants with higher income levels tend to have longer heating patterns and spend more energy than low-income families at homes.

Besides, the temperature set up on the thermostat will directly impact energy performance according to the research findings. The reason for this is that to maintain higher indoor temperatures will require higher energy consumption. Although 21°C is recommended as an appropriate indoor temperature, the majority of the occupants tend to set it higher. In the case study, 40% of the respondents set their wall thermostats higher than 21°C, which is relatively high, 30% of them set their wall thermostats less or equal to 21°C and 30% of them do not know what temperature they set it at. Besides, 24% of them set their wall thermostats equal or more than 24°C, which may imply to overheating problems and implications to their own health. On the other hand, occupants' perception of comfortable temperature is also influenced by the type of activities at home. Additionally, the research found that the use of windows, extractor fans and trickle vents also significantly impacts home energy performance. Moreover, occupants who have received energy advice tend to perform better on operating their homes in different aspects, such as 'using different types of heating controls' and 'energy conservation behaviour'.

Furthermore, the orientations of flats also have a significant impact on home energy performance concerning occurrences of housing issues, use of windows and temperature set on wall thermostats. Due to varying levels of solar radiation, south facing flats tend to be warmer and consequently may open the windows more frequently and set the indoor temperature lower. According to the research findings at the first stage, occupants did not have sufficient energy knowledge as they were not familiar with the heating systems and ventilation systems at their homes. They also tend to consume more energy than they should due to a variety of reasons. The key findings of the questionnaire survey were summarised in bullet points as follows:

- Majority of the respondents were social renters (84%). More than half of them (56%) have been living there for more than ten years.
- Thirty-eight per cent of participating households pay more than the national average level (£154.75) for their quarterly electricity bills and 36% of them pay more than the national average level (£157.5) for their quarterly gas bills (DBEIS, 2018).
- Thirty per cent of participating households do not have employed family members. Fifty-four per cent of them earn less than £12,000 per annum. The low-income level and high energy bills lead to significant fuel poverty problems.
- Respondents do not have sufficient knowledge of using heating controls, extractor fans and trickle vents to maintain a good living environment. They failed to follow a number of energy conservation approaches.
- The way that homes operated was found significant impact on energy performance. They include the use of wall thermostats, boiler thermostats, radiator valves, windows, extractor fans, trickle vents and indoor temperature.
- Occupants' energy consumption behaviour is correlated with a number of socio-demographic factors, such as family size, income levels, employment status and occupants' health conditions.
- The provision of energy advice was proven effective for promoting behavioural change.
- Housing problems, such as mould, condensation and cold, could cause more heating consumption. Flat orientations also significantly impact on occupants' energy consumption behaviour and energy performance.

A focus group interview was arranged to probe into details of the key issues. The barriers of energy conservation were explored. Besides, the implementation of smart meters and the likelihood of occupants using energy management applications are explored. The findings of the focus group interview are demonstrated in the next section.

References

Abanda, F. H., & Byers, L. (2016). An investigation of the impact of building orientation on energy consumption in a domestic building using emerging BIM (Building Information Modelling). *Energy, 97*, 517. https://doi.org/10.1016/j.energy.2015.12.135.

Abrahamse, W., & Steg, L. (2009). How do socio-demographic and psychological factors relate to households' direct and indirect energy use and savings? *Journal of Economic Psychology, 30*(5), 711. https://doi.org/10.1016/j.joep.2009.05.006.

Ahn, K., & Park, C. (2016). Correlation between occupants and energy consumption. *Energy and Buildings, 116*, 420. https://doi.org/10.1016/j.enbuild.2016.01.010.

Berneiser, J., Auerswald, S., Maier, D., Gölz, S., Carbonare, N., & Pflug, T. (2023). Feeling the breeze? Ventilation practices and occupant requirements for mechanical ventilation in residential buildings. *Energy and Buildings, 302*, 21–23. https://doi.org/10.1016/j.enbuild.2023.113702

Blackman, D. (2017). Why social landlords need to think seriously about smart meters. *Inside Housing*. [Online]. Availableat:https://www.insidehousing.co.uk/insight/insight/why-social-landlords-need-to-think-seriously-about-smart-meters2-53231. [Accessed 06 June 2018].

Bogin, D., Dorman, M., Kissinger, M., & Erell, E. (2024). *Beyond efficiency: evaluation of the electricity saving potential of green residential buildings*. Elsevier BV. https://doi.org/10.1016/j.enbuild.2024.114627.

Bone, A. (2018). Preventing avoidable deaths this winter – Public health matters. [Online]. Available at: https://publichealthmatters.blog.gov.uk/2014/10/21/preventing-avoidable-deaths-this-winter [Accessed 30 November 2017].

Bonett, D., & Wright, T. (2000). Sample size requirements for estimating Pearson, Kendall and Spearman correlations. *Psychometrika, 65*(1), 23–28.

Brounen, D., Kok, N., & Quigley, J. M. (2012). Residential energy use and conservation: Economics and demographics. *European Economic Review, 56*(5), 931. https://doi.org/10.1016/j.euroecorev.2012.02.007.

Colclough, S., & Salaris, C. (2023). Quantifying overheating in nZEB Irish residential buildings. An analysis of recorded interior temperatures of Irish newbuild and retrofit residential buildings against CIBSE, Passive House and WHO overheating criteria and recorded occupant satisfaction. *Energy and Buildings, 303*, 45–47. https://doi.org/10.1016/j.enbuild.2023.113571

Conan, G. (1981). *Domestic gas consumption, household behaviour pattern, and window opening*. Ph.D. Brunei University.

DBEIS. (2013). Smart meters: information for industry and other stakeholders. [Online]. Available at: https://www.gov.uk/guidance/smart-meters-information-for-industry-and-other-stakeholders [Accessed 14 April 2017].

De Castella, T. (2013). How do you prepare for a lifetime of renting? *BBC News*. [Online]. Available at: https://www.bbc.co.uk/news/magazine-22952667. [Accessed 06 June 2018].

DECC. (2013). *A framework for future action*. DECC.

Designing Buildings. (2015). Trickle ventilation in buildings. [Online]. Available at: https://www.designingbuildings.co.uk/wiki/Trickle_ventilation_in_buildings. [Accessed 21 March 2018].

Edalatnia, S., & Das, R. R. (2024). Building benchmarking and energy performance: analysis of social and affordable housing in British Columbia, Canada. *Energy and Buildings, 313*, 121–123. https://doi.org/10.1016/j.enbuild.2024.114259.

Ehrhardt-Martinez, K., Donnelly, K., & Laitner, J. (2010). *Advanced metering initiatives and residential feedback programs: a meta-review for household electricity-saving opportunities*. Energy Savings and Advanced Metering Meta-Analysis, ACEEE, United States.

Fabi, V., Andersen, R. V., Corgnati, S., & Olesen, B. W. (2012). Occupants' window opening behaviour: a literature review of factors influencing occupant behaviour and models. *Building and Environment, 58*, 188. https://doi.org/10.1016/j.buildenv.2012.07.009.

Gram-Hanssen, K. (2013). New needs for better understanding of household's energy consumption – behaviour, lifestyle or practices? *Architectural Engineering and Design Management, 10*(1–2), 91. https://doi.org/10.1080/17452007.2013.837251.

Guerra Santin, O. (2011). Behavioural patterns and user profiles related to energy consumption for heating. *Energy and Buildings, 43*(10), 2662. https://doi.org/10.1016/j.enbuild.2011.06.024.

Gupta, R., & Gregg, M. (2016). Do deep low carbon domestic retrofits actually work? *Energy and Buildings, 129*, 330. https://doi.org/10.1016/j.enbuild.2016.08.010.

Hargreaves, T., Nye, M., & Burgess, J. (2010). Making energy visible: a qualitative field study of how householders interact with feedback from smart energy monitors. *Energy Policy, 38*(10), 6111. https://doi.org/10.1016/j.enpol.2010.05.068.

Ho Kim, S., & Jun Moon, H. (2024). *A detailed occupant activity classification model in a residential environment using building monitoring data: considering occupant characteristics*. Elsevier BV. https://doi.org/10.1016/j.enbuild.2023.113867.

Jacobs, S., Van Minnebruggen, S., Matbouli, H., Ghane, S., Hellinckx, P., & Verhaert, I. (2024). Evaluating innovative collective heating and cooling concepts by incorporating occupants' preferences for conflicting performance indicators. *Energy and Buildings, 314*, 123–127. https://doi.org/10.1016/j.enbuild.2024.114264.

Kavousian, A., Rajagopal, R., & Fischer, M. (2013). Determinants of residential electricity consumption: using smart meter data to examine the effect of climate, building characteristics, appliance stock, and occupants' behavior. *Energy, 55*, 184. https://doi.org/10.1016/j.energy.2013.03.086.

Kontoleon, K. J., & Zenginis, D. G. (2017). Analysing heat flows through building zones in aspect of their orientation and glazing proportion, under varying conditions. *Procedia Environmental Sciences, 38*, 348. https://doi.org/10.1016/j.proenv.2017.03.100.

Leon, A. R., & Zhu, Y. (2008). ANOVA extensions for mixed discrete and continuous data. *Computational Statistics & Data Analysis, 52*(4), 2218–2227.

Neuhoff, K., Amecke, H., Novikova, A., & Stelmakh, K. (2011). *Thermal efficiency retrofit of residential buildings: the German experience CPI report*. Climate Policy Initiative.

Nicholls, L., & Strengers, Y. (2017). *Rising household energy and water bills: Case studies of health, wellbeing and financial impacts*. RMIT University.

Ozarisoy, B. (2022). Energy effectiveness of passive cooling design strategies to reduce the impact of long-term heatwaves on occupants' thermal comfort in Europe: Climate change and mitigation. *Journal of Cleaner Production, 330*, Article 129675. https://doi.org/10.1016/j.jclepro.2021.129675.

Ozarisoy, B., & Altan, H. (2021). Regression forecasting of 'neutral' adaptive thermal comfort: A field study investigation in the south-eastern Mediterranean climate of Cyprus. *Building and Environment, 202*. https://doi.org/10.1016/j.buildenv.2021.108013.

Ozarisoy, B., & Altan, H. (2022). Significance of occupancy patterns and habitual household adaptive behaviour on home-energy performance of post-war social-housing estate in the South-eastern Mediterranean climate: energy policy design. *Energy, 244*. https://doi.org/10.1016/j.energy.2021.122904.

Ozarisoy, B., & Altan, H. (2023a). A novel methodological framework in energy policy. Taylor & Francis Group, Routledge Explorations in Energy Studies, https://www.taylorfrancis.com/chapters/mono/10.4324/9781003393092-3/novel-methodological-framework-energy-policy-bertug-ozarisoy-hasim-altan.

Ozarisoy, B., & Altan, H. (2023b). *Home energy performance assessment*. Taylor & Francis Group, Routledge Explorations in Energy Studies, https://www.taylorfrancis.com/chapters/mono/10.4324/9781003393092-5/home-energy-performance-assessment-bertug-ozarisoy-hasim-altan.

Pajula, J., & Tohka, J. (2016). How many is enough? Effect of sample size in inter-subject correlation analysis of fMRI. *Computational Intelligence and Neuroscience – Special issue on Simulation and Validation in Brain Image Analysis, 2*, 1–10.

Saeki, K., Obayashi, K., Iwamoto, J., Tone, N., Okamoto, N., Tomioka, K., & Kurumatani, N. (2014). Stronger association of indoor temperature than outdoor temperature with blood pressure in colder months. *Journal of Hypertension, 32*(8), 1582–1589.

Seyedrezaei, M., Tak, A. N., & Becerik-Gerber, B. (2023). Consumption and conservation behaviors among affordable housing residents in Southern California. *Energy and Buildings, 304*, 27–29. https://doi.org/10.1016/j.enbuild.2023.113840.

Shortt, N., & Rugkåsa, J. (2007). "The walls were so damp and cold" fuel poverty and ill health in Northern Ireland: results from a housing intervention. *Health & Place, 13*(1), 99. https://doi.org/10.1016/j.healthplace.2005.10.004.

Stazi, F., Naspi, F., Ulpiani, G., & Di Perna, C. (2017). Indoor air quality and thermal comfort optimization in classrooms developing an automatic system for windows opening and closing. *Energy and Buildings, 139*, 732. https://doi.org/10.1016/j.enbuild.2017.01.017.

Weaver, B., & Koopman, R. (2014). An SPSS macro to compute confidence intervals for Pearson's correlation. *The Quantitative Methods for Psychology, 10*(1), 29–39.

World Health Organization. (2009). *WHO guidelines for indoor air quality: dampness and mould*. WHO Regional Office for Europe.

Yun, G. Y., & Steemers, K. (2011). Behavioural, physical and socio-economic factors in household cooling energy consumption. *Applied Energy, 88*(6), 2191. https://doi.org/10.1016/j.apenergy.2011.01.010.

11

ENVIRONMENTAL MONITORING AND ENERGY SIMULATION

Introduction

This chapter explains the findings from the monitoring study and provides explanations along with discussions. This chapter has mainly focused on the post-occupancy study. The post-occupancy study has considered two aspects, namely, the 'energy consumption study' and the 'thermal comfort study'. With regard to energy consumption influence factors, three factors have been studied, namely, the number of households, the internal temperatures during the heating season and the time spent at home. With regard to thermal comfort, CIBSE Guide A and EN15251 criteria were used to determine the thermal comfort for selected rooms during the summer and winter periods. More relevant findings were explained in the related subsections below. This analysis was based on the monitoring data which was recorded for two years. Within this research, the winter or 'heating season' is defined as November to February and the summer or 'cooling season' is defined as June to August. During the monitoring periods, energy consumption data for gas and electricity and internal data such as temperatures and Relative Humidity (RH) were monitored. Table 11.1 shows the period of monitoring, and the analysis study was taken from the periods monitored.

During energy monitoring for the case study homes, several problems of both technical and human influence were experienced. In some periods, there was a GSM line blackout so data on site was recorded by meters; however, they were not transmitted to the server to be saved. There was a meter battery supply run out, so for the period from September 2008 to March 2010 for gas metering in flats 1 and 2, and in the period from April 09 to March 10 for electricity metering in house 1, there were no recordings. Furthermore, some of the meters needed to be restarted, in these cases the meters were disconnected and/or removed by the user (in house 1).

Energy Consumption and CO_2 Emissions

Gas and Electricity Consumption

The case homes energy consumption data was recorded and the results are represented in Figure 11.1. For two years, September 2007 to August 09, gas usage and electricity consumption were monitored using the DYNAMATplus software. For some of the periods, no data was recorded due to technical problems with measurements, as mentioned above.

DOI: 10.1201/9781003490364-14

Table 11.1 Monitoring and analysing periods

	Monitoring period		*Analysing periods*	
	Energy data	*Internal data*	*Heating season*	*Cooling season*
Flat 1	September 2007 to March 2010	2007.08.17 to 2009.02.27	2007–2008 and 2008–2009	June to August 2008
Flat 2		2007.10.12 to 2009.02.27	2007–2008 and 2008–2009	June to August 2008
House 1		2007.05.04 to 2007.09.23	2007–2008	June to August 2007
House 2		2006.12.15 to 2009.02.27	2007–2008 and 2008–2009	June to August 2007 and 2008
Flat 3		2006.12.15 to 2008.03.07	2007–2008	June to August 2007
Flat 4		2006.12.15 to 2009.02.27	2007–2008 and 2008–2009	June to August 2007 and 2008

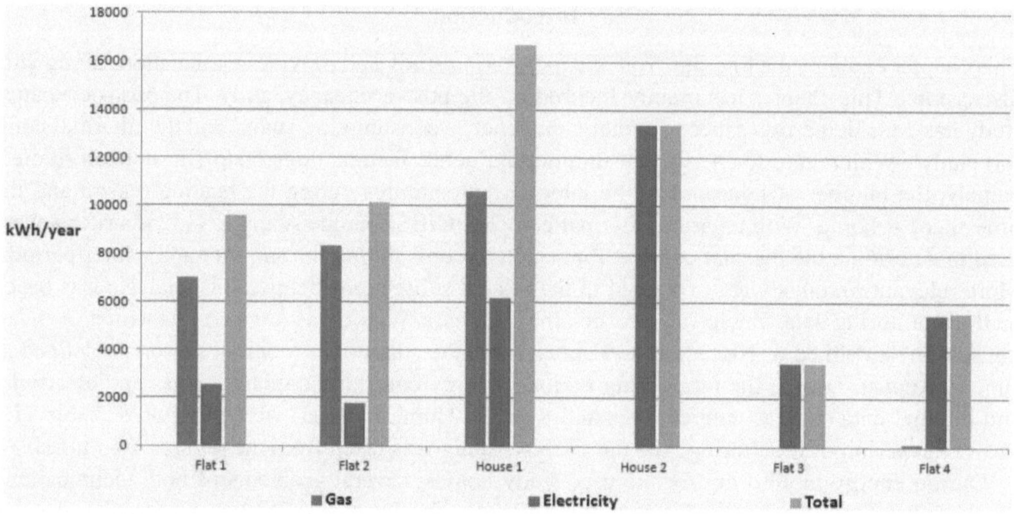

Figure 11.1 Energy consumption from case study homes.

In general, as the floor area and number of households increased, the energy consumption increased. In terms of the influence of the main energy sources, the case study homes' heating systems using electricity as a main energy source showed better performance than the gas central heating system.

This can be explained by the efficiency of the end-use energy source. Specifically, when heating systems with electricity, the efficiency is almost 100%. In contrast, heating systems with gas rely on the boiler's efficiency, which was around 90% in this research. Therefore, energy source for electricity system homes could perform better than homes with gas heating systems as shown by the efficiency of their heating systems (Calderon & Beltran, 2018).

Figure 11.1 shows each case study homes gas, electricity and total energy consumption. With regard to energy consumption, in house 1 the most energy was consumed followed by house 2, due to the size of the floor area and number in the household. House 1 and house 2 have an identical floor

Table 11.2 Comparison between house 1 and house 2

	House 1	*House 2*
Total floor area (m^2)	127	127
Number in the household	5 (one adult and four teenager)	5 (two adults and three kids)
Heating system	Gas radiator heating	Electric panel heating
Average internal temperature (during the heating season)	Living room: 22.5°C Peak Temp: 28.6°C	Living room: 21.4°C Peak Temp: 25.4°C
Total energy consumption	16,718 kWh	13,358 kWh

plan, floor area and size of household. However, their total energy consumption differs by 25% (see Figure 11.1 and Table 11.2). Table 11.2 explains why this energy consumption difference has occurred.

Heating systems energy efficiency difference: as mentioned above, electricity heating systems have better energy efficiency than gas central heating systems. Therefore, a difference of at least 10% in heating energy consumption occurred between these systems. An additional factor could account for this, which is heat loss through pipes and other parts of the gas heating systems. From this point of view, the difference between their heating systems' efficiency would be more than 10%. The average internal temperature difference, turning a room's thermostat down by just 1°C can reduce heating bills by 10% and by 10% of heating energy consumption. In line with this concept, the house 1 living room's average internal temperature during the heating season was 22.5°C and this is 1.1°C higher than the living temperature in house 2 and the peak temperature difference for the living room is more than 3°C. Thus, the internal temperature difference could influence the total energy consumption. Number of children/teenagers: the presence of children in a household is likely to affect the number of appliances per home and the resulting energy use, especially electricity consumption. The use of energy by children is that 58% of children aged between 12 and 14 have electric appliances and they use them most of their home time. In line with this trend, house 1 would consume more electricity than house 2, due to the difference between the number of children and their ages. Background: the occupants of house 1 originate from Africa and whereas the occupants of house 2 are typically British. Their thermal comfort temperature expectations during the winter were different from each other and the occupants of house 1 preferred a higher temperature than those in house 2, due to their country of origin. Therefore, the house 1's peak temperature for living room temperature was higher by 3°C and they may consume more energy than the house 2.

The least energy consumed is flat 3 in which the retired couple live. It was found that older people are more likely to consider the cost and bills more than other age groups and also to switch off lights that are no use. From these reasons, low energy consumption in flat 3 could be explained and when the author visited the flat 3, the author found out they do not have electrical appliances, as many as other case study homes. In terms of the energy efficiency within these case study homes, again the most energy efficiency home is flat 3 due to their energy consumption and in contrast, the least energy efficiency home is flat 1, which consumed three times more than flat 3 in kWh per m^2. More detail about the energy efficiency for the case study homes will be explained in the relevant section in below.

Carbone Dioxide Emissions

Figure 11.2 shows the results for carbon dioxide emissions from the case study homes. The figures were calculated based on the CO_2 emissions factor which is calculated on the CO_2 emissions rate for energy generation from different energy sources. The emissions factor is based on SAP

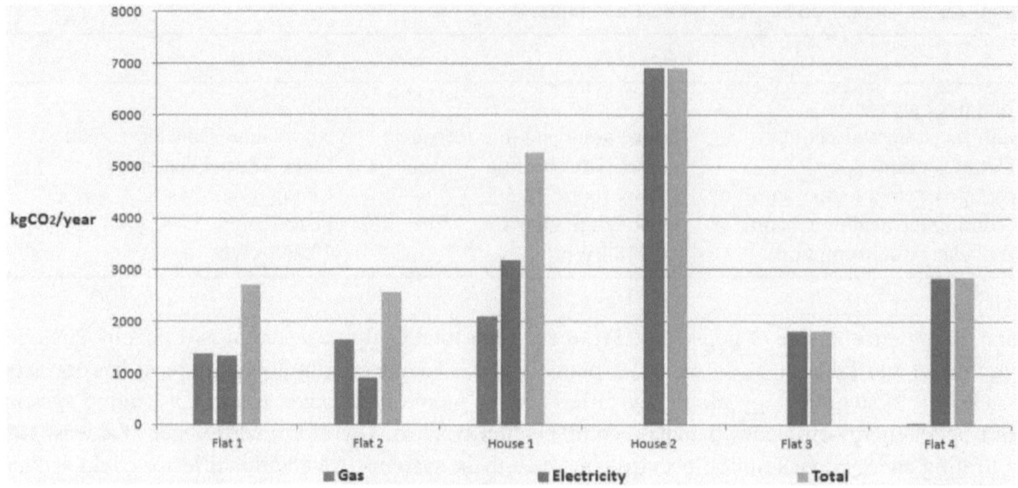

Figure 11.2 CO_2 emissions from case study homes.

emission factors which are gas for 0.198 kWh/kg CO_2 and electricity for 0.517 kWh/kg CO_2. In terms of CO_2 emissions from energy sources, energy generation from gas emits almost 2.5 times less than energy generation from electricity. Therefore, households that consume more electricity than others have a greater level of carbon dioxide emissions. In general, the CO_2 emissions from case study homes show almost the same trend as energy consumption in case study homes which means as the size of homes and number in the household increase, the CO_2 emissions increase (Gupta & Gregg, 2017). With regard to energy sources, it seems homes with more electricity consumption had a disadvantage in CO_2 emissions. From this point of view, electricity consumption could play a very important role in carbon dioxide emissions.

For example, in terms of energy consumption, house 1 consumed 25% more than house 2. With regard to CO_2 emissions, house 2 had the most CO_2 emissions among the case study homes and house 2 emitted almost 30% more than house 1, due to their main energy source. To compare the CO_2 emissions between flat 2 and 4, the total energy consumption for flat 4 was almost 50% less than flat 2. However, in terms of CO_2 emissions, flat 4 consumed 12% more than flat 2, again due to the main energy source. Flat 3 is still the home with the lowest CO_2 emissions among the case study homes due to its total energy consumption even though it uses electricity as its main energy source. However, the home with the least CO_2 emissions in a square metre case study home is flat 2, and again the worst case is flat 1 due to the high energy consumed with small floor area and more electricity consumption.

Factors Influence on Energy Consumption

User Influence

The energy consumption from the homes was very well matched with the size of the household and number of children. As mentioned earlier, when the monitoring study started, only one male was in flat 1 then his partner with a baby moved into this flat some time in 2008 (When the survey study was carried out for the case study homes, flat 1 and flat 2 interviews were not carried out. There is, therefore, no analysis for the survey study for these flats and it could be one of the limitations of this research. However, the surveyor had a chance to have a short talk with flat 1

Table 11.3 Number of households and children in case study homes

	Flat 1	*Flat 2*	*House 1*	*House 2*	*Flat 3*	*Flat 4*
Number of households	3	3	5	5	2	2
Children	1 (baby)	2 (child)	4 (teenager)	3 (child)	0	0
Bedroom	21.4°C	20.7°C	–*	19.2°C	18.8°C	19.8°C
Living room	21.6°C	19.4°C	22.5°C	21.4°C	18.4°C	18.3°C
Total energy (kWh/yr)	9,569	10,124	16,718	13,358	3,496	5,573

* No data recorded.

occupant to ask general questions about the flat and this information was used to analyze the post-occupancy study).

The correlation between the size of household and their age and the number of children with energy consumption shows that total energy demand increases with households' size, and number of children and their age (see Table 11.3). For example, houses 1 and 2 have the same number of households. But house 1 has 4 teenagers and their energy consumption was much higher than house 2. It is assumed that the electricity consumption of appliances for house 1 would be higher than other case homes, due to the teenagers having more electrical appliances and spending most of their time with them when they are at home. However, further research is needed to understand the influence of a child's age on electricity consumption. Flat 3 shows that they used less energy than other case homes. This may be because older people are more likely to show cost-conscious behaviour than other age groups because they have to spend within their pension and switch off appliances that are no longer needed (Mitchell & Natarajan, 2020). They also do not have so many electric goods compared to the case study homes (when the author visited the flat, there were not even any TVs and any Information Technology goods).

Internal Temperatures Influence

The study shows that the most influential factors for heating energy consumption are the size of dwelling and internal temperatures (O'brien et al., 2016). Table 11.4 represents the average internal temperatures for selected rooms in each case study homes from November 2007 to February 2008. It shows very similar results which means that case homes with children show higher average internal temperature than homes without children (see Figure 11.4) and homes with children consumed more energy, as mentioned above. During the monitoring period, house 1's bedroom data was missing due to a technical problem. However, from the average temperature of the living room and total energy consumption, it can be assumed that house 1's internal bedroom temperature would be higher than the case study home without children.

Figure 11.3 shows the internal temperature during the 2008–2009 heating season for flats 1, 2 and 4 bedrooms. Again, it clearly shows that homes with children, which are flats 1 and 2, have a higher average internal temperature and they consumed more heating energy.

Similarly, Figure 11.4 shows the internal temperatures of flat 1's bedroom for two heating seasons. After February 2008, which was moving the baby into flat 1, the internal temperature of flat 1's bedroom was clearly higher than before moving the baby into the flat. Due to a technical problem, the energy consumption data for 2008–2009 was not recorded. However, from the above findings, it can be assumed that the energy consumption for flat 1 after moving the baby would be higher than before moving the baby.

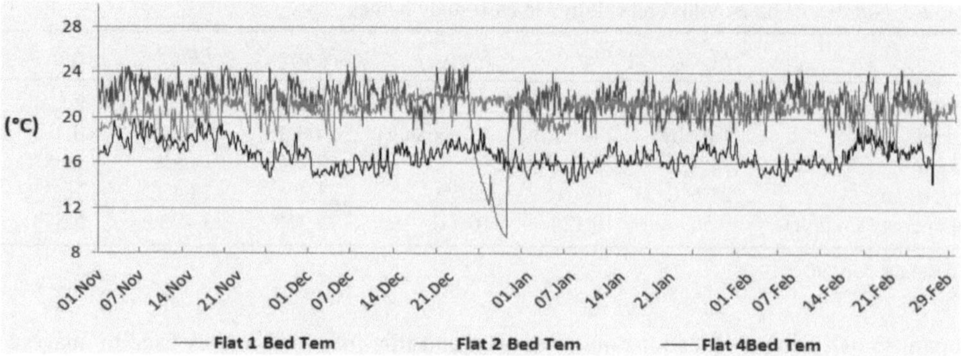

Figure 11.3 Bedroom temperatures in selected case study homes.

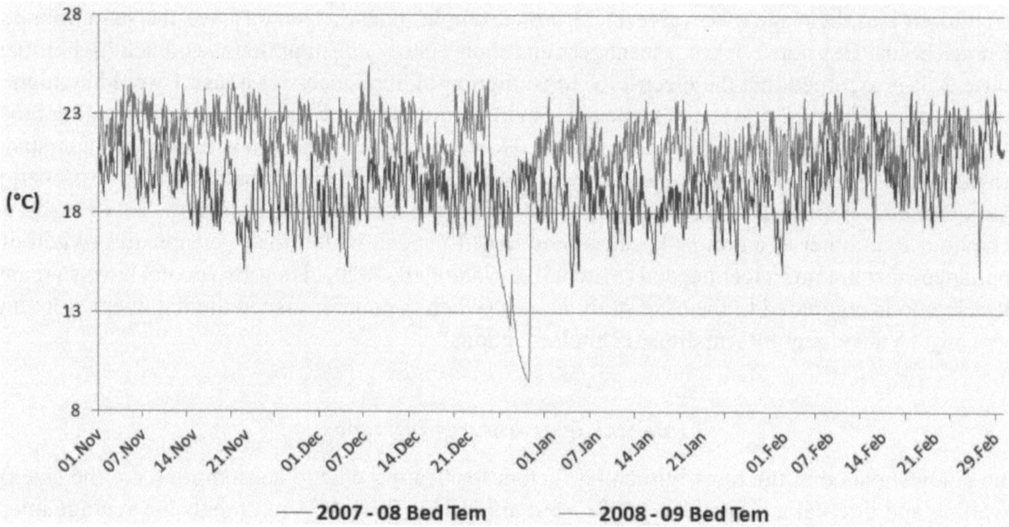

Figure 11.4 Flat 1's bedroom temperature for two heating seasons.

Time Spend at Homes

From the internal temperature and RH monitoring results, it was found that flat 3 was unoccupied for almost 27 days during the monitoring period, due to personal circumstances, such as holidays and medical situations (see Figure 11.5 assumed unoccupied households' periods are shown as grey colour). The energy consumption study pointed out that flat 3's energy consumption was unusually lower, and this could be explained by user influences such as age group and number of households. However, these influences were not enough to explain flat 3's energy consumption. Therefore, the author considered one more factor which is 'Time spent at home'. Flat 3's unusual energy consumption could be explained by the relationship between energy consumption and time spent at home. This could explain why flat 3's energy consumption was extremely lower than other case study homes, namely, that they spent less time in their home.

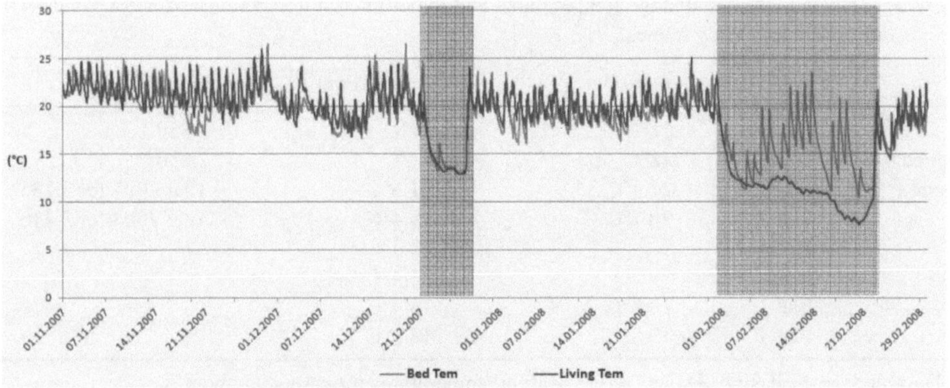

Figure 11.5 Internal temperature monitoring data for flat 3.

Table 11.4 Comparison of measured energy consumptions with SAP calculation

Case homes	*Measured (kWh/yr)*	*SAP (kWh/yr)*	*Difference (%)*
Flat 1	9569	5811	65
Flat 2	10124	9841	3
House 1	16718	9249	81
House 2	13358	8943	49
Flat 3	3496	5483	–36
Flat 4	5573	5981	–7

Energy Efficiency

SAP Assessment of the Case Study Homes

Yearly sets of measurements were compared with different energy consumption prediction methods, including SAP and represented in Table 11.4. In general, most SAP predictions performed better than the actual homes except flats 3 and 4 and both flats performed better than SAP predictions. With regard to SAP results, the most energy-consumed case study home is flat 2 and even the size of the home is smaller than houses 1 and 2.

Flats 2 and 4 show that the actual energy consumption and SAP prediction were quite close. These case homes were grouped as full-time working households and having a south-facing living room. From this, it could be said that these dwellings were operated as typical UK homes and/or in a way of energy consumption as SAP prediction and influence of solar gain on heating energy consumption as the SAP calculation method (Ozarisoy & Elsharkawy, 2019). The most underestimated case was house 1 with a difference of 81% followed by flat 1, which showed a 65% difference. The most energy efficient case home is flat 3, and it performed 36% better than the SAP estimation. A retired couple lived in this flat.

As mentioned before, most SAP calculation cases showed better performance of energy consumption than other cases, due to bedrooms' heating temperature being set at 18°C and the other cases' internal temperature being higher than the SAP (see Table 11.5). Table 11.5 shows the summary of internal temperatures and Heating Degree Days (HDD) for the case study homes and for the SAP calculation. It clearly shows that internal temperature and HDD for calculating the SAP

Table 11.5 Measured internal average temperatures and target internal temperatures for SAP and HDD for designed case study homes

		Bedroom	*Living room*	*HDD****
SAP		18°C	21°C	2,850
Designed*		18°C	21°C	2,850
Measured**	Flat 1	21.4°C	21.6°C	Year 2007 for 3,135
	Flat 2	20.7°C	19.4°C	Year 2008 for 3,410
	House 1	–	22.5°C	
	House 2	19.2C	21.4°C	
	Flat 3	18.8°C	18.4°C	
	Flat 4	19.8°C	18.3°C	

* Designed: in the designed stage, target heating temperatures for case study homes.
** Average internal temperatures during the heating season.
*** HDD temperature sets 18.5°C.

is lower than actual case study homes' average internal temperature during the heating period (Ozarisoy, 2022). The HDD difference between SAP and measured data was almost 10% in 2007 and 16% in 2008 and SAP calculation always underestimated the HDD. That is why most SAP predictions performed better than actual energy consumption from the case study homes.

From the results, it could be concluded that the SAP calculation method is a very useful tool to predict energy consumption in the design stage. However, the internal temperature for calculation would need to be changed. The UK home's preferable living room temperature is around 21°C and other space is 2°C lower (Botti et al., 2021). Therefore, the SAP's internal set temperature for predicting heating energy consumption, especially bedroom set temperature needs to be reset more realistically in line with current trends.

Energy Efficiency Gap

To make the comparison study, two different benchmarks were brought into this research, namely, 'typical UK home's energy consumption and CO_2 emission data' and 'energy' consumption and CO_2 emissions by built flat and mid-terraced home with 2006 Building Regulation standards. Figure 11.6 represents the results for comparing case study homes and typical UK homes. The purple dash line represents typical UK home's energy consumption, and the green dash line shows the CO_2 emissions from typical UK home. In terms of energy consumption, all case study homes show better performance than the typical UK home. House 1 was the home with the greatest energy consumption consuming 42% less than a typical UK home's energy consumption and flat 3 has only consumed one-sixth of a typical UK home. With regard to CO_2 emissions, most case study homes show better performance than typical UK homes except house 2 due to the fact that house 2 mainly uses energy from electricity and when calculating the typical UK home energy consumption, the main heating energy source is gas. As mentioned above, electricity generation results in almost 2.5 times more CO_2 emissions than gas generation.

Figure 11.7 shows the result of a comparison study for case homes' energy consumption against typical UK homes which have adapted to 2006 Building Regulations. It was not possible to compare directly due to the fact that the case study homes' floor area was not likely to be the same as that of typical UK homes. Therefore, this research used the energy consumption per square meter (kWh/m²) in order to make a comparison study. In general, a case study home with gas heating

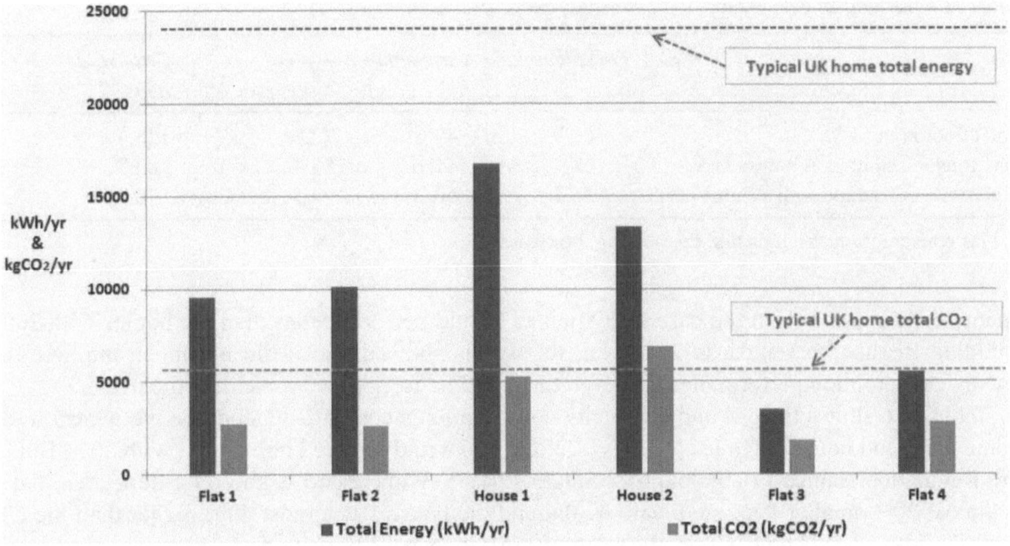

Figure 11.6 Comparison between case study homes and typical UK homes.

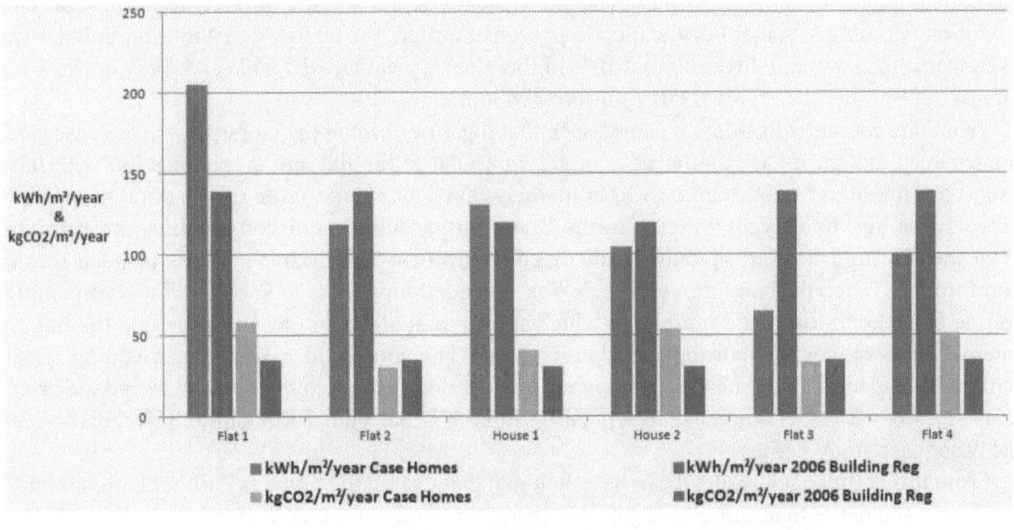

Figure 11.7 Comparison of case study homes with 2006 Building regulation homes.

uses more energy than a 2006 Building Regulation standard home with the expectation of flat 2. In contrast, the case home with electricity heating showed better performance than homes with 2006 Building Regulation and the best performed case study home was flat 3 which consumed less than half of the energy against the 2006 Building Regulation adapted flat. The worst case for energy consumption per square metre (kWh/m²) is flat 1, which consumed almost a third more than the typical UK flat with adapted 2006 Building Regulation standard (Cole et al., 2023).

As can be seen from the building fabric thermal properties, the case study homes' building fabric standards were better than the 2006 Building Regulation, and according to this, the overall

Table 11.6 Reference data against the case study home (flat 1 and house 1)

	Typical flat	*Case study flat 1*	*Typical mid-terraced*	*Case study house 1*
Total floor area (m^2)	60.9	49.6	78.8	127.6
Gas consumption* ($kWh/m^2/year$)	102.7	141.1	85.3	89.3
Electricity consumption ($kWh/m^2/year$)	36.1	60.2	34.5	48.2

* Gas consumption assumes that the heating, hot water and cooking.

energy consumption from the case study homes should perform better than the homes with 2006 Building Regulation standards. However, the results showed that it did not much improve the energy consumption performance and even flat 1 and house 1 were worse than the others.

Table 11.6 shows the gas and electricity consumption's comparison study between case study home flat 1 and house 1 against a UK typical flat and a mid-terraced home built with 2006 Building Regulation standard (Colclough & Salaris, 2023). With regard to the total floor area, flat 1 is almost 20% smaller than a typical UK flat and the house 1 is almost 40% bigger than the UK typical mid-terraced home. Given the relationship between the floor area and energy consumption, flat 1 should consume less heating energy than a typical UK flat. However, the overall energy consumption of flat 1 is almost 40% higher than a typical UK flat. In terms of gas consumption, flat 1's gas consumption was almost 30% higher than typical UK flat and the mid-terrace case's gas consumption was almost same. For the electricity consumption, the energy consumption in flat 1 was even worse than gas and it was almost 40% higher than typical UK flat and case study house 1 was almost 30% higher than typical UK mid-terraced home.

From this comparison study, it can be seen that flat 1 performed the worst among the case study homes even though it has smaller floor area. On a visit to the flat and a conversation with flat 1 user, very important issues came to light, namely, flat 1 has some issue of the draught problem, especially a big 'door-type' window in the living room, due to poor construction and finishing, even the flat used heating system, it still needs extra heat from other systems to keep the flat comfortable. When the baby moved in this flat, it needed more heat to keep the flat warm enough for the baby's comfort (see Figure 11.4 which shows that, after moving the baby into the flat, the internal temperatures for bedroom clearly increased). The householder uses his gas heating system with mobile electric panel heater for topping up heat and that is why his gas and even electricity consumption was much higher than typical both the UK flat with 2006 Building Regulation and the other case study homes.

From this comparison study, it can be concluded that even if the home is built with higher building fabric standards than those at a national level, the total energy consumption does not always show better performance and the most important factor to influence on energy consumption can be user behaviour (Gupta et al., 2018).

Winter Thermal Comfort

There are two different comfort terms which consider thermal and health (Ozarisoy & Altan, 2023a). These two comfort criteria have a different temperature range for comfort. This research considers only the thermal comfort which is based on CIBSE Guide A Recommendation (see Table 11.7). The section below shows the finding from winter comfort study from the monitoring data for case study homes and these considered occupied periods only.

Table 11.7 Recommended comfort criteria for dwelling during heating season

	Bedroom	*Living room*
Temperature (°C)*	17–19	22–23
RH (%)*	40–70	

* CIBSE comfort levels only apply to occupied period.

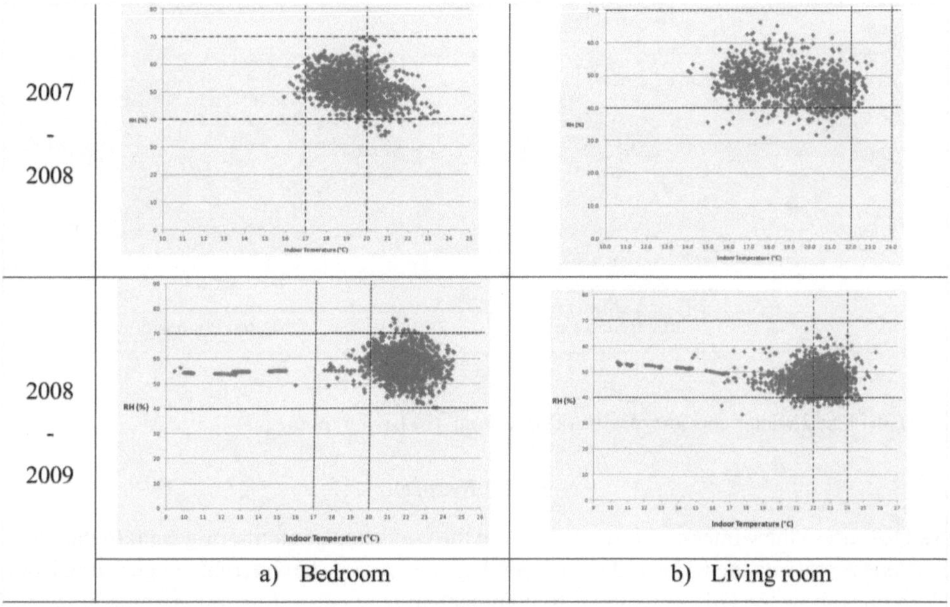

Figure 11.8 Flat 1 winter comfort results. (a) Bedroom. (b) Living room.

Flat 1 Results

Figure 11.8 shows the winter comfort results from the two years monitoring data. In the figure, the vertical axis represents the RH and the horizontal axis represents the internal temperature. Both axes and dash lines show the CIBSE Guide A Recommendation ranges for internal temperature and RH.

In the 2007–2008 winter, 57% of bedroom's temperatures were higher than CIBSE Guide A recommendation and 6% of its temperatures were under heated. Most of the RH for the bedroom was within the comfort range, and 5% of bedroom's RH was under 40%. In the living room, 96% of the temperatures were below the CIBSE Guide A recommendation, and 93% of RH was within the comfort range.

In 2008–2009 winter, the trend has changed. 88% of bedroom's temperatures were higher than CIBSE Guide A recommendation and 4% of its temperatures were under heated. Most of RH was within the comfort range and only 1% was over 70% of RH. In the living room, almost 50% of temperatures were below the CIBSE Guide A recommendation range and 4% of the temperatures were recorded as overheated. The RH trend was almost the same as the previous winter period. The bedroom and living room show scatter plotted lines due to unoccupied periods and especially during the Christmas holidays.

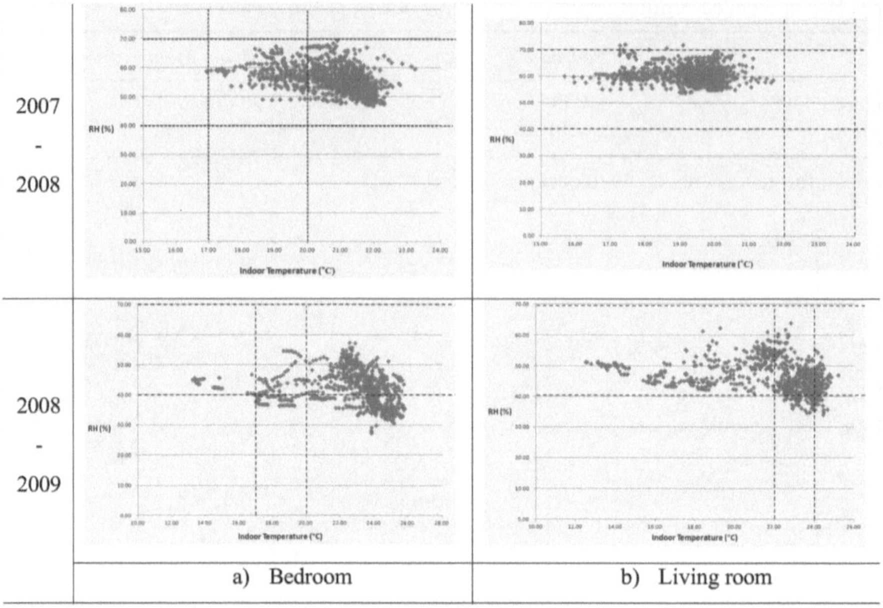

Figure 11.9 Flat 2 winter comfort results. (a) Bedroom. (b) Living room.

Flat 2 Results

Figure 11.9 shows the winter comfort results from the two years of monitoring data. In the figure, the vertical axis represents the RH and the horizontal axis represents the internal temperature. Both axes and dash lines show the CIBSE Guide A Recommendation ranges for internal temperature and RH.

In 2007–2008 winter, 80% of bedroom's temperatures were higher than the CIBSE recommendation and 20% of its temperatures were within the comfort temperature range. Most of the RH for the bedroom was within the comfort range, and less than 1% of time was highly humid. In the living room, almost 100% of the temperatures were below the comfort temperature and 99% of the RH for the living room was within the CIBSE recommendation range.

In 2008–2009 winter, the thermal comfort for bedroom's trend was almost same as 2007–2008 winter. 84% of the bedroom's temperatures were overheated and 4% of its temperatures were colder than CIBSE recommendation. 42% of RH for the bedroom was under the recommended range. In the living room, almost 40% of temperatures were below the CIBSE comfort temperature range and 14% of its temperatures were recorded as overheated. With regard to the living room's RH, 7% of its RH was under 40%.

House 1 Results

Figure 11.10 shows the winter comfort results from the one year's monitoring data. In the figure, the vertical axis represents the RH and the horizontal axis represents the internal temperature. Both axes and dash lines show the CIBSE Guide A Recommendation ranges for internal temperature and RH. As mentioned before, the house 1's monitoring data were missing due to the technical problems. Therefore, house 1, during 2007–2008, 44% of the living room's temperatures were higher than the CIBSE comfort temperature range, and 24% of its temperatures were under the

2007 - 2008	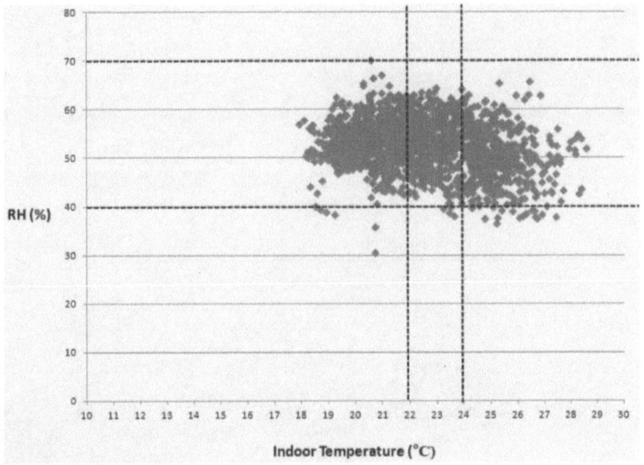
	Living room

Figure 11.10 House 1 winter comfort results.

comfort range. Most of RH for the living room was within the comfort range and 3% of time was under the comfort range.

House 2 Results

Figure 11.11 shows the winter comfort results from the two years of monitoring data for house 2. In the figure, the vertical axis represents the RH and the horizontal axis represents the internal temperature. Both axes and dash lines show the CIBSE Guide A Recommendation ranges for internal temperature and RH. In 2007–2008 winter, 24% of bedroom's temperatures were higher than the CIBSE Guide A recommendation and 4% of its temperatures were underheated. Most of the RH for the bedroom was within the comfort range and 1% of bedroom's RH was under 40%. In the living room, almost 70% of the temperatures were below the CIBSE Guide A recommendation. Additionally, 77% of the time, the living room's RH was within the comfort range and 24% of its RH was under 40%. In 2008–2009 winter, 33% of bedroom's temperatures were higher than the CIBSE Guide A recommendation and 60% of its temperatures were within the comfort temperature range. In addition, 8% of bedroom's RH was under 40% of RH. In the living room, almost 80% of temperatures were below the CIBSE Guide A recommendation range and 1% of the temperatures were recorded as overheated. Additionally, 38% of its RH was under the comfort RH range.

Flat 3 Results

Figure 11.12 shows the winter comfort results from the one years monitoring data for flat 3. In the figure, the vertical axis represents the RH and the horizontal axis represents the internal temperature. Both axes and dash lines show the CIBSE Guide A Recommendation ranges for internal temperature and RH. 2008–2009 winter data for the bedroom and living room were missing due to technical problems, therefore, flat 3 only has the monitoring data for 2007–2008.

In addition, 36% of bedroom's temperatures were higher than CIBSE Guide A recommendation and 21% of its temperatures were below the recommended temperature range. Most of the RH for the

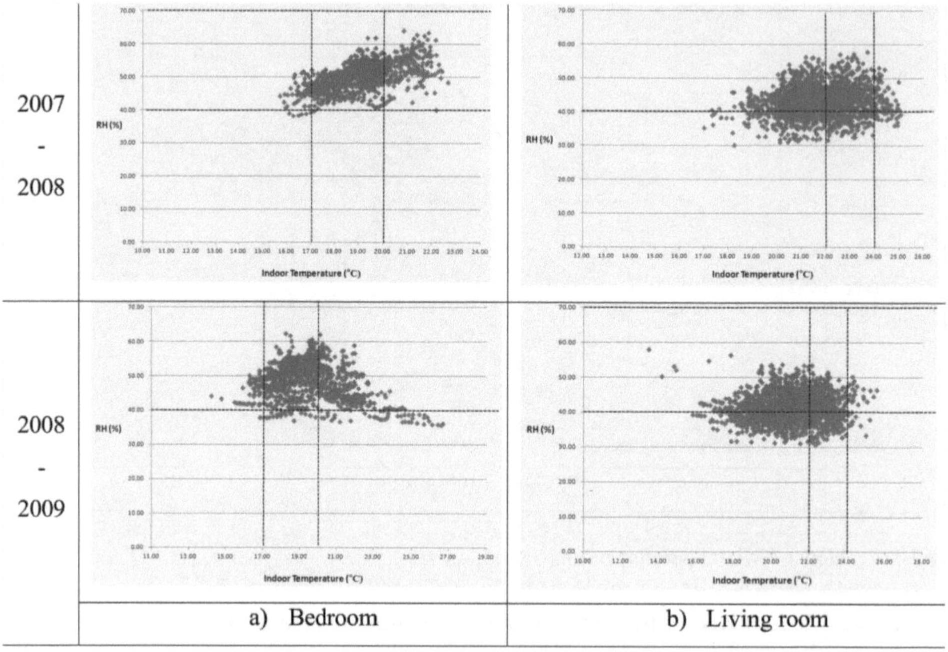

Figure 11.11 House 2 winter comfort results. (a) Bedroom. (b) Living room.

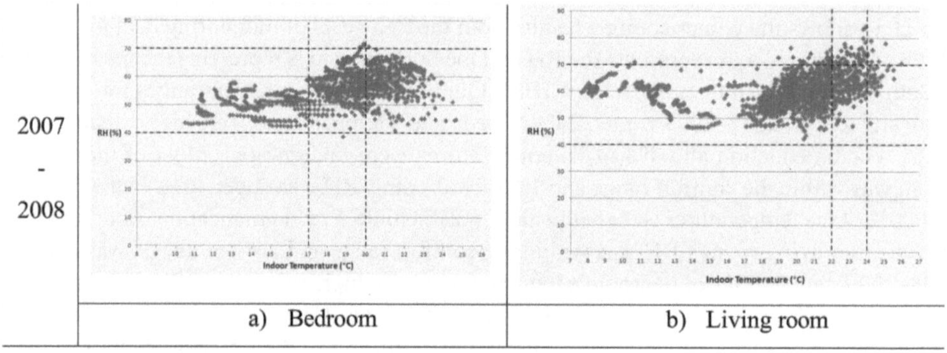

Figure 11.12 Flat 3 winter comfort results. (a) Bedroom. (b) Living room.

bedroom was within the comfort range. To consider the living room's thermal comfort, almost 85% of the temperatures were below the CIBSE Guide A recommendation. Almost 90% of the time, the living room's RH was within the comfort range and the rest of the time RH levels were above 70%.

Flat 4 Results

Figure 11.13 shows the winter comfort results from the two years monitoring results for flat 4. In the figure, the vertical axis represents the RH and the horizontal axis represents the internal

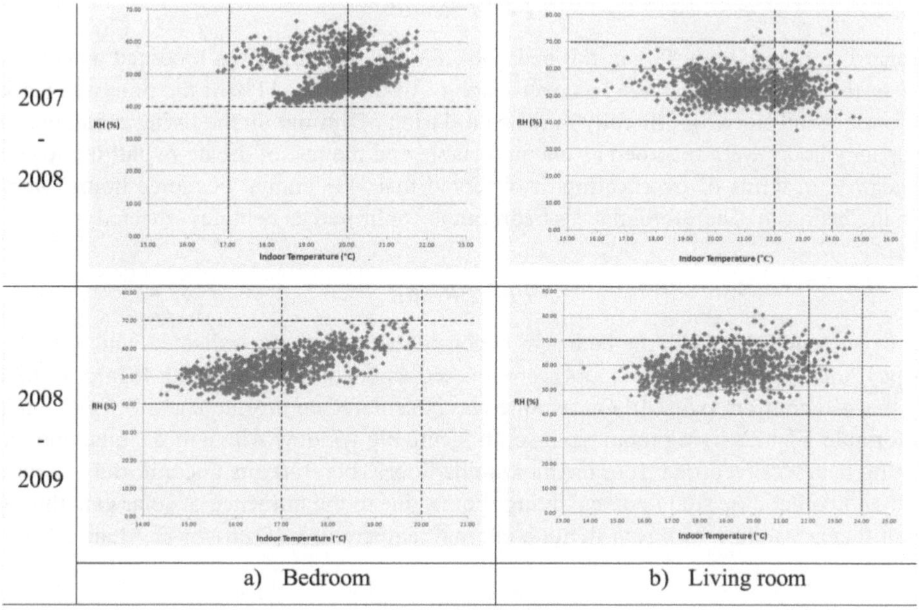

Figure 11.13 Flat 4 winter comfort results. (a) Bedroom. (b) Living room.

temperature. Both axes and dash lines show the CIBSE Guide A Recommendation ranges for internal temperature and RH. In 2007–2008 winter, more than 40% of bedroom's temperatures were higher than CIBSE Guide A recommendation and 57% of its temperatures were within the CIBSE recommended temperature range. Most of the RH for the bedroom was within the comfort range. In the living room, almost 85% of the temperatures were below the CIBSE Guide A recommendation. Most of the RH for the living room was within the comfort range. In the 2008–2009 winter, 33% of bedroom's temperatures were higher than CIBSE Guide A recommendation and 60% of its temperatures were within the comfort temperature range. Additionally, 8% of bedroom's RH was under 40% of RH. In the living room, almost 80% of temperatures were below the CIBSE Guide A recommendation range and 1% of the temperatures were recorded as overheated. In addition, 38% of its RH was under the comfort RH range.

Summer Thermal Comfort

To consider overheating for case study homes, two different methods were adopted within this research, and these are namely 'CIBSE Guide A summer benchmark' and 'EN15251'. The sections below show the results from the monitoring study.

CIBSE Guide A Recommendation

Table 11.8 shows the thermal comfort analysis results from the case study homes' monitoring study based on the CIBSE Guide A Summer Benchmark temperature criteria and these considered occupied periods only.

Flat 1 Result

With regard to thermal comfort in the bedroom, 88% of the time was recorded as comfortable and 1% of the time was considered as overheating. Additionally, 11% of the time was marked as uncomfortable and sleeping difficulty may occur during nighttime. In the living room, only 1% of its occupancy hours were recorded as uncomfortable and the rest of the hours fall in the comfortable category. In terms of overheating, it observed that '1% annual occupied hours' were over 26°C in the bedroom. Therefore, flat 1's bedroom is confirmed as being overheated.

Flat 2 Result

As can be seen in Table 11.8, the bedroom is considered as being overheated which is 6% of its occupancy hours and 37% of its hours were marked as uncomfortable. With regard to the living room's thermal comfort, over 90% of its time was considered comfortable and 7% was marked as uncomfortable. Flat 2's living room has a south facing big window with a mezzanine floor and the monitoring data was recorded from the mezzanine floor. Therefore, its uncomfortable hours were higher than in other case study homes' living rooms, due to the influence of solar gain through the big south facing window with roof light on internal temperatures (Ozarisoy & Altan, 2021).

House 1 Result

In the bedroom, uncomfortable hours were higher than comfortable hours and its overheating hours were 7% of its summer period. This bedroom has big south facing window and solar gain

Table 11.8 CIBSE thermal comfort results for case study homes

	*Bedroom**		*Living room***	
	2007	*2008*	*2007*	*2008*
Flat 1	–	Comfort: 88%	–	Comfort: 99%
	–	Discomfort: 11%	–	Discomfort: 1%
	–	Overheating: 1%	–	Overheating: 0%
Flat 2	–	Comfort: 57%	–	Comfort: 93%
	–	Discomfort: 37%	–	Discomfort: 7%
	–	Overheating: 6%	–	Overheating: 0%
House 1	Comfort: 45%	–	Comfort: 98%	–
	Discomfort: 48%	–	Discomfort: 2%	–
	Overheating: 7%	–	Overheating: 0%	–
House 2	Comfort: 71%	Comfort: 81%	Comfort: 99%	Comfort: 99%
	Discomfort: 27%	Discomfort: 19%	Discomfort: 1%	Discomfort: 1%
	Overheating: 2%	Overheating: 0%	Overheating: 0%	Overheating: 0%
Flat 3	Comfort: 78%	–	Comfort: 100%	–
	Discomfort: 20%	–	Discomfort: 0%	–
	Overheating: 2%	–	Overheating: 0%	–
Flat 4	Comfort: 96%	Comfort: 94%	Comfort: 100%	Comfort: 98%
	Discomfort: 4%	Discomfort: 6%	Discomfort: 0%	Discomfort: 2%
	Overheating: 0%	Overheating: 0%	Overheating: 0%	Overheating: 0%

* Bedroom temperature: comfort (18–23°C), discomfort (24–26°C), overheating (>26°C).
** Living room temperature: comfort (18–25°C), discomfort (26–27°C), overheating (>28°C).

through the window may influence its overheating and such a high level of discomfort. With regard to the living room's thermal comfort, only 2% of its time was recorded as uncomfortable and the rest was marked as comfortable.

House 2 Result

As can be seen in Table 11.8, 99% of the living room was thermally comfortable and only 1% was recorded as uncomfortable. According to the result, 71% of bedroom's hours were determined as comfortable and 27% and 2% were marked as uncomfortable and overheated, respectively. With regard to living room's thermal comfort in 2008, the result was same as the 2007 result. In the bedroom, the situation was improved. The comfortable hours for bedrooms in 2008 were improved by 10% comparing with 2007 and eliminated the overheating.

Flat 3 Result

With regard to bedroom's thermal comfort, the result shows that 78% of its hours were within the comfort temperature and 2% of it was considered as overheated. In terms of CIBSE Guide A overheating criteria, the bedroom could be described as overheated. In the living room, it shows that almost 100% of its time was marked as comfortable.

Flat 4 Result

With regard to bedroom's thermal comfort in 2007, more than 95% of its monitoring period was determined as comfortable and the rest time was in uncomfortable. In the living room, it was marked as 100% comfortable. Compare with 2007 and 2008 for bedroom's thermal comfort, its thermal comfort was decreased by 2% and the discomfort was increased by 2% and there was no record for overheating. With regard to living room's thermal comfort and comparison with 2007 and 2008 results, it shows the same trend as the bedroom result. The thermal comfort was decreased, and discomfort was increased by 2%.

EN 15251 Upper Limit Temperatures

Table 11.9 shows the thermal comfort analysis results from the case study homes' monitoring study based on the EN15251Upper Limit temperature criteria.

Flat 1 Result

With regard to thermal comfort in the bedroom, 87% of total hours were within the comfortable range and 1% of hours were marked as overheating. In the living room, 95% of its time was within the comfort temperature range and again 1% of hours were considered as overheating.

Flat 2 Result

With regard to thermal comfort for the bedroom, 12% of monitoring hours were considered as overheating and 70% of its hours were within the thermal comfort temperature range. In the living room, 9% of its monitoring hours were considered as overheating and 73% of the monitoring hours were within the thermal comfort temperature range based on the EN15251 calculation.

Table 11.9 EN15251 thermal comfort results for case study homes

	Bedroom		*Living room*	
	2007	*2008*	*2007*	*2008*
Flat 1	–	Comfort: 87%	–	Comfort: 95%
	–	Discomfort: 12%	–	Discomfort: 4%
	–	Overheating: 1%	–	Overheating: 1%
Flat 2	–	Comfort: 70%	–	Comfort: 73%
	–	Discomfort: 18%	–	Discomfort: 18%
	–	Overheating: 12%	–	Overheating: 9%
House 1	Comfort: 83%	–	Comfort: 92%	–
	Discomfort: 0%	–	Discomfort: 3%	–
	Overheating: 13%	–	Overheating: 5%	–
House 2	Comfort: 94%	Comfort: 94%	Comfort: 98%	Comfort: 99%
	Discomfort: 4%	Discomfort: 6%	Discomfort: 2%	Discomfort: 1%
	Overheating: 2%	Overheating: 0%	Overheating: 0%	Overheating: 0%
Flat 3	Comfort: 95%	–	Comfort: 94%	–
	Discomfort: 2%	–	Discomfort: 5%	–
	Overheating: 3%	–	Overheating: 1%	–
Flat 4	Comfort: 86%	Comfort: 88%	Comfort: 83%	Comfort: 90%
	Discomfort: 14%	Discomfort: 12%	Discomfort: 16%	Discomfort: 9%
	Overheating: 0%	Overheating: 0%	Overheating: 1%	Overheating: 1%

House 1 Result

The bedroom results show that 17% of the monitoring period was above the thermal comfort range and the rest of its hours were within the comfort temperature range. With regard to the living room thermal comfort, 92% of the monitoring hours were within the EN15251's suggested comfort temperature range and 5% of its hours were considered as overheating (Ozarisoy & Altan, 2023b).

House 2 Result

As can be seen in Table 11.9 in 2007, 2% of the monitoring hours for the bedroom were considered as overheating and 94% of its hours were within the thermal comfort temperature range. With regard to living room's thermal comfort result, 98% of its monitoring hours were within the EN15251's thermal comfort temperature range and the rest hours were below the temperature range (Symonds et al., 2016).

As can be seen in Table 11.9 in 2008, overheating was not observed during the monitoring period for the bedroom and living room. In the bedroom, 94% of its hours were within the comfort temperature range and 6% of it was considered underheating, which means that temperatures were below the comfort temperature range. With regard to the living room, almost 100% of the monitoring hours were marked as comfortable.

Flat 3 Result

With regard to the bedroom's thermal comfort result, 3% of its monitoring hours were considered overheating and 95% of the monitoring period was within the comfort temperature range. In the living room, 94% of its monitoring hours were within the thermal comfort temperature rage and 1% its hours were considered as overheating.

Flat 4 Result

Table 11.9 shows flat 4's thermal comfort analysis result in 2007. In the bedroom, there were no recorded hours for overheating, 86% of its hours were within the comfort temperature range and 14% of the monitoring hours were considered underheating, which was below the lower limit of comfort temperature. With regard to the living room's thermal comfort, only 1% of its monitoring hours were considered as overheating and 83% of the monitoring hours were within the comfort temperature range.

Again in 2008, in the bedroom, there were no recorded hours for overheating and 88% of the monitoring hours were within the comfort temperature range and the rest of it was below the low limit of comfort temperature. With regard to the living room's thermal comfort, 1% of its monitoring hours were considered overheating and 90% of the monitoring hours were within the comfort temperature range and the rest of its hours were below the low limit of comfort temperature.

Discussion

Energy Consumption and CO_2 Emissions

In general, case study homes show much better energy consumption performance against that of typical UK housing, and most case study homes show better CO_2 emissions performance than the national average except in the case of study house 2 (Taylor et al., 2021). The case study homes' building fabric standard level is almost the same as the Best Practice in which energy consumption is 25% less than the 2006 Building Regulation. This means that the case study homes should be more energy efficient than homes built with 2006 Building Regulation standard. However, flat 1 and house 1 performed worse than 2006 Building Regulation home. From this result, it could be concluded that even in dwellings that have achieved specific standards, energy consumption may be dramatically different depending on the occupant's energy use behaviour. Post-occupancy evaluations, therefore, are important to gain an increased understanding of occupant behaviour in new homes in order to optimise the energy-saving measures (Tian & de Wilde, 2011). More research is needed into the effect of occupant behaviour and lifestyles on energy use; this will allow for more targeted interventions to be applied.

Research also has shown the main factors that affect energy consumption are:

- Type of dwelling and size of floor
- Number of persons with number of children in household and their occupancy patterns
- Internal temperature
- Fabric heat loss characteristics
- Main energy source for the heating system

With regard to heating energy consumption, the most important influence factors are the heating system efficiency, the main energy source for the heating system, and internal temperature. From the case study homes energy consumption result, the electricity heating system shows better performance than the gas heating system. However, considering the carbon emissions, using electricity as the main energy source has a disadvantage (Evins et al., 2018).

Table 11.10 Ranking for case study homes

	Energy consumption	*CO_2 emissions*	*Energy efficiency*	*CO_2 efficiency*
Flat 1	4	4	1	1
Flat 2	3	5	3	6
House 1	1	2	2	4
House 2	2	1	5	2
Flat 3	6	6	6	5
Flat 4	5	3	4	3

* Small number represents worse performance than other case study homes in each category.

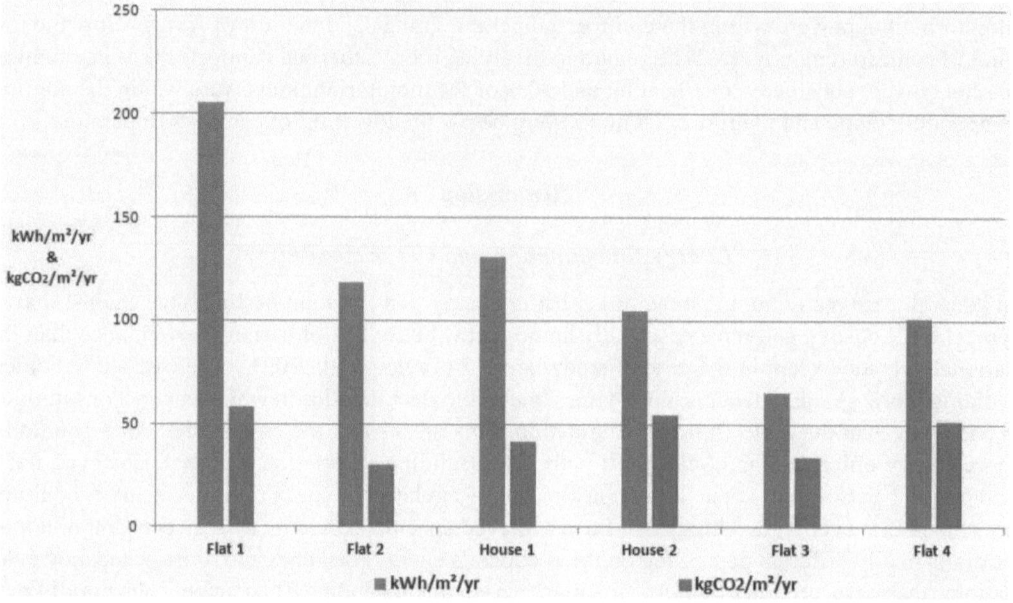

Figure 11.14 Energy and CO_2 efficiency results.

Furthermore, the out date for the target internal temperature for SAP calculation needs to be addressed. From the comparison study results, it is seen that SAP's target temperature to calculate the prediction of energy consumption for dwellings could be lower than the occupants' preferable internal temperature. Most SAP predictions show better than the actual energy consumption monitored (Laouadi et al., 2020a).

Time spent at home is an important issue as the more time people spend at home the more likely they are to have their heating on, use lights, and so on. Lifestyle factors that influence the time spent at home include retirement, ill health and home parenting.

Table 11.10 and Figure 11.14 show the summary of the energy consumption and CO_2 emissions results from the case study homes. Energy efficiency means how much energy consumption made kWh per m² per year and a small number represents better performance in energy efficiency. The meaning of CO_2 efficiency is how much of CO_2 is emitted for case study homes kg CO_2 per m² per year and a small number represents better performance in CO_2 emissions.

Table 11.11 Summary of winter comfort results

		Bedroom		*Living room*	
Case homes	Monitoring years	Overheating (%)	Internal temperature	Underheating (%)	Internal temperature
Flat 1	07–08	57	19.3	96	18.7
	08–09	88	21.4	49	21.6
Flat 2	07–08	80	20.7	100	19.4
	08–09	84	22.6	39	21.7
House 1	07–08	no data		44	22.5
House 2	07–08	24	19.2	67	21.4
	08–09	33	19.5	79	20.7
Flat 3	07–08	36	18.8	86	18.4
Flat 4	07–08	43	19.8	84	20.5
	08–09	–	16.7	98	18.3

As mentioned above, the energy consumption is increased in line with the size of the dwelling and the number in the household, and the energy efficiency of the case study homes relies upon the energy sources and the size of the case study homes. The CO_2 emissions from the case study homes show CO_2 emissions results were highly dependent on the main energy sources. It can be seen that having electricity as a main energy source has a disadvantage for carbon emissions.

Thermal Comfort Studies

Winter Comfort Results

In general, most of the case study homes' RH was within the acceptable range which is 40%–70% and most of bedroom's average temperature was higher than the CIBSE Guide A recommendation and most living rooms' average temperatures were lower than the CIBSE recommendation. Table 11.11 shows the summary of winter comfort results for the case study homes. As can be seen, the average temperature of bedrooms was higher by 0.2°C–3.6°C than the CIBSE recommended temperature which is 17°C–19°C (Attia et al., 2023). From this result, it might be necessary to consider reviewing the CIBSE comfort temperature for bedrooms which means the CIBSE recommended temperature could be outdated or be lower than the occupant's expected internal temperature for the bedroom during the heating season.

With regard to the living room's thermal comfort, most case study homes' living room average temperatures were recorded lower than CIBSE Guide A recommended temperature except for house 2 and every living room was considered as being underheated. In general belief, new homes have warmer lounges and would expect it warmer than bedrooms. However, the results show some different trends when the living room is located in the north and depending on the pattern of occupancy. For instance, flat 4 has south facing living room, and living room's internal temperature is higher than the bedroom. House 1 and house 2 are defined as all-day occupancy which means the house would be heated all day at heating set temperature. In these cases, the internal temperature of the living room is close to the CIBSE recommendation, and underheating has been decreased. Again, the CIBSE recommended living room temperature could be outdate or not in line with user expectations. With regard to winter recommendation temperature, the CIBSE Guide A recommendation would need to be reviewed with consideration for occupant expectations.

Summer Comfort Results

Figure 11.15 shows the thermal comfort results based on CIBSE Guide A summer benchmark temperature.

According to CIBSE overheating definition, house 1, house 2 and flat 3 bedrooms were considered as being overheated in the 2007 summer period and all living rooms recorded no overheating in the same period. The number of overheating hours for house 1's bedroom was higher than in other case study homes and could be explained by the influence of room orientation and the size of the window. House 1's bedroom has a big window is south facing. In 2008, flat 1 and flat 2 bedrooms were recorded as being overheated and only flat 2's living room was considered as being overheated which has a big window with a roof light in south facing.

From these findings, it could be said that room orientation and the size of the window can affect the thermal comfort of occupancy. However, this could not be concluded from this research, due to the quantity of case study homes and monitoring data. Therefore, if the author could have had all case study homes for the same period of monitoring data and the larger number of case study homes, the author would have been able to provide the influence of factors on overheating for such well-insulated new homes. The size of strong evidence of the case study homes is one of the significant limitations of this research.

Table 11.12 shows the summary of thermal comfort results for case study homes based on EN15251 upper limit comfort temperature. To compare the CIBSE results and EN15251 result, the EN15251 is overestimated than the CIBSE result due to the upper limit temperature difference between them which are 28°C and about 27°C for CIBSE and EN15251, respectively. In 2007, house 1's bedroom showed the worst performance over the monitoring period and the best-performed case study home was flat 4. Again, south facing bedroom shows worse performance for overheating. In 2008, the flat 2 was the worst performance among the case study homes and again flat 4 shows the best.

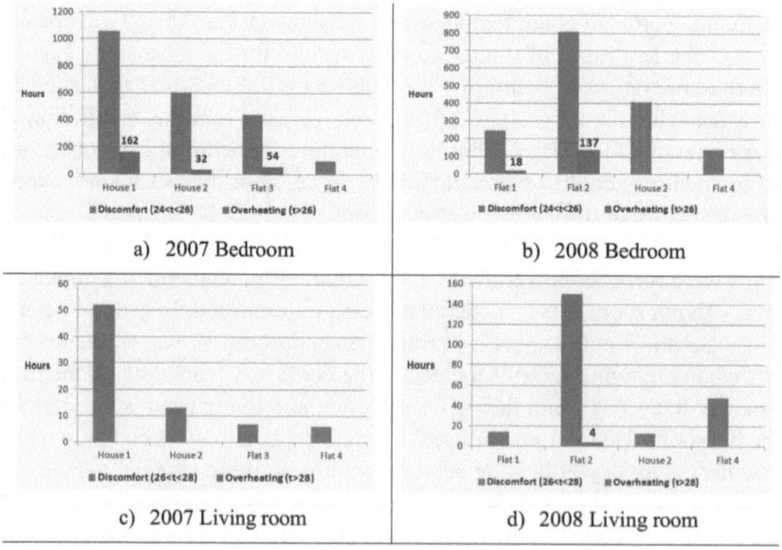

Figure 11.15 CIBSE Guide A overheating results. (a) 2007 Bedroom. (b) 2008 Bedroom. (c) 2007 Living room. (d) 2008 Living room.

Table 11.12 EN15251 overheating results

	Bedroom		*Livingroom*	
	2007	*2008*	*2007*	*2008*
Flat 1	No data	15	No data	10
Flat 2	No data	260	No data	212
House 1	369	No data	163	No data
House 2	55	0	3	12
Flat 3	100	No data	70	No data
Flat 4	1	0	10	28

Summary

In general, as the floor area and number in the household increased, the energy consumption and CO_2 emissions increased. With regard to the influence of the main energy source, the case study homes' heating systems using electricity as a main energy source showed better performance than those using gas central heating systems. However, in terms of CO_2 emissions and the influence of the main energy source, it seems that homes with more electricity consumption were at a disadvantage. In terms of the correlation between internal temperature and number of children with energy consumption, homes with children showed higher internal temperatures and higher energy consumption than homes with no children.

The applied building standard to case study homes is better than '2006 Building Regulation'. However, two of the case study homes performed worse than the 2006 Building Regulation homes. From this result, it could be concluded that it is the occupants who determine how energy efficient a dwelling will be; even if the building is well insulated and the dwelling has an efficient energy source. Therefore, reducing energy use depends not only upon improving building performance but also on understanding the relationship between how occupants use the property and their particular personal circumstances. For example, house 1 and house 2 homes have an identical floor plan, floor area and size of household, the energy consumption can show large variations due to user behaviour impact.

As a result of the winter comfort study, it shows that the occupants in the case study homes preferred higher internal temperature for bedrooms than 'CIBSE Guide A' recommendation. With regard to the overheating, some of the case study homes show overheating during the monitoring periods due to the fact that they had south facing living rooms and bedrooms with large windows.

Validation Study Result

Energy Consumption

This section shows the results from the simulation study which was based on the simulation interests and scenarios. It also shows the validation study results. To carry out simulation studies, model validation is an essential starting point. With regard to the validation of the simulation model, the accuracy of the simulation model was checked. To do this, the results obtained by the DesignBuilder and the monitoring data were compared to obtain the percentage differences between the measurements. In terms of the acceptable percentage difference, it is a maximum of

Table 11.13 Validation study result

	Actual consumption (kWh)	*Simulation prediction (kWh)*	*Difference (%)*
Flat 1*	6,998	6,888.9	−1.5
Flat 2*	8,351	8,012.5	−4
House 1*	11,392	10,573.5	−7.2
House 2**	13,358	13,043.6	−2.4
Flat 3*	3,553	4,258.2	19.8
Flat 4*	5,573	5,859.7	5.2

* Total gas consumption (heating + DHW) and **Total energy consumption.

Table 11.14 HDD difference in 2007, 2008 and simulation weather file

	Year 2007		*Year 2008*		*Simulation file*
	HDD	*%*	*HDD*	*%*	*HDD*
18.5°C	3,135	1.9	3,411	9.8	3,076
15.5°C	2,100	0.9	2,370	12.2	2,082
13.5°C	1,527	1.7	1,776	15.5	1,501

15% as mentioned earlier. Table 11.13 shows the summary of the validation study results. Most case study homes are successfully validated and fall within the acceptable percentage difference with the exception of flat 3. However, from the post-occupancy study, flat 3 has almost 25% less proportion of days occupied than other case study homes during the heating season. The simulation prediction, therefore, may need to have a 25% deduction from its prediction to incorporate the actual occupancy pattern. After deduction, its simulation prediction will be 3,193.7 kWh and the difference between actual consumption and simulation prediction is – 10.1% and the difference is within the acceptable percentage difference. As a result, flat 3 simulation model could be confirmed as a valid model.

The summary of the possible sources of error from the simulation study is explained as how the author tried controlling the uncertainty to minimise it for the simulation model. Even when controlled, the same difference between actual consumption and simulation prediction is still shown and the possible remaining sources of errors are differences between the actual weather at the site and the weather data used for simulation, due to the fact that weather data is represented as average weather not the actual weather for monitoring.

Table 11.14 shows the HDD difference between actual weather and simulation weather. The difference between the year 2007 and simulation weather was relatively close and is within 2%. However, the difference between the year 2008 and simulation weather is greater than the year before and ranges from 9.8% to 15.5%. As a result, the difference between the actual energy consumption and the simulation prediction could differ from the difference between the actual occupant behaviour and assumed occupancy input data (Jenkins et al., 2011). For example, during the monitoring period, the occupier of flat 3 was in the hospital for almost one month, which is why total energy consumption for flat 3 shows such a difference; the difference between the NCM lighting and appliances input data and the actual lighting and appliances from case study homes; the difference between the actual construction data and the simulation construction data and thermal bridge impact.

Table 11.15 Average internal temperature for actual and simulation data

	Actual monitoring data		*Simulation data*	
	Bedroom	*Living room*	*Bedroom*	*Living room*
Flat 1	22.7°C	23.0°C	23.7°C	24.5°C
Flat 2	23.5°C	23.2°C	23.7°C	24.9°C
House 1	24.2°C	23.5°C	23.2°C	23.2°C
House 2	23.1°C	23.6°C	22.7°C	22.9°C
Flat 3	23.4°C	23.0°C	23.4°C	23.2°C
Flat 4	22.4°C	23.0°C	24.5°C	25.3°C

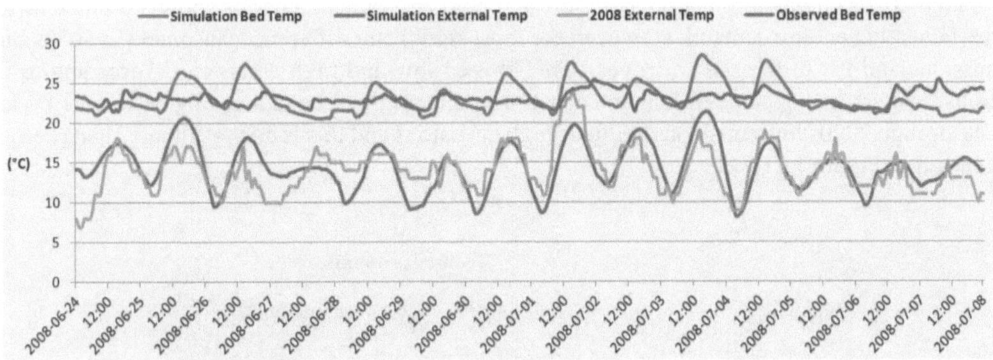

Figure 11.16 Comparison Flat 1's bedroom temperature for six days.

Internal Temperatures

Table 11.15 shows the average internal temperature for actual monitoring data and simulation results. In the bedroom, the temperature difference between actual data and simulation data ranges from 0°C to 2.1°C. The difference between them is up to 8.6%. In the living room, the temperature difference is from 0.2°C to 2.3°C and there is a 9.1% maximum difference between them.

Figure 11.16 shows the example for internal temperature comparison results and for all case homes' summer internal temperature. It shows when the external temperature is below 16°C both simulation and monitoring data are well matched. When the external temperature went over 16°C, simulation data shows peak temperature in simulation results due to solar gain from the window.

In general, the north facing room's monitoring data when compared with simulation data shows better matching than the south facing room's data (Laouadi et al., 2020b). It could be said, therefore, that solar gain through the window could play a very important role in internal temperature for case homes' data and simulation data. However, further study is needed with a reasonable number of case study homes.

Scenario A Results: Climate Sensitivity Study

Scenario A considered the effect of climate change scenarios and differences on energy consumption and overheating. The case study homes were simulated with low and high emissions scenarios through 2080s future climatic projections and the results are shown below as subsections.

Heating and Cooling Energy Consumption

In general trend, heating demand falls, and cooling demand increases through the climate scenarios. Therefore, global warming has an impact on heating and cooling demand, and Figure 11.17 represents the overall results for energy consumption influenced by two different climate scenarios. The figure shows the most significant difference was indicated in cooling energy demand, which is flat 2 in the 2080s high emissions scenario (Amaripadath et al., 2023). However, the greatest reduction was indicated in heating energy consumption. In terms of heating demand, the high emissions scenario shows better performance than the low emissions scenario. This result is very obvious due to the increasing average global temperature of 'high emission scenario' being greater than the 'low emissions scenario' throughout the time slices.

In the 2020s climate projection, both low and high emissions scenarios show not much difference in heating energy demand. However, the most significant difference was made in 2080s high emissions and the difference was over 10% between low and high emissions. According to the results, winters will become milder and demand for heating will be reduced by between 14% and 56% through 2020s low emissions to 2080s high emissions and this is due to climate change alone, as shown in Figure 11.18(a).

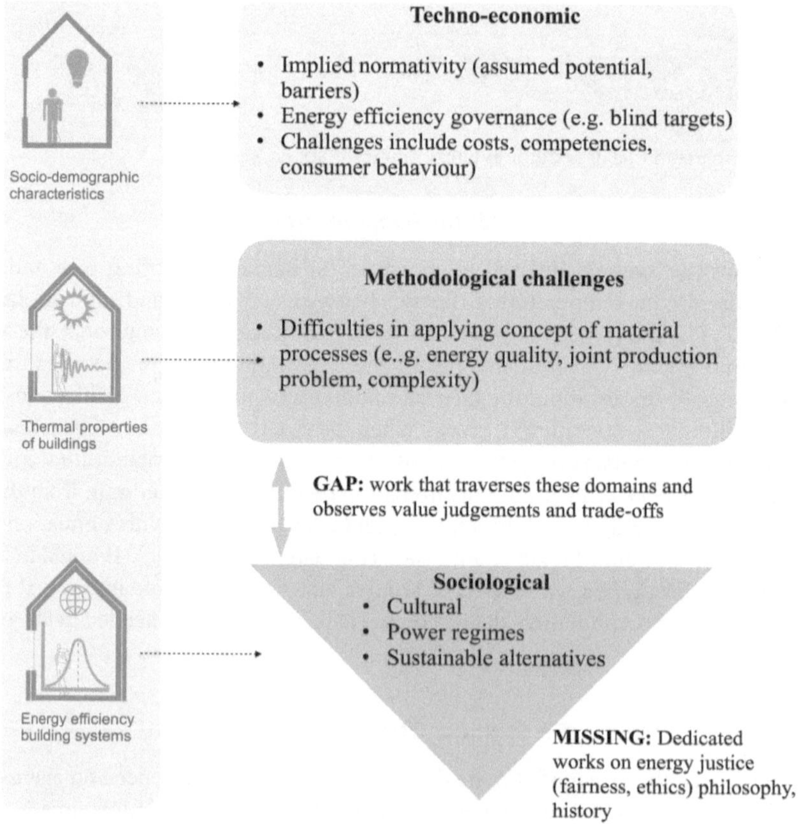

Figure 11.17 Heating and cooling energy consumption.

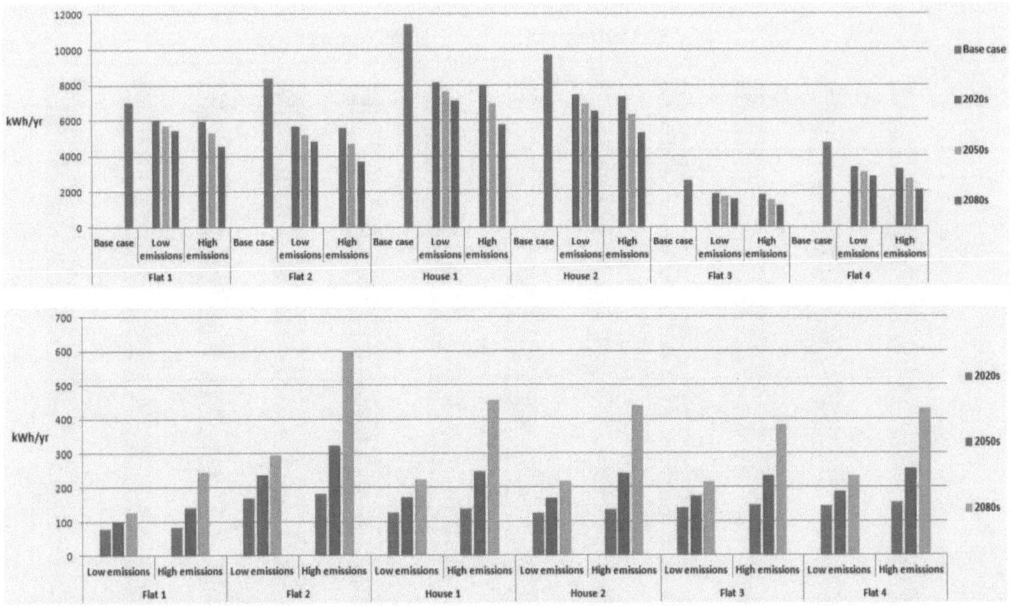

Figure 11.18 (a) Heating energy consumption reduction by 2080. (b) Cooling energy consumption reduction by 2080.

In contrast, this advantage gives an opposite effect on cooling energy consumption. Therefore, the most remarkable difference between low and high emissions was predicted in the 2080s again and the difference was almost double, as shown in Figure 11.18(b).

Figure 11.19 shows the energy consumption distribution by heating and cooling energy consumption. It shows more energy-efficient homes' heating energy reduction was greater than less energy efficiency ones. Again, this advantage could be seen as a disadvantage in cooling energy demand. Under climate change scenarios, for all case study homes, the energy consumption is still dominated by heating and in most cases; cooling energy consumption is less than 20% of overall energy consumption except flat 3 under the 2080s climate scenario, which is predicted over 20%.

CO_2 Emissions

In general trend, CO_2 emissions from heating fall and from cooling increase through the climate scenarios. Therefore, again global warming has an influence on CO_2 emissions from heating and cooling. Figure 11.20 represents the overall results of CO_2 emissions studies for climate sensitivity.

With regard to CO_2 emissions reduction by decreased heating energy demand, the 209 most significant difference was indicated in house 2 whose main energy source is electricity in 2080s high emissions scenario and the influence of main energy source's carbon emissions rates were explained. Therefore, having electricity as the main heating energy source for case study homes shows better CO_2 emissions reduction performance than gas ones (Jenkins et al., 2012). However, this energy source benefit would be diminished in CO_2 emissions from cooling energy demand due to them all using electricity for the main cooling energy source. CO_2 emissions by increased cooling energy demand are not sufficient to offset a reduction in heating energy consumption through the climate change scenarios (Jentsch et al., 2008).

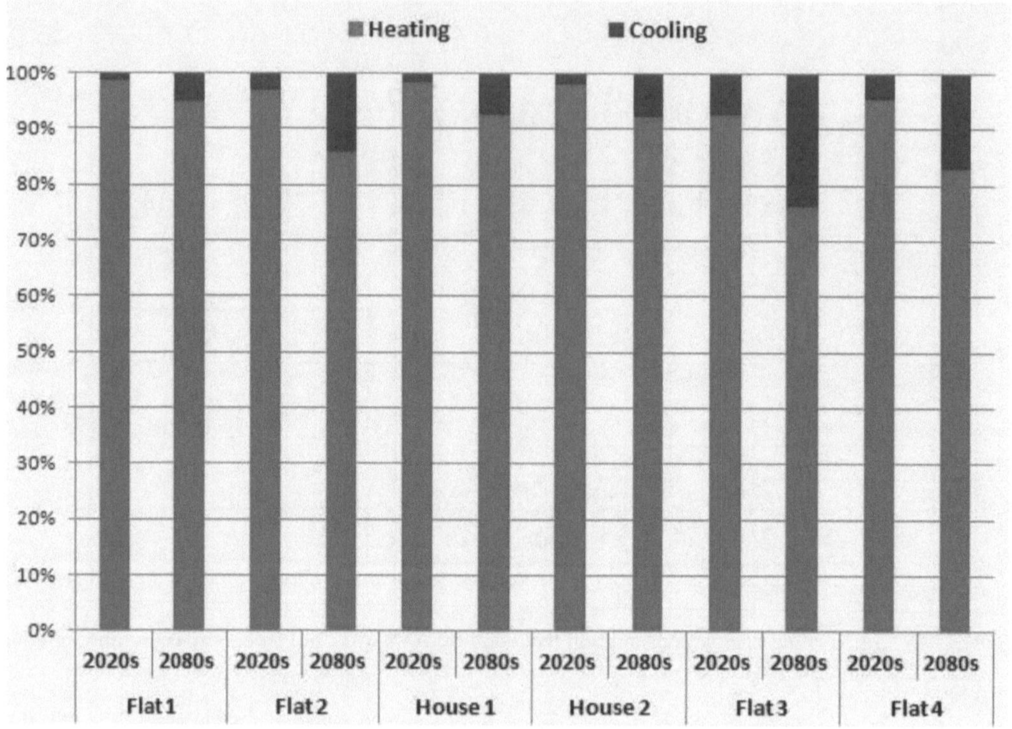

Figure 11.19 Energy consumption distribution by heating and cooling consumption.

Overheating Study

In general, overheating hours increase through the climate change scenarios. As a result, global warming has an impact on thermal comfort, and high emissions scenario shows a greater impact than low emissions scenario due to the temperature-increasing difference (Porritt et al., 2012). This overheating study was only considered from June to August, which is confirmed as 'summer' within this research. Figure 11.21 shows the results of overheating hours under CIBSE Guide A summer overheating criteria. Bedroom's overheating hours are higher than in the living rooms in both low and high emission scenarios due to security reasons the bedroom windows could not be opened during the night and bedroom's overheating temperature is 2°C lower than the living room's overheating temperature. Furthermore, the hours of the external temperatures exceed the bedroom's overheating temperatures, which are also more than those of the living room (see Table 11.16 and Figure 11.22). Figure 11.22 shows the thermal comfort temperatures and external temperature through high emissions scenarios. It clearly shows that in the future sometimes during the summer the natural ventilation may not deal with cooling without air conditioning systems.

Figure 11.23 shows the results of the comparison CIBSE Guide A summer overheating and EN15251 upper limit temperatures. This comparison study only considered the high emissions scenarios. In the bedroom, CIBSE Guide A results were overestimated than EN15251 and in the living room, this trend was reversed. The reason can be found in Table 11.17. Table 11.17 shows the percentage of EN15251's upper limit hours above the CIBSE Guide A overheating temperatures. In the 2080s, more than 90% of the bedroom's CIBSE temperature was below the EN15251

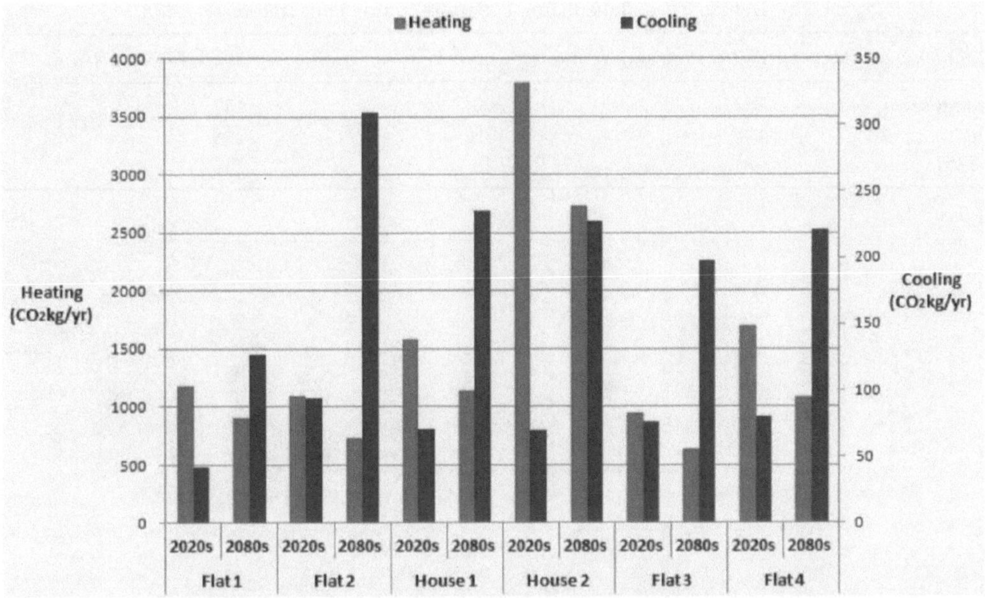

a) Carbon emissions from heating and cooling

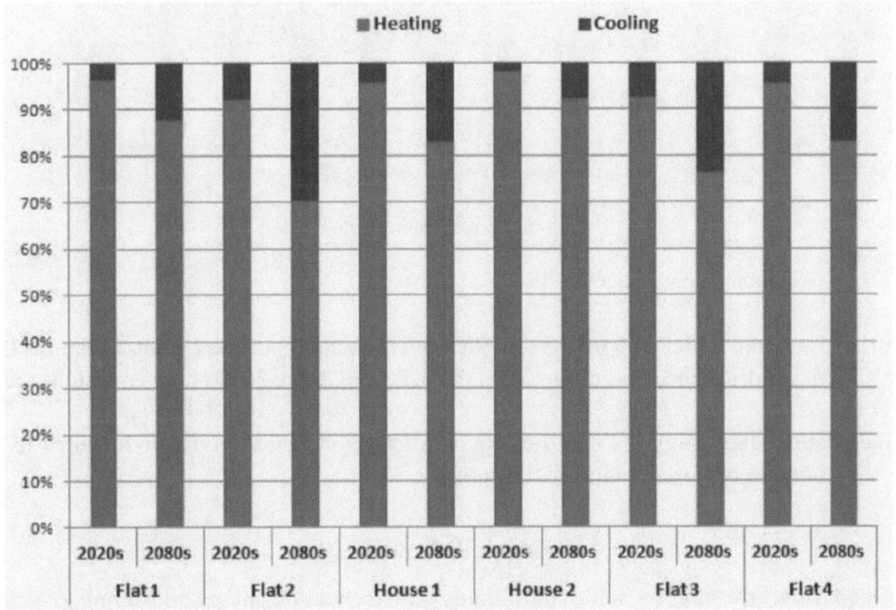

b) CO2 emissions contribution by heating and cooling energy consumption

Figure 11.20 CO_2 emissions influence by climate change scenarios. (a) Carbon emissions from heating and cooling. (b) CO_2 emissions contribution by heating and cooling energy consumption.

Table 11.16 External temperatures exceed the thermal comfort criteria temperatures

Criteria temperature	*2020s high emission*	*2050s high emission*	*2080s high emission*
Bedroom (26°C)	181	209	447
Living room (28°C)	108	175	298
EN15251	98	146	243

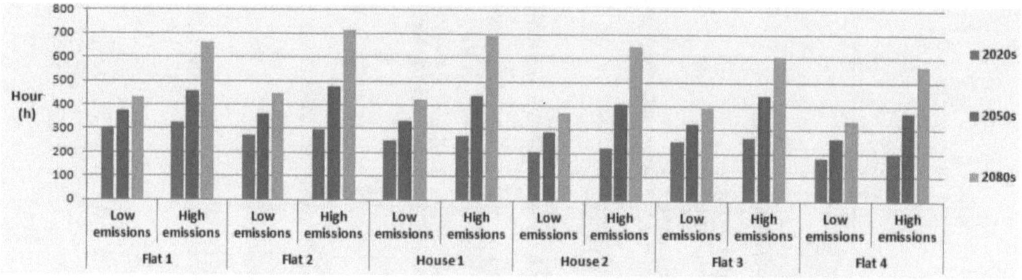

Bedroom overheating hours

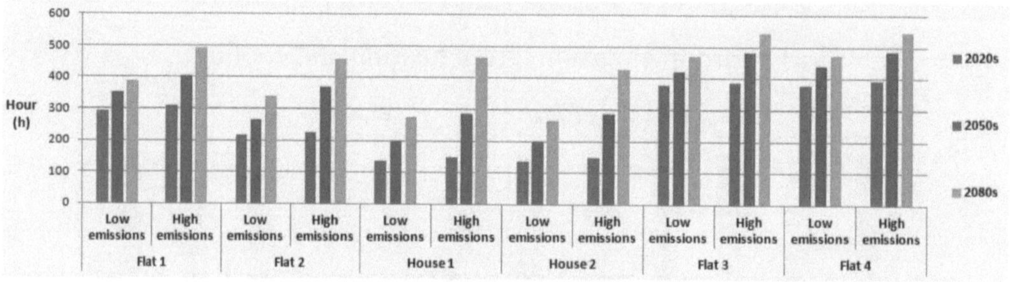

Living room overheating hours

Figure 11.21 Overheating hours under CIBSE Guide A criteria.

upper limit temperature which is why the CIBSE's overheating result was greater than EN15251. The most significant difference was observed in the bedroom in the 2080s high emissions scenario (Patidar et al., 2012).

The overheating study suggests that the air conditioning systems should be installed to make the case study homes comfortable during the summer.

Scenario B Results

Scenario B considers the effect of house morphology differences on energy consumption and overheating. The case study homes simulated different types of housing and only considered the use of high emissions climate weather for the simulations.

Energy Consumption

Figure 11.24 shows the result of morphology influence on heating and cooling energy consumption in case study homes. To predict climate change's effect on energy consumption, this study

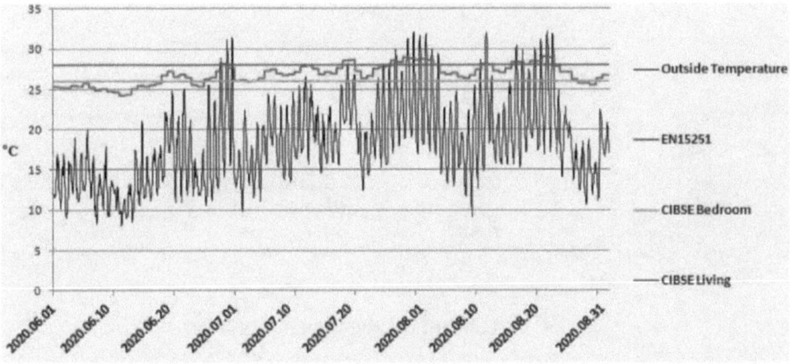

2020s thermal comfort temperatures and external temperatures

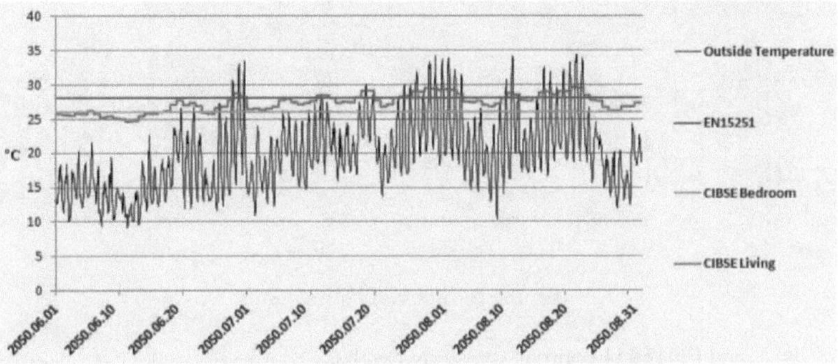

2050s thermal comfort temperatures and external temperatures

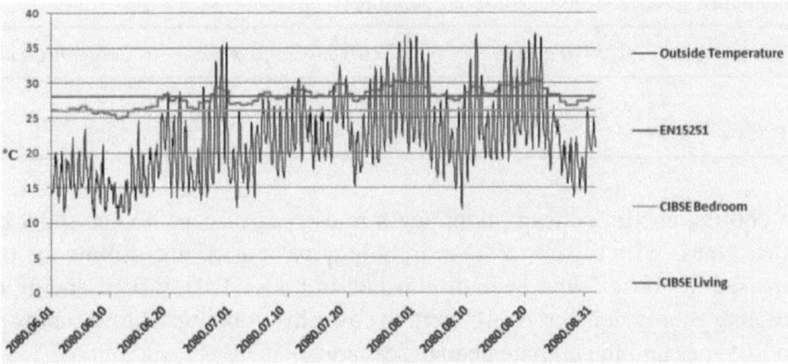

2080s thermal comfort temperatures and external temperatures

Figure 11.22 Thermal comfort temperatures and external temperatures.

only considered high emission scenarios. As a general trend, heating demand falls and cooling demand increases through the climate scenarios. The most significant difference was indicated in heating energy consumption, and the most significant change, however, was found in cooling energy demand. In terms of heating energy consumption, mid-floor flats and mid-terraced houses show better performance than other types of homes. This could be explained that less exposed surface to ambient means less heat losses through exposed walls and windows than less heating energy demand (Kendrick et al., 2012).

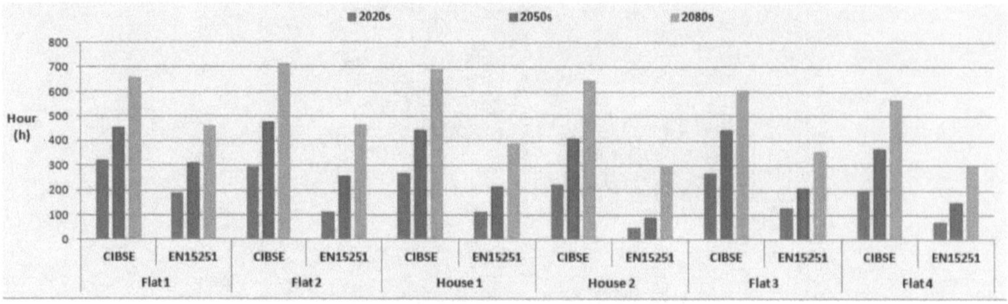

Bedroom overheating hours

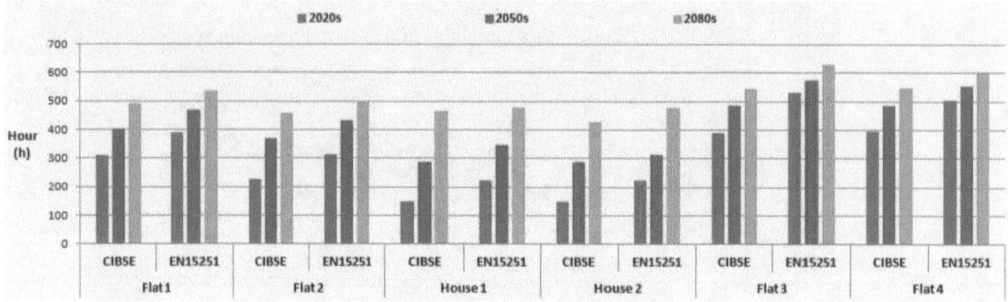

Living room overheating hours

Figure 11.23 CIBSE and EN15251 comparison study results.

Table 11.17 Time below EN15251 upper limit temperatures

	2020s high emission	*2050s high emission*	*2080s high emission*
Bedroom (26°C)	72%	80%	91%
Living room (28°C)	19%	29%	48%

In terms of cooling energy consumption, again mid-types of case homes show better performance than other types. With regard to the morphology of houses, flats follow the trend, but this trend was decreased in house 2 and even diminished in house 1. However, energy consumption increased by cooling energy demand is sufficient to cover by reducing heating energy consumption through the house types and the climate change scenarios.

Overheating Study

In general, overheating in bedrooms and living rooms increased throughout the climate scenarios and where most energy was demanded in heating and cooling also predicted the worst performance in overheating (Rostami & Bucking, 2024). The result is summarised in Table 11.18, which also shows the comparison study CIBSE Guide A against EN15251. The comparison results show the same trend in Scenario A, meaning that the CIBSE results overestimate in the bedroom temperature and underestimate in the living room temperature.

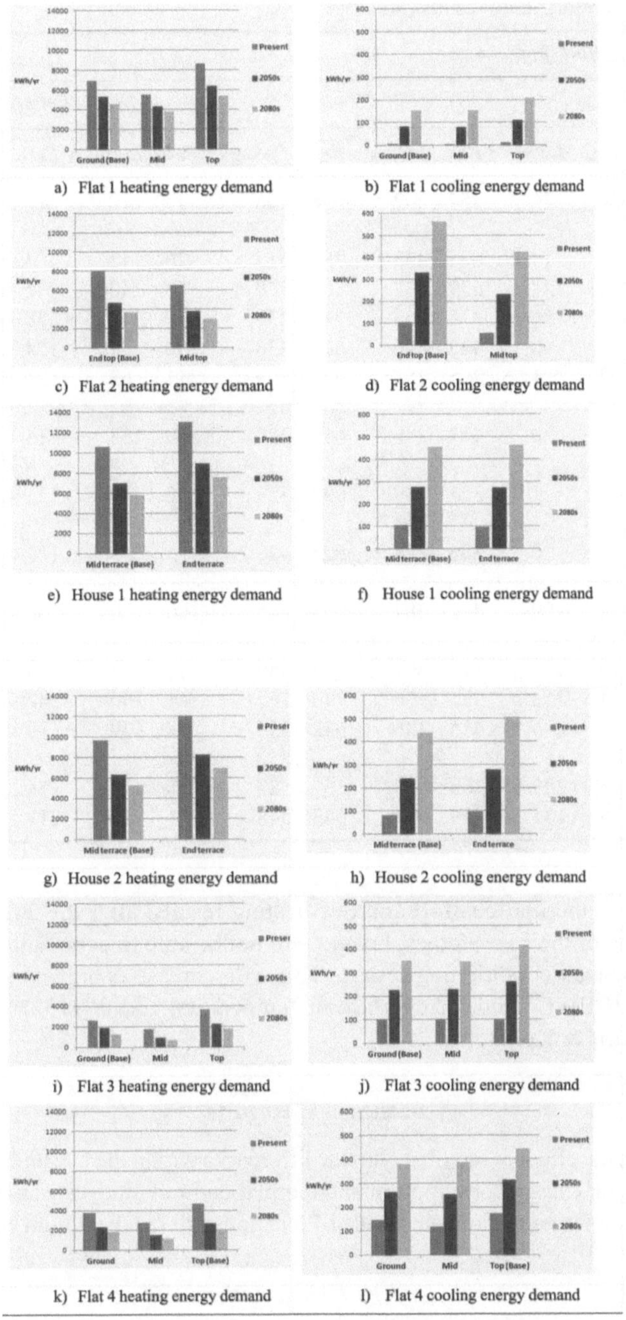

Figure 11.24 Heating and cooling energy consumption results. (a) Flat 1 heating energy demand. (b) Flat 1 cooling energy demand. (c) Flat 2 heating energy demand. (d) Flat 2 cooling energy demand. (e) House 1 heating energy demand. (f) House 1 cooling energy demand. (g) House 2 heating energy demand. (h) House 2 cooling energy demand. (i) Flat 3 heating energy demand. (j) Flat 3 cooling energy demand. (k) Flat 4 heating energy demand. (l) Flat 4 cooling energy demand.

Table 11.18 Thermal comfort study results (hours)

		CIBSE guide A						*EN15251*					
		Base case		*2050s*		*2080s*		*Base case*		*2050s*		*2080s*	
		Bed	*Living*	*Bed*	*Living*	*Bed*	*Living*	*Bed*	*Living*	*Bed*	*Living*	*Bed*	*Living*
Flat 1	End ground	30	76	454	402	658	492	15	10	310	469	461	536
	Mid	17	14.3	605	433	787	517	20	142	404	523	599	564
	End top	90	124	639	438	805	520	85	240	471	518	640	586
Flat 2	End top	1	20	475	368	714	459	260	212	258	432	465	501
	Mid Top	0	0	425	306	677	438	0	57	174	370	383	473
House 1	End terraced	2	1	255	181	432	269	11	63	121	205	239	285
	Mid terraced	1	0	441	286	659	428	369	163	215	346	391	476
House 2	End terraced	5	1	397	289	637	429	0	64	133	351	317	480
	Mid terraced	2	0	385	286	625	428	55	3	122	345	303	476
Flat 3	End ground	24	189	422	484	607	545	100	70	206	572	353	630
	Mid	16	134	511	494	714	559	16	304	336	590	503	643
	End top	51	213	535	506	743	567	41	360	347	600	523	647
Flat 4	End ground	13	236	415	486	622	554	18	379	180	560	333	620
	Mid	15	289	368	485	567	582	18	430	154	552	300	600
	End top	8	135	429	528	643	548	1	10	211	616	374	661

According to these thermal comfort and overheating results, after the 2050s, no matter what the house type in any of the case studies, houses will not be able to cope with overheating during the summer without an air conditioning system. This will increase as higher external temperatures rise, surpassing the 'CIBSE Guide A' overheating temperature and 'EN15251's' upper limit temperatures (Santamouris & Kolokotsa, 2014).

Scenario C Results

Scenario C considers the effect of regional climate differences within the UK and attempts to establish the heating and cooling energy consumption and thermal comfort study for case study homes. Two selected case study homes were simulated with 14 locations of UKCP 09 high emissions scenarios.

Energy Consumption

Figure 11.25 shows the heating and cooling energy consumption influence on three selected locations for case study homes. In general, heating energy increases and cooling energy decreases from London to Edinburgh. With regards to heating energy consumption, London shows better performance than other cities and Edinburgh consumes more heating energy than the base case, due to London's milder winter than Edinburgh's colder winter (see Figure 11.26). The energy

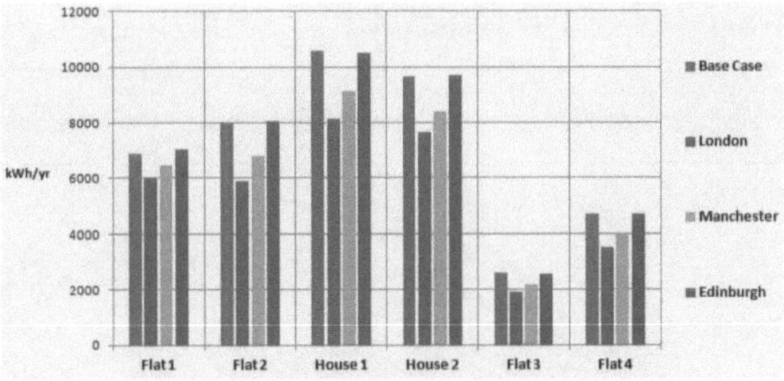

Heating energy consumption locations

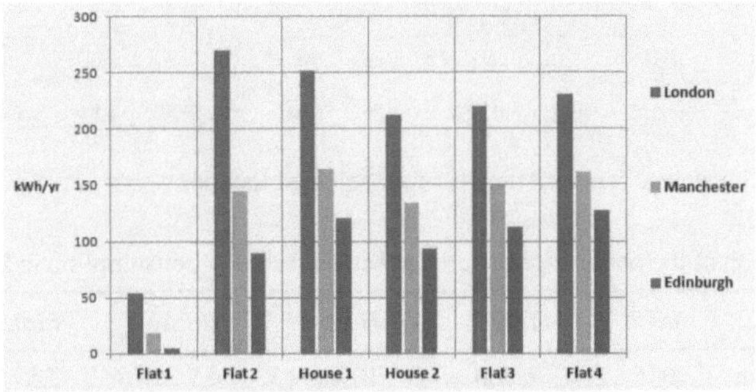

Cooling energy consumption by locations

Figure 11.25 Heating and cooling energy consumption by locations.

consumption between Bradford which is the case study homes' site and Edinburgh was almost the same even if there is quite a distance between them.

As can be seen in Figure 11.26, monthly external temperatures for both cities were almost the same and that is why their heating energy consumption was also almost the same. Two factors could affect the external temperature in this case which are sea level and distance from the sea. The sea level of Edinburgh is around 41 m and that of Bradford is around 60 m, i.e., 20 m higher than Edinburgh. Height above sea level means increased chance of more wind and cooler air temperature. The second is the location. The location of Edinburgh is much closer to the sea than Bradford which is almost central within the UK and also almost 90 km from Blackpool. Changes in ocean temperatures take longer than those in land due to thermal capacity differences between soil and water. The temperature range over the year will therefore be less for a coastal city than for an inland city. As a result, cities near oceans are warmer than cities that are far from the ocean. These two factors could affect the external temperature for both cities and for heating energy consumption.

In contrast, the cooling demand shows the opposite trend against heating energy consumption. Therefore, it shows very obvious results, i.e., cities with lower heating demand clearly consume more cooling energy, as their climate is warmer than others.

The difference between London and Edinburgh is represented in Table 11.19. With regard to heating energy consumption in Edinburgh and London, Edinburgh consumes more than London and the variation can be between 15% and 27% with most flats showing better performance than

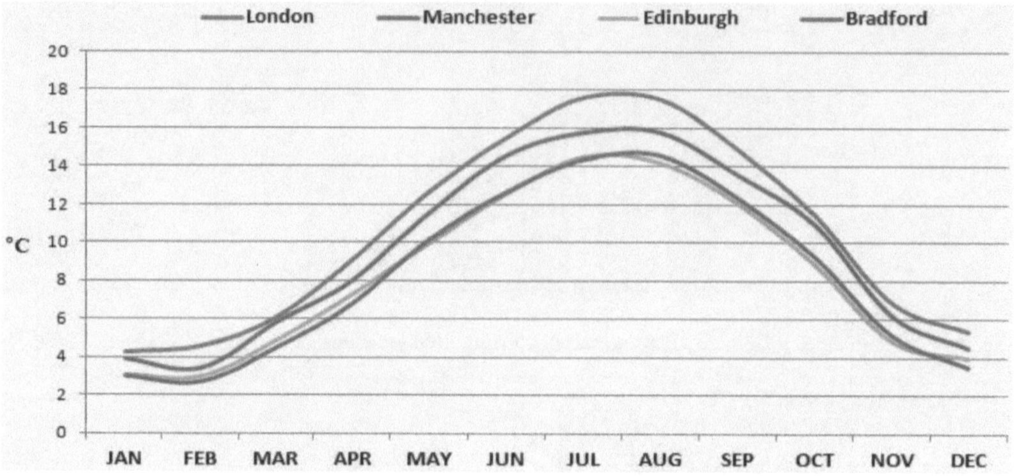

Figure 11.26 Monthly external temperatures for four selected locations.

Table 11.19 Comparison of the percentage difference between energy consumption in London and Edinburgh

		Flat 1	*Flat 2*	*House 1*	*House 2*	*Flat 3*	*Flat 4*
Heating	Edinburgh	7,025	8,066	10,499	9,710	2,574	4,717
(kWh/yr)	London	5,982	5,873	8,138	7,631	1,904	3,508
	Difference	15%	27%	22%	21%	26%	27%
Cooling	Edinburgh	5	89	120	93	112	127
(kWh/yr)	London	54	270	252	213	221	231
	Difference	91%	67%	52%	52%	49%	45%

houses with the exception of flat 1. With regard to cooling energy demand in London and Edinburgh, London case homes demand more cooling energy than those of Edinburgh and the variation can be between 45% and 91%. The results for case study homes show better heating efficiency but worse performance in cooling demand. The average energy consumption difference between London and Edinburgh is 23% and 53% of heating and cooling, respectively.

Figure 11.27 shows that heating demand falls and cooling demand increases between results for present and throughout climate scenarios. However, although cooling demand increases gradually, it starts at such a low baseline that is dwarfed by reducing heating demand, even in 2080s. Proportionally, all locations present similar reactions of heating energy reduction and cooling energy increase. The biggest reduction was in London 2020s, which is around 22% reduction in house 1.

Again Figure 11.28 shows how the heating and cooling energy consumption could be affected by regional differences which are ordered by latitude. With regard to the heating energy consumption trend, northern cities consumed more heating energy than southern cities. For example, flat 1 in Glasgow could consume 16% more heating energy than flat 1 in London and house 1 in London could consume 28% less heating energy than house 1 in Glasgow. As a result, London has the least heating energy consumption. With regard to cooling energy consumption, it shows the opposite trend. Southern cities are likely to consume more cooling energy than northern cities. For example, flat 1 in London could consume 90% more cooling energy than flat 1 in Glasgow and house 1 in London could demand 52% more cooling energy than house 1 in Glasgow. Considering energy

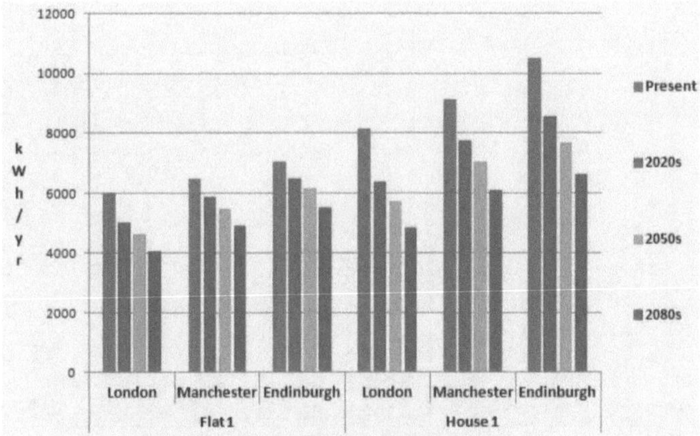

Heating energy consumption prediction

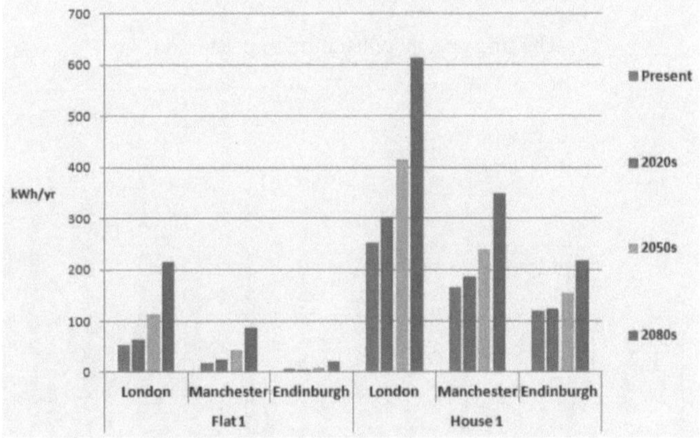

Cooling energy consumption prediction

Figure 11.27 Energy consumption through climate scenarios.

consumption influenced by locations, it could be concluded that the northern part of the UK may need to improve heating efficiency and the southern part of the UK would consider increasing cooling energy demand (Jones et al., 2016).

CO_2 Emissions

In terms of CO_2 emissions, the general trend shows as same as energy consumption trends. As increased temperature through the climate change scenarios, savings by heating energy reduction and emissions from cooling are increased.

Figure 11.29 shows that the CO_2 emissions from cooling and saving by heating reduction through the climate change scenarios. House 1 in London shows almost 50% of carbon emissions saving from heating reduction was required to cover the CO_2 emissions from cooling demand increased through the climate scenarios. In flat 1, CO_2 emissions contributed by cooling energy demand is almost 25% through climate scenarios in London. Therefore, the southern part of the

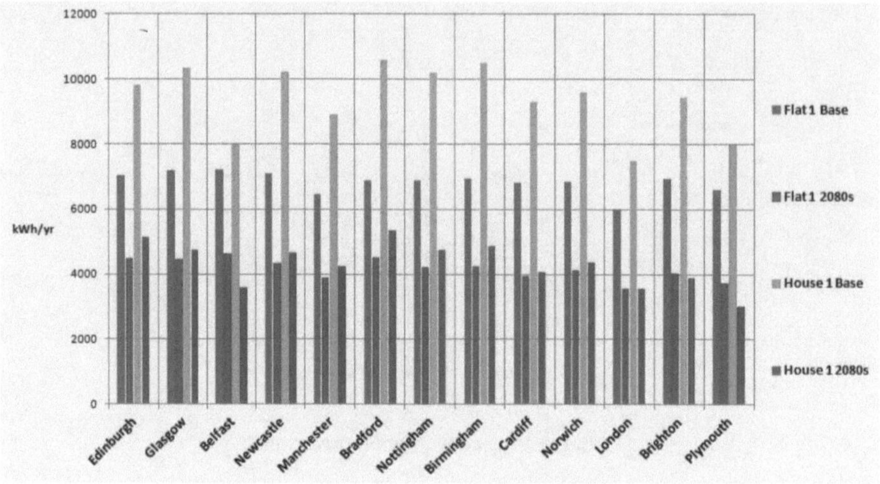

Heating energy consumption prediction

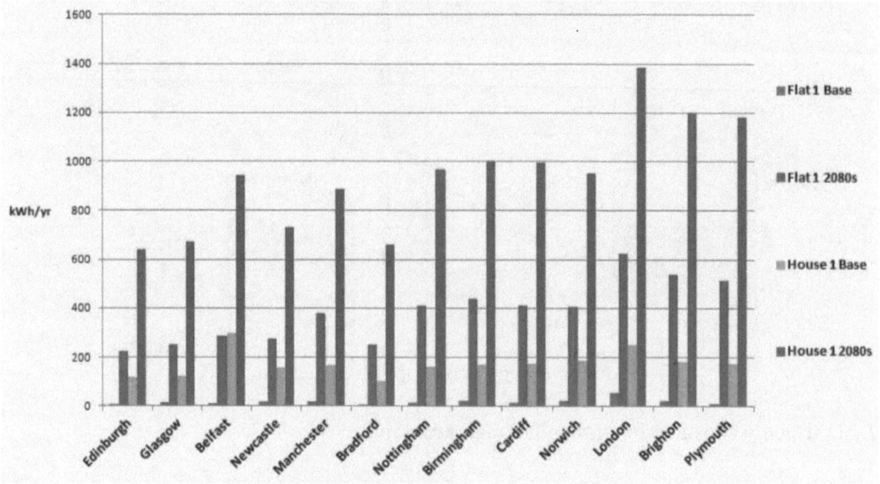

Cooling energy consumption prediction

Figure 11.28 Heating and cooling energy demand by 13 national locations.

UK in areas such as London should be aware that CO_2 emissions will increase as demand for cooling increases in line with climate change scenarios and this should be offset by a reduction in demand for heating energy.

Overheating Study

Figure 11.30 shows a comparison of the results for the bedroom of flat 1 and house 1 for London, Manchester and Edinburgh under high emissions scenario. For Manchester, the hours exceeding 26°C do not become greater than those for London through the climate scenarios and fall below those for London present and 2050s. For Edinburgh, the hours exceeding 26°C do not become

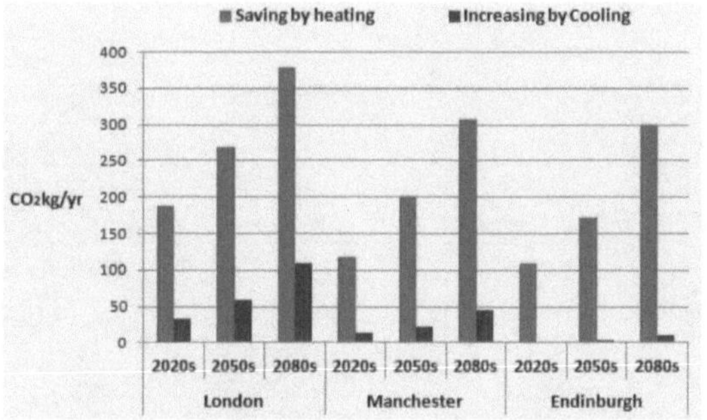

CO_2 emissions by cooling and saving from heating for flat 1

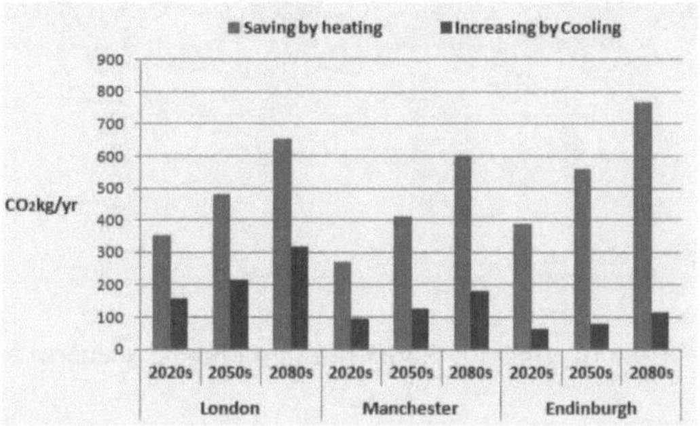

CO_2 emissions by cooling and saving from heating for house 1

Figure 11.29 CO_2 emissions by cooling and saving from heating.

larger than those for Manchester and for London until the 2080s, and then are comparable to those for Manchester 2020s. These results indicate that levels of overheating in Manchester and Edinburgh by the 2080s under high emissions scenario would be comparable or slightly larger to those experienced in the 2020s in London and Manchester (Pathan et al., 2017).

Scenario D Results

Scenario D studies the effect of various construction standards such as 'No insulation', '2006 Building Regulation' and 'Best Practice' with two different construction methods such as 'light-weight' and 'heavy-weight' constructions for attempting to establish the trend of heating and cooling energy consumption and overheating in case study homes. The effect of climate change on this relationship was also examined.

Energy Consumption

In general trend, heating energy consumption decreased, and cooling demand increased as improving building standards and this trend is greater through the climate change scenarios (Murano et al., 2018). Through the results, the heavy-weight construction method shows better performance in heating and cooling energy consumption. With the improved building fabric standards, the heating energy demand decreased, and the most significant reduction was made 'No insulation' case

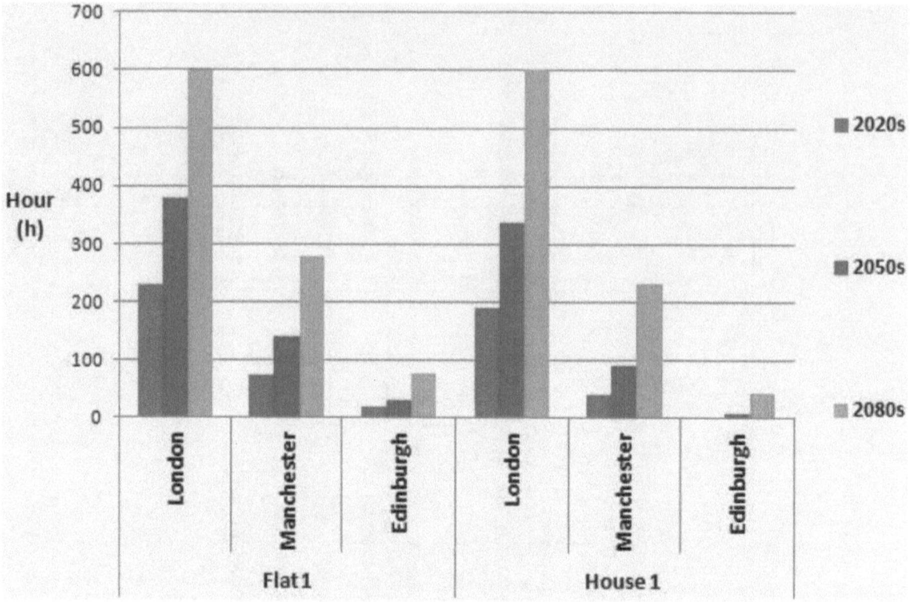

Figure 11.30 Comparison of the overheating hours through climate scenario for London, Manchester and Edinburgh.

to the '2006 Building Regulation' case. The most efficient construction method for heating energy consumption is 'Best Practice' with a 'heavy-weight' structure. With regard to cooling energy consumption, having no insulation in heavy-weight construction shows the best performance. Certainly, the most important factor for these trends is made by level of insulation and construction methods (Tomrukcu & Ashrafian, 2024).

Figure 11.31 shows heating and cooling energy consumption for flat 1 and house 1 through the climate change scenarios with construction method and standard consideration. With regards to heating energy consumption, heavy-weight construction consumed less heating energy than light-weight structure. However, flat 1's 'Best Practice' with heavy-weight construction shows more heating energy demand than light-weight ones.

The heating energy consumption difference between light-weight and heavy-weight is fixed at around 18%, 9% and 17% in 2020s, 2050s and 2080s respectively. With regard to heating energy consumption reduction through the building standards for the case study home, the heating demand difference between 'No insulation' and '2006 Building Regulation' is fixed at around 80% and the difference between 'No insulation' and 'Best Practice' is fixed around 75% and around 30% difference is fixed for the difference between '2006 Building Regulation' and 'Best Practice'. However, flat 1 shows less heating reduction than other case study homes.

As can be seen in Figure 11.31, the cooling energy consumption is less in heavy-weight construction with no insulation standard and the cooling energy demand is gradually increased as improved insulation levels through the climate scenarios. For example, the cooling energy difference between heavy-weight and light-weight in the 2020s with no insulation is around 55% and 74% in house 1 and flat 1, respectively. However, the difference in 2080s Best Practice is around 4% in all case study homes.

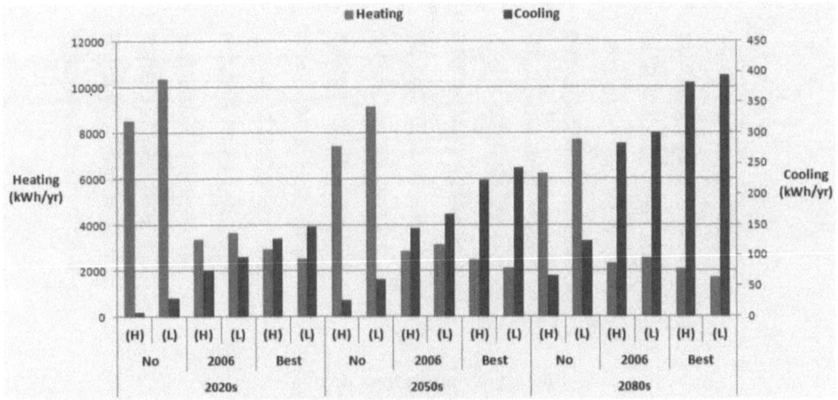

Flat 1 energy consumption results

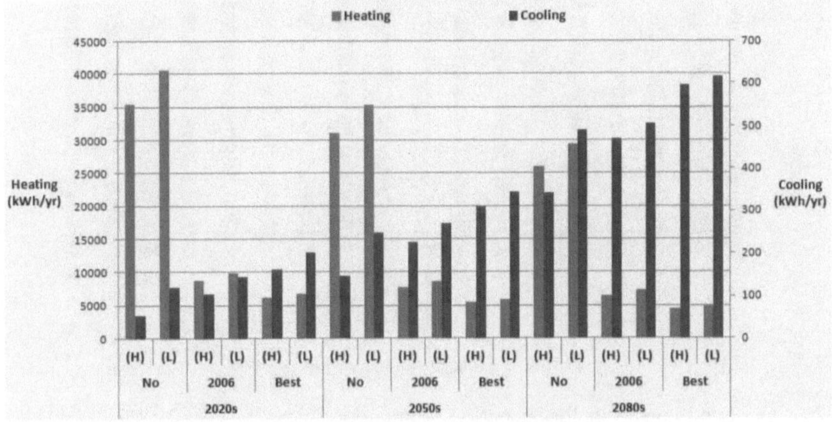

House 1 energy consumption results

Figure 11.31 Energy consumption results for scenario D simulation study.

In terms of CO_2 emissions, the general trend shows as same as energy consumption trends. As increased temperatures through the climate change scenarios, in all cases dramatic CO_2 emissions reductions are achieved from heating energy reduction by improving the level of insulation and this has happened in both construction methods. With regards to comparison construction methods, heavy-weight construction shows better performance in CO_2 emissions from heating and cooling through the climate change scenarios (Parker, 2020). Figure 11.32 shows the CO_2 distribution from heating and cooling while considering building standards and construction methods. As can be seen in Figure 11.32, most CO_2 emissions come from heating and CO_2 emissions are gradually increased as increased temperatures throughout the climate change scenarios improved insulation levels. In terms of construction methods sensitivity, the heavy-weight structure shows less emissions than the light-weight structure as mentioned above.

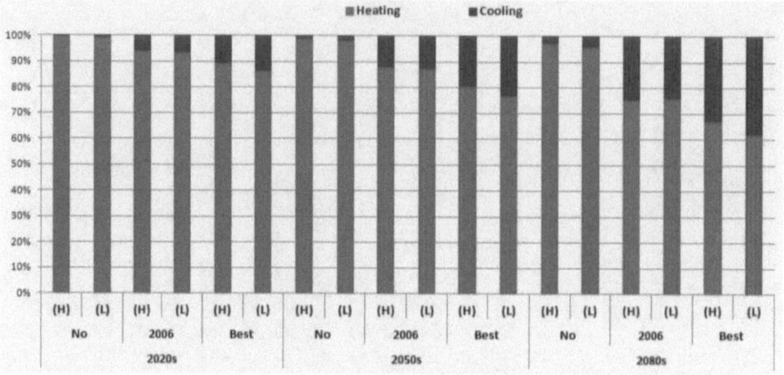

Flat 1 CO_2 emissions results

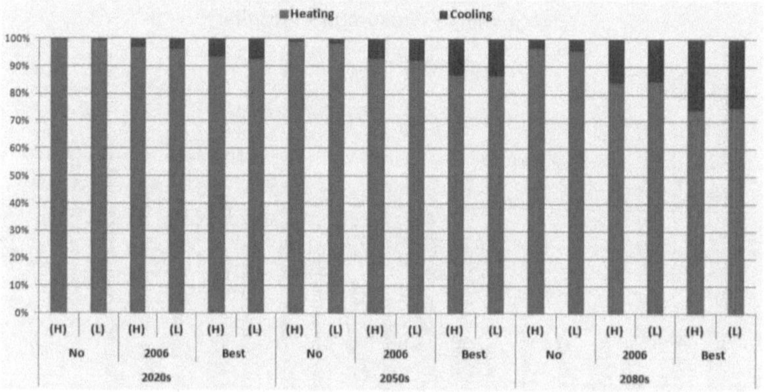

House 1 CO_2 emissions results

Figure 11.32 CO_2 emissions results for Scenario D simulation study.

Overheating Study

Figure 11.33 shows the results of the overheating study for flat 1 and house 1's bedroom and the comparison study between CIBSE Guide A and EN15251. In general trend, overheating hours were observed more in heavy-weight structures than light-weight structures through the climate change scenarios. With regard to building standard influence on overheating hours, 'No insulation' shows much fewer overheating hours throughout the building standards (Gunawardena & Steemers, 2019; Carlucci et al., 2021). Therefore, the most effective home in the overheating study is 'No insulation' with a light-weight structure throughout the climate scenarios. Overheating is significantly developed after improving the level of insulation for case study homes. As improved insulation level with increased temperatures, the overheating difference between heavy-weight and light-weight is gradually reduced. In 2080s 'Best Practice', the overheating hours difference between Best Practice heavy-weight and light-weight is about 1%.

Figure 11.33 also shows the results of the comparison CIBSE Guide A and EN15251 upper limit temperatures. This comparison study only considered the bedroom during occupied hours. The overheating hours based on EN15251 show the same trend as CIBSE Guide A, which means the light-weight structure shows better performance than the heavy-weight structure.

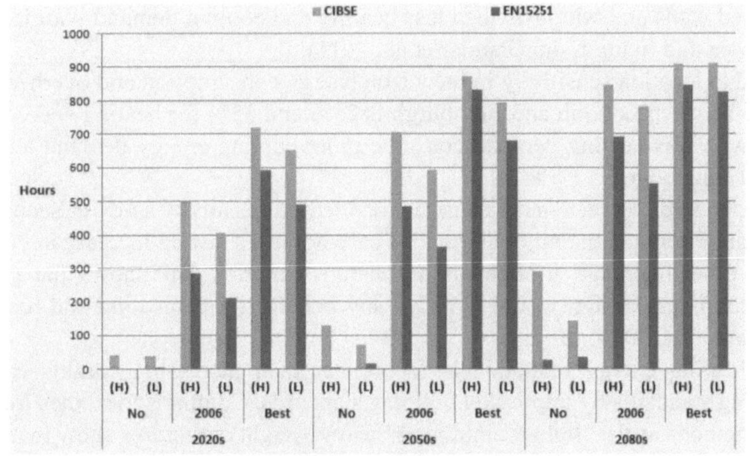

Overheating results for flat 1's bedroom

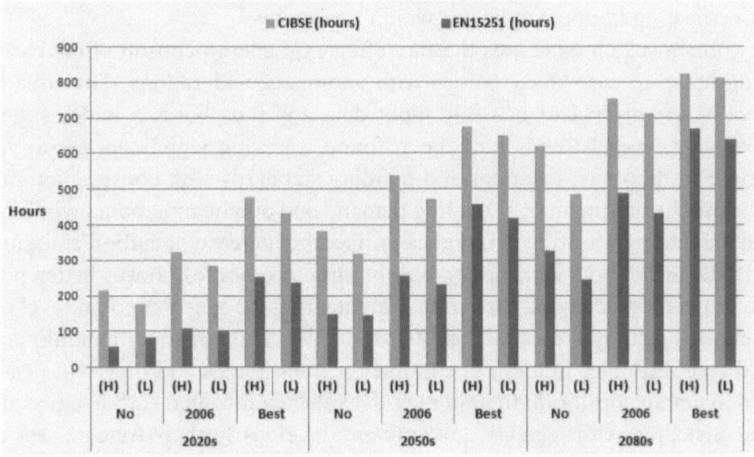

Overheating results for house 1's living room

Figure 11.33 Overheating results and comparison study for adapting climate.

Conclusions

Despite testing a pessimistic scenario with high cooling demands, the size of the heating demand still far outweighs the effects of an increase in cooling load. Although the climate is predicted to warm up, case study homes still have a significant heating load to bear. As climate change occurs, winters will become milder and demand for heating will reduce between 14% and 56% through 2020s low emissions to 2080s high emission scenarios. This is due to climate change alone, without any improvement in existing case study homes.

With regard to morphology influence on energy consumption and overheating, mid-floor flat and mid-terraced home show better performance than top-floor flat and end-terraced home. From these results, it is pointed out that less exposed surface to ambient means less heat losses and gains through exposed walls and windows then less heating and cooling demand with less overheating hours in bedroom and living room (Patidar et al., 2011).

With regard to location sensitivity influence on energy consumption and overheating, the average difference between London and Edinburgh is 23% and 53% for heating and cooling, respectively. Cities with less heating demand consume more cooling energy demand and increase the level of overheating hours.

The better standards, as seen in the construction method sensitivity study in Scenario D, such as the 2006 Building Regulation and Best Practice case homes, show an increase in cooling demand and level of overheating hours through climate change scenarios. This shows that a well-insulated with better airtightness design would not give any benefit of overheating and reducing cooling demand (Santamouris, 2019).

However, heating energy reduction is far greater than increasing cooling energy demand through building standards by improving building standards. Comparison studies for two different construction methods as the 'light-weight' and 'heavy-weight' structures show that heavy-weight structure has better performance of heating and cooling energy consumption than light-weight structures. In contrast, the level of overheating in case study homes with light-weight structures shows fewer overheating hours than heavy-weight structures.

In summary, this research examines the like effects on energy consumption, carbon emissions and overheating hours in case study homes with various considerations. This area of research is highly sensitive to the myriad of possible input data and thus holds a wide range of predicted outcomes. However, general trends have been found, showing significant sensitivity to climate change scenarios, morphology, locations and building standards with construction methods. Heating demand will still be dominant over cooling demand and overheating hours will increase through the 2080s. The reduction in heating demand can offset the increase in carbon emissions from cooling systems. Better insulation with heavy-weight structure homes shows better performance of heating and cooling energy demand than light-weight structure ones but not in level of overheating hours. The overheating study indicated that after the 2050s without an air conditioning system, the occupant could face difficulty in thermal comfort for their homes and they will need to install air conditioning to maintain comfort in their homes. Therefore, unwanted carbon emissions from cooling should be considered while the UK government develops further strategies for carbon reduction targets in the future, due to the fact that present reduction targets are only considering heating energy consumption.

References

Amaripadath, D., Joshi, M. Y., Hamdy, M., Petersen, S., Stone, J., Brian, & Attia, S. (2023). Thermal resilience in a renovated nearly zero-energy dwelling during intense heat waves. *Journal of Building Performance Simulation, 1*, 11–12. https://doi.org/10.1080/19401493.2023.2253460.

Attia, S., Benzidane, C., Rahif, R., Amaripadath, D., Hamdy, M., Holzer, P., Koch, A., Maas, A., Moosberger, S., Petersen, S., Mavrogianni, A., Maria Hidalgo-Betanzos, J., Almeida, M., Akander, J., Khosravi Bakhtiari, H., Kinnane, O., Kosonen, R., & Carlucci, S. (2023). Overheating calculation methods, criteria, and indicators in European regulation for residential buildings. *Energy and Buildings, 292*, 132–134. https://doi.org/10.1016/j.enbuild.2023.113170.

Botti, A., Leach, M., Lawson, M., & Hadjidimitriou, N. S. (2021). Developing a meta-model for early-stage overheating risk assessment for new apartments in London. *Energy and Buildings, 254*. https://doi.org/10.1016/j.enbuild.2021.111586.

Calderón, C., & Beltrán, M. R. (2018). Effects of fabric retrofit insulation in a UK high-rise social housing building on temperature take-back. *Energy and Buildings, 173*, 470. https://doi.org/10.1016/j.enbuild.2018.05.046

Carlucci, S., Erba, S., Pagliano, L., & De Dear, R. (2021). ASHRAE likelihood of dissatisfaction: a new right-here and right-now thermal comfort index for assessing the Likelihood of dissatisfaction according to the ASHRAE adaptive comfort model. *Energy and Buildings, 250*, 32–34. https://doi.org/10.1016/j.enbuild.2021.111286.

Colclough, S., & Salaris, C. (2023). Quantifying overheating in nZEB Irish residential buildings. An analysis of recorded interior temperatures of Irish newbuild and retrofit residential buildings against CIBSE, Passive House and WHO overheating criteria and recorded occupant satisfaction. *Energy and Buildings, 303*, 132–134. https://doi.org/10.1016/j.enbuild.2023.113571.

Cole, R., Evins, R., & Eames, M. (2023). An empirical review of methods to assess overheating in buildings in the context of changes to extreme heat events. *Journal of Building Performance Simulation, 17*(3), 322. https://doi.org/10.1080/19401493.2023.2276711.

Eames, M., Kershaw, T., & Coley, D. (2011). The appropriate spatial resolution of future weather files for building simulation. *Journal of Building Performance Simulation, 5*(6), 347. https://doi.org/10.1080/19401493.2011.608133.

Evins, R., Alexandra, R., Wiebe, E., Wood, M., & Eames, M. (2018). The impact of local variations in a temperate maritime climate on building energy use. *Journal of Building Performance Simulation, 13*(2), 167. https://doi.org/10.1080/19401493.2018.1536167.

Gunawardena, K., & Steemers, K. (2019). Adaptive comfort assessments in urban neighbourhoods: simulations of a residential case study from London. *Energy and Buildings, 202*, 145–146. https://doi.org/10.1016/j.enbuild.2019.07.039.

Gupta, R., & Gregg, M. (2017). Assessing energy use and overheating risk in net zero energy dwellings in UK. *Energy and Buildings, 158*, 897. https://doi.org/10.1016/j.enbuild.2017.10.061.

Gupta, R., Kapsali, M., & Howard, A. (2018). Evaluating the influence of building fabric, services and occupant related factors on the actual performance of low energy social housing dwellings in UK. *Energy and Buildings, 174*, 548. https://doi.org/10.1016/j.enbuild.2018.06.057.

Jenkins, D. P., Gul, M., Patidar, S., Banfill, P. F. G., Gibson, G., & Menzies, G. (2012). Designing a methodology for integrating industry practice into a probabilistic overheating tool for future building performance. *Energy and Buildings, 54*, 73. https://doi.org/10.1016/j.enbuild.2012.07.001.

Jenkins, D. P., Patidar, S., Banfill, P. F. G., & Gibson, G. J. (2011). *Probabilistic climate projections with dynamic building simulation: predicting overheating in dwellings*. Elsevier BV. https://doi.org/10.1016/j.enbuild.2011.03.016.

Jentsch, M. F., Bahaj, A. S., & James, P. A. B. (2008). Climate change future proofing of buildings—generation and assessment of building simulation weather files. *Energy and Buildings, 40*(12), 2148. https://doi.org/10.1016/j.enbuild.2008.06.005.

Jones, R. V., Fuertes, A., Boomsma, C., & Pahl, S. (2016). Space heating preferences in UK social housing: a socio-technical household survey combined with building audits. *Energy and Buildings, 127*, 382. https://doi.org/10.1016/j.enbuild.2016.06.006.

Kendrick, C., Ogden, R., Wang, X., & Baiche, B. (2012). Thermal mass in new build UK housing: a comparison of structural systems in a future weather scenario. *Energy and Buildings, 48*, 40. https://doi.org/10.1016/j.enbuild.2012.01.009.

Laouadi, A., Bartko, M., & Lacasse, M. A. (2020a). A new methodology of evaluation of overheating in buildings. *Energy and Buildings, 226*. https://doi.org/10.1016/j.enbuild.2020.110360.

Laouadi, A., Gaur, A., Lacasse, M. A., Bartko, M., & Armstrong, M. (2020b). Development of reference summer weather years for analysis of overheating risk in buildings. *Journal of Building Performance Simulation, 13*(3), 301. https://doi.org/10.1080/19401493.2020.1727954.

Mitchell, R., & Natarajan, S. (2020). UK Passivhaus and the energy performance gap. *Energy and Buildings, 224*, 79–81. https://doi.org/10.1016/j.enbuild.2020.110240.

Murano, G., Dirutigliano, D., & Corrado, V. (2018). Improved procedure for the construction of a typical meteorological year for assessing the energy need of a residential building. *Journal of Building Performance Simulation, 13*(2), 139. https://doi.org/10.1080/19401493.2018.1479774.

O'brien, W., Gaetani, I., Gilani, S., Carlucci, S., Hoes, P., & Hensen, J. (2016). International survey on current occupant modelling approaches in building performance simulation. *Journal of Building Performance Simulation, 10*(5–6), 653. https://doi.org/10.1080/19401493.2016.1243731.

Ozarisoy, B. (2022). Energy effectiveness of passive cooling design strategies to reduce the impact of long-term heatwaves on occupants' thermal comfort in Europe: Climate change and mitigation. *Journal of Cleaner Production, 330*, Article 129675. https://doi.org/10.1016/j.jclepro.2021.129675.

Ozarisoy, B., & Altan, H. (2021). Regression forecasting of 'neutral' adaptive thermal comfort: a field study investigation in the south-eastern Mediterranean climate of Cyprus. *Building and Environment, 202*. https://doi.org/10.1016/j.buildenv.2021.108013.

Ozarisoy, B., & Altan, H. (2023a). Thermal comfort survey I: a field study investigation to assess on households' thermal discomfort and overheating risk of european buildings. In: *Handbook of retrofitting high density residential buildings*. Springer (456–458). https://doi.org/10.1007/978-3-031-11854-8_6.

Ozarisoy, B., & Altan, H. (2023b). Thermal comfort survey II: a field study investigation on the regression forecasting of neutral adaptive thermal comfort. In: *Handbook of retrofitting high density residential buildings*. Springer (547–549). https://doi.org/10.1007/978-3-031-11854-8_7.

Ozarisoy, B., & Elsharkawy, H. (15 March 2019). Assessing overheating risk and thermal comfort in state-of-the-art prototype houses that combat exacerbated climate change in UK. *Energy and Buildings, 187*, 201–217. https://doi.org/10.1016/j.enbuild.2019.01.030.

Parker, J. (2020). The Leeds urban heat island and its implications for energy use and thermal comfort. *Energy and Buildings, 235*. https://doi.org/10.1016/j.enbuild.2020.110636.

Pathan, A., Mavrogianni, A., Summerfield, A., Oreszczyn, T., & Davies, M. (2017). *Monitoring summer indoor overheating in the London housing stock*. Elsevier BV (23–25). https://doi.org/10.1016/j.enbuild.2017.02.049.

Patidar, S., Jenkins, D. P., Gibson, G. J., & Banfill, P. F. G. (2011). Statistical techniques to emulate dynamic building simulations for overheating analyses in future probabilistic climates. *Journal of Building Performance Simulation, 4*(3), 271. https://doi.org/10.1080/19401493.2010.531144.

Patidar, S., Jenkins, D. P., Gibson, G. J., & Banfill, P. F. G. (2012). Analysis of probabilistic climate projections: heat wave, overheating and adaptation. *Journal of Building Performance Simulation, 6*(1), 65. https://doi.org/10.1080/19401493.2012.684447.

Porritt, S. M., Cropper, P. C., Shao, L., & Goodier, C. I. (2012). Ranking of interventions to reduce dwelling overheating during heat waves. Paper presented at the, *55* 16, 1–7. https://doi.org/10.1016/j.enbuild.2012.01.043.

Rostami, M., & Bucking, S. (2024). Adaptation to extreme weather events using pre-conditioning: a model-based testing of novel resilience algorithms on a residential case study. *Journal of Building Performance Simulation, 1*. https://doi.org/10.1080/19401493.2024.2307636.

Santamouris, M. (2019). Recent progress on urban overheating and heat island research. Integrated assessment of the energy, environmental, vulnerability and health impact. Synergies with the global climate change. *Energy and Buildings, 207*, 23–25. https://doi.org/10.1016/j.enbuild.2019.109482.

Santamouris, M., & Kolokotsa, D. (2014). On the impact of urban overheating and extreme climatic conditions on housing, energy, comfort and environmental quality of vulnerable population in Europe. *Energy and Buildings, 98*, 125. https://doi.org/10.1016/j.enbuild.2014.08.050.

Symonds, P., Taylor, J., Chalabi, Z., Mavrogianni, A., Davies, M., Hamilton, I., Vardoulakis, S., Heaviside, C., & Macintyre, H. (2016). *Development of an England-wide indoor overheating and air pollution model using artificial neural networks*. Informa UK Limited. https://doi.org/10.1080/19401493.2016.1166265.

Taylor, J., Symonds, P., Heaviside, C., Chalabi, Z., Davies, M., & Wilkinson, P. (2021). Projecting the impacts of housing on temperature-related mortality in London during typical future years. *Energy and Buildings, 249*, 35–38. https://doi.org/10.1016/j.enbuild.2021.111233.

Tian, W., & De Wilde, P. (2011). Thermal building simulation using the UKCP09 probabilistic climate projections. *Journal of Building Performance Simulation, 4*(2), 105. https://doi.org/10.1080/19401493.2010.502246.

Tomrukcu, G., & Ashrafian, T. (2024). Climate-resilient building energy efficiency retrofit: evaluating climate change impacts on residential buildings. *Energy and Buildings, 316*, 123–125. https://doi.org/10.1016/j.enbuild.2024.114315.

12
CONCLUSION

Introduction

This final chapter presents a short summary of the main findings of this research and a discussion of both its original contribution to the body of knowledge and its limitations. It also compared findings from this research with other's results on energy consumption under climate change. This is followed by a discussion on future work arising out of this thesis.

Research Findings

This research sets out with the aim of assessing the importance of building morphology and building standards to energy consumption and overheating in the case study homes, which represent modern methods of dwelling construction, under current and future climate projections. This research started with two important trends, namely, 'Climate Change and Global Warming' and 'Energy Efficiency Homes'.

> The scientific consensus is that for the UK to play its part in helping the world avoid a rise of more than 2°C, we must reduce our carbon emissions by 80 per cent by 2050. The household sector represents 27 per cent of our total emissions and achieving deep cuts here is an imperative. The low-carbon revolutions start at home.
>
> University of Oxford's Environmental Change Institute (Jentsch et al., 2008)

> Changes in atmospheric composition resulting from human activity have enhanced the natural greenhouse effect, ……. The IPCC's best estimate was that globally averaged surface temperatures would be between 2.5 ~ 4.7°C higher by 2100 compared to pre-industrial levels. The full range of projected temperature increases by 2100 was found to be 1.8–7.1°C based on the various scenarios and uncertainties in climate sensitivity.
>
> The Royal Society (Collins et al., 2010)

> Household energy use is still growing at a rate of 1.5 per cent per year. Household energy use will need to fall by 1.8 per cent per year to deliver a 60 per cent fall in emissions by 2050.–
>
> NERA Economic Consulting (Din & Brotas, 2017)

DOI: 10.1201/9781003490364-15

> As a result of poor insulation, £1 in every £3 spent heating homes in the UK is wasted – the current trend of focusing on the standards of new build housing fails to address this issue
>
> British Gas (Gupta & Gregg, 2012)

As a result of climate change studies, predictions contained in the IPCC's recent assessment report show that current emission trends imply that the actual temperature increase could be far higher than we expected. Modelling by the Headley Centre shows that, by the 2040s, a 2003 heat wave type of summer is predicted to be typical, and this will clearly have a great impact on energy consumption for cooling and thermal comfort (De Wilde et al., 2008).

Climate change is caused by the emissions of GHGs into the atmosphere and the UK is responsible for contributing around 1.5% to the global level. According to the Kyoto Protocol, the UK government has set the very ambitious target of reducing GHG emissions by 80% by 2050 from the 1990 level (Kendrick et al., 2012). The extent to which these savings will be achieved is to some extent predicted by the response of the domestic sector, which is currently responsible for approximately 27% of all UK carbon emissions (Lindley et al., 2007). In line with this reduction target, the UK government is actively encouraging fast construction, together with better standards of construction using off-site construction techniques. In addition, Part L of the Building Regulation requirements on energy efficiency and carbon, demands increasingly stringent energy performance for new construction (Ozarisoy, 2022). These factors have led to an increased interest in lightweight timber and steel frame construction as it can be easier to engineer these forms of construction to achieve high standards of thermal insulation and air tightness.

However, a move to light-weight construction raises concerns about internal thermal comfort due to a perceived reduction in available thermal mass (Ozarisoy & Elsharkawy, 2019). This can result in rapid swings of internal temperatures. The concern is increased by projected climate change and global warming. Without optimised design of new buildings, many of the benefits accruing from reduced heating consumption may be lost as air conditioning systems are installed to maintain comfortable internal temperatures.

Findings from the Post-Occupancy Study

In general, as the floor area and number in the household increased, energy consumption and CO_2 emissions increased (Salem et al., 2019). With regard to the influence of the main energy source, the case study homes' heating systems using electricity as a main energy source showed better performance than those using gas central heating systems. However, in terms of CO_2 emissions and the influence of the main energy source, it seems that homes with more electricity consumption were at a disadvantage. For example, comparing House 1 and House 2, they have identical floor plans, floor areas and sizes of households. However, House 1 total energy consumption differs by 25% due to the heating system energy efficiency difference.

In terms of the correlation between internal temperature and number of children for energy consumption, homes with children showed higher internal temperatures and higher energy consumption than homes with no children. The home with children's average internal temperature during the heating period is almost 1.6°C higher than the home with no children. Even if homes have an identical floor plan, floor area and size of household, the energy consumption can show large variations due to user behaviour impact. Again, house 1 total energy consumption is almost 25% higher than house 2 and they have an identical floor plan and size of household due to user behaviour differences, children's age differences and heating system differences. The applied building standard to case study homes is better than '2006 Building Regulation', which should be almost 25% less energy consumption expected due to the case study homes building standard being very close to 'Best Practice standard'. However, two of the case study homes performed worse than the 2006 building regulation homes (Williams et al., 2012). From this result, it could be concluded that it is the occupants who determine how energy efficient a dwelling will be; even if the building is well insulated and the dwelling has an efficient energy source. Therefore, reducing energy use depends not only upon improving building performance but also on understanding the relationship between how occupants use the property and their particular personal circumstances.

With regard to energy efficiency for case study homes, it is seen that SAP prediction shows better performance than case study homes and the difference between SAP prediction and actual energy consumption is from 3% to 81%. The case study house shows more energy consumption than SAP prediction due to the fact that the heating temperature set for SAP is lower than that for the case study homes' internal temperature, especially in the bedroom. The SAP's internal set temperature for predicting heating energy consumption needs to be reflected more realistically in line with current trends which means the preferable temperature for rooms in the UK homes is around 21°C and for other areas with the home, it is 2°C lower.

A result of the winter comfort study shows that the occupants in the case study homes preferred higher internal temperature for bedrooms than the 'CIBSE Guide A' recommendation (Wright & Venskunas, 2022). The CIBSE recommended bedroom temperature is from 17°C to 19°C. However, the case study homes' preferred internal temperature range is from 18.8°C to 22.5°C and the average bedroom temperature across the case study home is 20.4°C, which is almost 1.5°C higher than CIBSE recommended bedroom upper limit temperature. With regard to the overheating, some of the case study homes show overheating during the monitoring periods due to the fact that they had south facing living rooms and bedrooms with large windows. It is also seen that solar gains through windows and exposed walls have an effect on thermal comfort. To minimise energy demand in the heating season, case study homes were designed increasing airtightness as well as fabrics' thermal insulation levels (Xie et al., 2023). Combined with increasing summer temperatures raise and with more airtight and energy-efficient homes could face the potential for overheating. It is important to look at the potential risk for hotter homes and their effect on occupant health.

Findings from Simulation Study

Input data from the monitoring study and the survey helped to reduce any uncertainty regarding the simulation model and also improved the accuracy of the simulation results. The difference between actual consumption and simulation results is from 1.5% to 10% and the difference is within the acceptable percentage which is 15%.

Although the climate is predicted to warm up, case study homes still have a significant heating load when compared to cooling demand, and overheating hours will increase through the 2080s (Moazami et al., 2019). Depending on the climate change scenarios, space heating energy demand in the present case study homes fell by 14% in the 2020s and demand in the 2080s level was 54% lower than at present. The energy consumption in cooling is less than 20% of overall energy consumption in the 2080s. Overheating hours in the bedroom increased by 19%–31% in 2020s and overheating in the 2080s level increased by almost 70% more than at present. In the case of overheating in the living room in the 2080s level was increased by 27%–62% than at present.

Carbon saving by the reduction in heating demand can offset the increase in carbon emissions from cooling systems (Kelly, 2009). In the 2080s, the carbon emissions contribution by cooling demand, however, could be more than one-third of total carbon emissions from case study homes. Well-insulated homes with better airtightness design would not result in any benefit to the level of overheating and cooling demand. The impact of improving building standards will reduce demand for heating energy by 30%–80%. It will also easily offset the carbon emissions from the demand for cooling energy. As a result of this, improving insulation with better airtightness is recommended to reduce overall energy demand throughout climate change scenarios (Holmes & Hacker, 2007).

With regard to the influence of regional location on energy consumption and overheating, cities in the southern part of the UK will need to prepare for increasing cooling demand and high levels of overheating, whereas northern cities within the UK will need to focus on dealing with a reduction in demand for heating energy (Gupta et al., 2018). Home in London would consume less heating energy by 15%–27% than home in Edinburgh and home in London would demand more cooling energy by 45%–91% than home in Edinburgh. As a result of this, it is suggested that separate building standards and construction methods need to develop to reflect regional differences (Kelly et al., 2012). Better insulation with heavy-weight structure homes shows better performance of energy consumption than those with a light-weight structure by 8%, however, this does not apply to the level of overheating hours. In terms of overheating in bedrooms, heavy-weight structure shows almost 7% greater than the light-weight structure in the 2020s and only 1% in the 2080s. After the 2050s, the external temperature projections show that the external temperature will be far higher than the thermal comfort criteria temperatures (Tian & De Wilde, 2011). It will, therefore, be very difficult for occupants to deal with their comfort using only passive cooling without air conditioning systems, irrespective of what type of structure their homes were built with heavy-weight or light-weight. Therefore, unwanted carbon emissions from cooling should be considered while the UK government develops further strategies for carbon reduction targets in the future.

Comparison of Relevant Research Results on Internal Temperatures

Last year, very interesting research results were published by Tyndall Centre for Climate Change Research who developed a new method for future building stock models to incorporate the dynamics of internal temperature demand. Moreover, the method was adapted to quantify the take-back effect to estimate changing internal temperature by combinations of different energy efficiency measures. This research also shows how this method allows modellers to include behavioural and socio-demographic factors to estimate their effect on internal temperatures.

> For example, if a child under five is present in the dwelling then the mean internal daily temperature is expected to be ~0.5°C warmer when compared to a home without a child, ceteris paribus.
>
> (Vardoulakis et al., 2015)

Within this research, it agreed that homes with children show a higher average internal temperature than homes without a child. However, this research shows the bedroom's mean internal temperature is 1°C higher and by 1.1°C–4.3°C in the living room. However, there will be a wide variation in the population as this is only a statistical model.

> The number of people living in a dwelling represents one of the most important determinants for explaining dwelling energy consumption. The results presented in this analysis support this finding. For each additional person living in a dwelling the mean daily internal

temperature increase, ~0.25°C on average. Thus, a dwelling with a family of five would be ~1.25°C warmer than a single person household.

(Gaterell & Mcevoy, 2005)

The results from this research support this finding. The greater the number of people living in the dwellings show higher the average internal temperature than where fewer people are living in the dwellings. For instance, for two people, three people and five people living in the case study homes', the average internal temperature is 18.4°C, 20.5°C and 22°C, respectively. The presence of a child under five is shown to increase internal temperatures by an average of ~0.5°C while each child under the age of 18 increases mean internal temperature by ~0.22°C for each additional child. A home where the oldest person is aged 64–74 will be ~0.37°C warmer than a home where everyone is under 60. The presence of an elderly person over the age of 75 increases internal temperature by ~0.56°C more than a home where the oldest person is under 64.

The home with a child under five shows the highest average internal temperature during the heating period which is 21.6°C, and next is the case home with children under the age of 18 and its internal temperature during the heating season, which is 21.1°C. One of the case study homes was occupied by a retired couple and their age group is between 60 and under 75. This case study home should be recording higher internal temperature. However, its average internal temperature is around 19°C. It could be explained by the time spent at home because the retired couples' home was unoccupied for more than three weeks during the monitoring period and during the unoccupied period, the daily internal temperature dropped below 12°C and even 10°C.

Comparison of Relevant Research Results on Energy Consumption

This section shows a comparison of relevant research results on energy consumption under climate change scenarios, and it is summarised below: The Gaterell and Mcevoy (2005) study shows the impact of climate change on the performance of heating energy consumption in UK dwellings. Their study confirmed that delivered energy for space heating is estimated by adapting to climate change and for the low emissions case, a space heating energy reduction by 25%, and that of 74% for the high emission case in the 2050s. Day et al. (2009) show the forecasting of future cooling demand in London and the results suggest around a 40% increase in cooling energy demand on base case levels in the 2030s. They estimated trends in London's temperature and cooling degree-day (CDD), and forecast future climate change scenarios. From the results, London's CDD projecting trend forward to 2010, 2020 and 2030 gives 66, 87 and 110 CDD, respectively.

The Collins et al. (2010) study shows climate change and future energy consumption in UK housing stock and a nationwide worst-case scenario for the 2050s UK dwelling stocks show a 20% reduction in gas consumption and a 26% in the 2080s. For a detached house, the study suggests CO_2 emissions from heating energy consumption could be reduced by 40% in the 2080s and for a purpose-built flat reduction by 37%. In terms of cooling energy consumption, the results suggest around a 60% increase in cooling energy demand on base case levels in the 2080s. The Kendrick et al. (2012) study estimated that space heating loads decreased by approximately 25% between 1990 and the 2050s, as a result of the medium-high emissions scenario. Heavy-weight structure construction case heating load is higher by 5–8% than light-weight structure construction due to a result of the extra thermal storage capacity in the structure. Gupta and Gregg (2012) suggested that the impact of climate change on the heating load reduction in UK dwellings will be around 60% in the 2080s.

Most research shows heating energy consumption reduced by adapting to climate change by 20–25% in the 2050s (Shrubsole et al., 2014). However, in this study, the reduction could be around 38% by the 2050s. The heating energy reduction difference between this research and that of others is around 13–15%. This difference could be explained by the fact that most relevant researchers' case study homes simulated typical UK homes, whereas this research case study homes were built close to 'Best Practice standard' and were more energy efficient thus saving more heating energy consumption by adapting to climate change. In terms of construction method sensitivity, heavy- weight structure construction homes consumed 5–8% more than light-weight structure construction (Kendrick et al., 2012), a result that is well matched with this research.

Contribution to the Body of Knowledge and the Limitations of the Research

Contribution to the Body of Knowledge

The work that has been undertaken within this research could make an important contribution to the body of knowledge in a number of different ways, these include: It represents a detailed attempt that has been undertaken to project the energy consumption and overheating consideration in the well-insulated recently built UK homes in the climate leading up to the 2080s. It shows the way a simulation study can be built and how collaboration can take place with post-occupancy study, in order to control the uncertainty of the model input data. Results will give insights into a better understanding of construction methods for the reduction of energy consumption and overheating with the UK housing stock. These insights will enable rational and considered responses to problems, such as overheating and necessitating the installation of air conditioning systems with the resultant increase in energy consumption problems that are likely to be faced in the future due to climate change.

It should have value to referenced studies and to the direction of government policy or strategies for UK housing, in their efforts to reduce energy consumption and overheating under a range of future climate change scenarios. In terms of cooling and overheating, this research also provides explanations for the disadvantageous effect of using modern construction methods for the UK homes, such as insulation with better airtightness and light-weight timber framed structures. This study could also be used to explain the advantages and disadvantages of having heavyweight and light-weight structures in terms of heating load.

Limitations of the Research

Size of case study homes: The total number of case study homes was not representative of all types of UK housing stock. The amount of data available from the case study homes was limited due to the fact that each case study house has only two internal monitoring data devices. More data could have provided more evidence to support the results and eliminate the uncertainty of input data in order to improve the accuracy of simulation models and simulation results. Monitoring data was lost during the monitoring period: some technical problems arise during the monitoring period, such as power-outages for the data logger and the monitoring system; breakdown of the data logger; and the removal and loss of the data logger by an occupant. This loss of data had an effect on the analysis and built simulation model.

Future Work

The limitation of this work would suggest that further work is needed to improve the method of monitoring. Some technical problems arose, and these issues had an effect on this research. It is necessary to improve the way data is monitored and gathered from loggers such as real-time upload of data to the server, backup power storage for loggers and the installation of a warning system when the logger detects any fault. The performance of low-carbon homes is very much dependent on occupant behaviour. As a result, it is suggested that the full benefits of low-carbon homes are not being realised because occupants do not fully understand how to control and to use and live in their homes more efficiently. Thus, there is a need to better understand occupant behaviour and the way in which occupants interact with buildings, as this can have a massive impact on the energy used and the comfort levels achieved.

The findings of this work would suggest that further work is needed to provide a detailed analysis of adaptations which are currently a part of every retrofit under 'Green Deal' and new build, for example, various insulation types and the effectiveness of positioning, investigating even a combination of exterior, interior insulation and exposed thermal mass on different wall orientations. Even if the combinations available for analysis are seemingly endless, it needs to be done in order to find better solutions for reducing heating and cooling energy demand and the risk of overheating in climate change scenarios. Further work is needed to provide a detailed study for defining the upper limits of thermal comfort in domestic dwellings. If the behavioural response of UK householders to a warming climate is akin to that found in the US, then it is predicted that 18% of homes in the South of England will have installed air conditioning systems by 2030. This would suggest that extant climate change will pass through a 'tipping point' in the next two decades creating a domestic cooling load demand. However, defining the upper limits of thermal comfort in domestic dwellings is a complex and under-researched pathway.

Most thermal comfort research has been conducted in office buildings and then the same assumptions were made for domestic buildings. A key aspect of thermal comfort is that it is a result of physiological and mental responses to a climate condition. Our mental state at home and the range of adaptive behaviour possible is distinct from that found in the office and, therefore, perceptions of comfort are likely to be quite different. Thus, defining upper limits of thermal comfort in our homes, especially living area and bedroom, needs to be carried out. In line with the findings of this work, further work is needed to provide a detailed methodology for making simulation models based on real buildings. Predicting building energy usage will be a central part of the design process, but research has shown the discrepancy between simulation energy consumption and real building energy usage. This is a serious problem which can significantly impact on the potential to achieve carbon reduction targets. Methodological improvements need to be based on sufficient empirical data rather than assumptions and simulations. It is necessary to properly understand energy use and to provide better solutions for building designers, decision makers, constructors and even occupants.

The findings of this work would suggest that further work is needed to provide a detailed study of construction methods and materials. Light-weight timber and steel framed buildings with insulation tend to have significantly less thermal mass than dwellings with heavier masonry construction incorporating physically heavy building elements on the internal side of the insulation. Heavyweight construction will generally reduce the speed at which internal temperatures increase as a result of external solar radiation or internal heat emissions. The corollary of this, however, is that the same thermal mass may reduce the speed at which internal temperature reduces when heating effects are removed. A thermally heavy building may therefore heat up more slowly but may retain its heat for longer, with the effect that it is cooler during the heating period but hotter during the cooling period. Optimum levels of thermal mass are therefore dependent upon the duration and magnitude of excess heating loads, and the periods of building usage. For instance, it may be that living spaces in a dwelling might most advantageously be designed to avoid daytime overheating and therefore have relatively high thermal mass, whilst sleeping spaces may be designed to have low thermal mass such that daytime temperature gains be lost as soon as possible to provide a cooler outside temperature during the night. Furthermore, heavyweight materials, like solid concrete, do not have the ability to absorb and emit heat quickly and also buffer moisture, other heavy-weight but low environmental impact materials, such as unfired earth cob or straw-clay, are able to provide a good thermal performance. Therefore, more a detailed study of materials and how they comply with Building Regulation is needed.

References

Collins, L., Natarajan, S., & Levermore, G. (2010). Climate change and future energy consumption in UK housing stock. *Building Services Engineering Research and Technology, 31*(1), 75. https://doi.org/10.1177/0143624409354972.

Day, A. R., Jones, P. G., & Maidment, G. G. (2009). Forecasting future cooling demand in London. *Energy and Buildings, 41*(9), 942. https://doi.org/10.1016/j.enbuild.2009.04.001.

De Wilde, P., Rafiq, Y., & Beck, M. (2008). Uncertainties in predicting the impact of climate change on thermal performance of domestic buildings in the UK. *Building Services Engineering Research and Technology, 29*(1), 7. https://doi.org/10.1177/0143624407087261.

Din, A., & Brotas, L. (2017). Assessment of climate change on UK dwelling indoor comfort. *Energy Procedia, 122*, 21. https://doi.org/10.1016/j.egypro.2017.07.296.

Gaterell, M. R., & Mcevoy, M. E. (2005). The impact of climate change uncertainties on the performance of energy efficiency measures applied to dwellings. *Energy and Buildings, 37*(9), 982. https://doi.org/10.1016/j.enbuild.2004.12.015.

Gupta, R., & Gregg, M. (2012). Using UK climate change projections to adapt existing English homes for a warming climate. *Building and Environment, 55*, 20. https://doi.org/10.1016/j.buildenv.2012.01.014.

Gupta, R., Kapsali, M., & Howard, A. (2018). Evaluating the influence of building fabric, services and occupant related factors on the actual performance of low energy social housing dwellings in UK. *Energy and Buildings, 174*, 548. https://doi.org/10.1016/j.enbuild.2018.06.057.

Holmes, M. J., & Hacker, J. N. (2007). Climate change, thermal comfort and energy: meeting the design challenges of the 21st century. *Energy and Buildings, 39*(7), 802. https://doi.org/10.1016/j.enbuild.2007.02.009.

Jentsch, M. F., Bahaj, A. S., & James, P. A. B. (2008). Climate change future proofing of buildings—generation and assessment of building simulation weather files. *Energy and Buildings, 40*(12), 2148. https://doi.org/10.1016/j.enbuild.2008.06.005.

Kelly, M. J. (2009). Retrofitting the existing UK building stock. *Building Research & Information*, 37(2), 196. https://doi.org/10.1080/09613210802645924.

Kelly, S., Crawford-Brown, D., & Pollitt, M. G. (2012). Building performance evaluation and certification in the UK: Is SAP fit for purpose? *Renewable and Sustainable Energy Reviews, 16*(9), 6861. https://doi.org/10.1016/j.rser.2012.07.018.

Kendrick, C., Ogden, R., Wang, X., & Baiche, B. (2012). Thermal mass in new build UK housing: a comparison of structural systems in a future weather scenario. *Energy and, Buildings, 48*, 40. https://doi.org/10.1016/j.enbuild.2012.01.009.

Lindley, S. J., Handley, J. F., Theuray, N., Peet, E., & Mcevoy, D. (2007). *Adaptation strategies for climate change in the urban environment: assessing climate change related risk in UK urban areas*. Informa UK Limited. https://doi.org/10.1080/13669870600798020.

Moazami, A., Nik, V. M., Carlucci, S., & Geving, S. (2019). Impacts of future weather data typology on building energy performance – investigating long-term patterns of climate change and extreme weather conditions. *Applied Energy, 238*, 696. https://doi.org/10.1016/j.apenergy.2019.01.085.

Ozarisoy, B. (2022). Energy effectiveness of passive cooling design strategies to reduce the impact of long-term heatwaves on occupants' thermal comfort in Europe: climate change and mitigation. https://doi.org/10.1016/j.jclepro.2021.129675.

Ozarisoy, B., & Elsharkawy, H. (2019). Assessing overheating risk and thermal comfort in state-of-the-art prototype houses that combat exacerbated climate change in UK. Energy and Buildings, Volume 187, 15 March 2019, pp. 201–217, https://doi.org/10.1016/j.enbuild.2019.01.030.

Salem, R., Bahadori-Jahromi, A., & Mylona, A. (2019). Investigating the impacts of a changing climate on the risk of overheating and energy performance for a UK retirement village adapted to the nZEB standards. *Building Services Engineering Research and Technology, 40*(4), 470. https://doi.org/10.1177/0143624419844753.

Shrubsole, C., Macmillan, A., Davies, M., & May, N. (2014). *100 Unintended consequences of policies to improve the energy efficiency of the UK housing stock.* SAGE Publications. https://doi.org/10.1177/1420326x14524586.

Tian, W., & De Wilde, P. (2011). Uncertainty and sensitivity analysis of building performance using probabilistic climate projections: a UK case study. *Automation in Construction, 20*(8), 1096. https://doi.org/10.1016/j.autcon.2011.04.011.

Vardoulakis, S., Dimitroulopoulou, C., Thornes, J., Lai, K., Taylor, J., Myers, I., Heaviside, C., Mavrogianni, A., Shrubsole, C., Chalabi, Z., Davies, M., & Wilkinson, P. (2015). Impact of climate change on the domestic indoor environment and associated health risks in the UK. *Environment International, 85*, 299. https://doi.org/10.1016/j.envint.2015.09.010.

Williams, D., Elghali, L., Wheeler, R., & France, C. (2012). Climate change influence on building lifecycle greenhouse gas emissions: case study of a UK mixed-use development. *Energy and Buildings, 48*, 112. https://doi.org/10.1016/j.enbuild.2012.01.016.

Wright, A., & Venskunas, E. (2022). Effects of future climate change and adaptation measures on summer comfort of modern homes across the regions of the UK. *Energies, 15*(2). https://doi.org/10.3390/en15020512.

Xie, H., Eames, M., Mylona, A., Davies, H., & Challenor, P. (2023). Creating granular climate zones for future-proof building design in the UK. *Applied Energy, 357*. https://doi.org/10.1016/j.apenergy.2023.122549.

INDEX

Note: **Bold** page numbers refer to tables and *italic* page numbers refer to figures.